Ethnicity in Modern Africa

Other Titles in This Series

Apartheid and International Organizations, Richard E. Bissell

Zambia's Foreign Policy: Studies in Diplomacy and Dependence, Douglas G. Anglin and Timothy M. Shaw

Botswana: An African Growth Economy, Penelope Hartland-Thunberg

Westview Special Studies on Africa

Ethnicity in Modern Africa
edited by Brian M. du Toit

The fifteen essays written for this volume reflect the increasing importance for social scientists of ethnic, rather than physical or tribal, criteria for classifying modern population groups. The authors—from South Africa, the United States, South West Africa (Namibia), Nigeria, and Scotland—cover most of Africa south of the Sahara. They consider the range from large national population groupings to small-scale societies attempting to maintain their social boundaries, and discuss such topics as emergent nationalism, ethnic divisiveness, social distance, voluntary association, and the role of women.

The first section is concerned with particular communities, peoples, and ethnic groups, and treats traditional tribal groupings as well as communities delineated on phenotypic grounds. In the second section, the focus turns to modern situations of interaction; the two major themes discussed here are situational ethnicity and situational realignment. The third section deals with color, one of the physical criteria of ethnic identification; here the authors discuss the political and legal implications of a system based on color. The last essay reports on current changes in attitude and organization within the countries of white-ruled southern Africa.

Brian M. du Toit, professor of anthropology at the University of Florida, is a native of South Africa. The author of many books on African culture, his most recent publications are *Configurations of Cultural Continuity* and *Drugs, Rituals, and Altered States of Consciousness.*

Ethnicity in Modern Africa
edited by Brian M. du Toit

Westview Press / Boulder, Colorado

Westview Special Studies on Africa

Copyright © 1978 by Westview Press, Inc.

Published in 1978 in the United States of America by
Westview Press, Inc.
5500 Central Avenue
Boulder, Colorado 80301
Frederick A. Praeger, Publisher

Library of Congress Cataloging in Publication Data
Main entry under title:
Ethnicity in modern Africa.
(Westview special studies on Africa)
Bibliography: p.
1. Africa, Sub-Saharan—Race relations—Addresses, essays, lectures. 2. Blacks—Africa, Sub-Saharan—Race identity—Addresses, essays, lectures. 3. Ethnology—Africa, Sub-Saharan—Addresses, essays, lectures. 4. Africa, Southern—Race relations—Addresses, essays, lectures. 5. Africa, Southern—Social conditions—Addresses, essays, lectures. I. Du Toit, Brian M., 1935-
DT352.42.E86   301.45'1'0968   78-58295
ISBN 0-89158-314-9

*This book is dedicated to a new generation
of Africans of different ethnic groups
who seek an equitable and peaceful future*

Contents

ix

Ethnicity in Modern Africa

1
Introduction

Brian M. du Toit

During the last four decades, social scientists, politicians, and common folk have increasingly been using the concept of "ethnic" and its derivative "ethnicity." In a country that has been called the ethnic melting pot, we have politicians referring to "ethnic" neighborhoods, eating "ethnic" foods, and telling "ethnic" jokes. It seems that in many cases these terms are used as a substitute for more precise and descriptive terms, and in other cases they are used as catchall phrases because the speaker is too lazy to clarify his referent. After forty years of academic use, these terms are still as obscure in meaning as they were two and three centuries ago.

This discussion will trace the development of these terms and document their introduction to the academic jargon. Then we will explore the use of these terms in the African context.

I

The word *ethnic* is derived from the original Greek word ἐθνικος, meaning "heathen." This was based on the root form ἐθνος ("ethnos"), which is translated as *nation* and was applied in particular to the "non-Israelitish nations or Gentiles." Thus *The Shorter Oxford English Dictionary on Historical Principles* lists two meanings for the word *ethnic*: "A. Pertaining to nations not Christian or Jewish; Gentile, heathen, pagan. B. A Gentile heathen, pagan."[1] Soon, however, *ethnical* was used to refer to heathenish or pagan, while *ethnicism* indicated heathenism or paganism.

Derivative forms that were first recognized included *ethnological*, as when The Ethnological Society was founded in 1843 in London. The link between this name and the Société Ethnologique de Paris, founded five years earlier, is quite clear.[2] In 1842 the American Ethnological Society was founded in New York. As seen in the following statement, it originally had a somewhat unique concentration. When the society incorporated in 1916, it restated its purpose of earlier years: "To promote inquiries into the origin, progress, and characteristics of the *races of man*; by publishing and

1

distributing documents; by arranging scientific meetings and public lectures; and by other means adapted to the ends for which this corporation is organized" (emphasis added).

In a certain sense, and with particular reference to Europe, *ethnic* was also used to refer to race or phenotype. Thus *ethnic stock* had a physical referent. At this stage, however, it was not used as substitute for the term *race* or in avoidance of that term.

Except for occasional uses in this last sense, *ethnic* was clearly employed to designate differences in religion, behavioral forms, and life-style. Speakers tended to use it with reference to European subgroupings, while it was also used to contrast "others" with one's own group. Generally speaking, social groups and subgroups in the Americas, Africa, and further afield were designated by other terms, frequently *tribe.*

One of the first scientific or objective uses of the term *ethnic* was by Huxley and Haddon. In their study *We Europeans*, they lament the confusion between the "ideas of race, culture and nation" and recommend that the term *race*, as applied to human groups, be dropped from the vocabulary of science. Acting upon this recommendation, they state: "In what follows the word race will be deliberately avoided, and the term (ethnic) group or people employed for all general purposes."[3] In a later chapter in their book, they talk about "ethnic classifications" and refer to phenotypical groups and subgroups marked by obvious physical differences such as "black woolly hair, dark brown or black skin, and a broad nose" as contrasted with "wavy or curly hair of any colour from black to flaxen, dark brown or white skin, and typically a medium or narrow nose with usually a high bridge."[4] In fact, we even have reference to roundheaded and broad-faced peoples as representing ethnic groups. When Haddon reviewed Griffith Taylor's *Environment, Race and Migration* two years later, he did not use the word *race* but discussed the author's description of "widely separated *ethnic* types."[5]

On the other side of the Atlantic, the term *ethnic* was also gaining recognition, but it appears that it applied more specifically to sociocultural than to racial criteria. Though trained in Sweden, Gunnar Myrdal had some influence from American academics, including twelve months' residence some years before, prior to his preparation of the monumental *An American Dilemma*. In this book he does not skirt the issue of race, spending one whole chapter on the topic, but he does point out that in "modern biological or ethnological research 'race' as a scientific concept has lost sharpness of meaning."[6] He then regularly substitutes the terms *Negro people*, the *Negro group*, or the *Negro population*. But he also uses the word *ethnic*. One reason for residential concentrations of black Americans, he suggested, was "ethnic attachment." This resulted in clusters "for convenience and mutual protection." Nor is this tendency unique to American Negroes, says Myrdal, for "ethnic cohesion" marked most of the

European migrants, who in time, however, lost their "ethnic affiliation" and "ethnic background."[7]

Though published a year after Myrdal's study, the Yankee City research, under the direction of Lloyd Warner, employed the concept "ethnic" as a central theme. The third volume in fact uses that word in its title. The authors clarify that the term *ethnic* "refers to any individual who considers himself, or is considered to be, a member of a group with a foreign culture and who participates in the activities of the group. Ethnics may be either of foreign or of native birth."[8] It is quite clear that Warner and his associates had a completely different frame of reference for this term, namely, a sociocultural rather than racial context. In a final section of this same study, the authors differentiate between ethnic group, racial group, and ethno-racial group, and they hypothesize that "when the combined cultural and biological traits are highly divergent from those of the host society the subordination of the group will be very great, their subsystem strong, the period of assimilation long, and the processes slow and usually painful."[9] There are occasional references in American social sciences to the term *ethnic*,[10] but the handbook of the day, *General Anthropology* (1938), does not index the term, while Boas's chapter on "Race" treats just that.[11]

From this sketchy overview, it seems abundantly clear that while British scientists, including A. C. Haddon, who had earlier written *The Races of Man* (1925), were using *ethnic* to refer to race, Americans were giving it a special sociocultural meaning. These orientations are even reflected in modern dictionaries. The *Oxford English Dictionary* lists a secondary meaning under *ethnic* as "Pertaining to race; peculiar to a race or nation; ethnological,"[12] while the *Webster's New World Dictionary* lists a secondary meaning under *ethnic* as "designating or of any of the basic divisions or groups of mankind as distinguished by custom, characteristics, language, etc.; ethnological."[13]

In the decades that followed the use of the term *ethnic* by Warner and Srole, a few other authors employed it. Lowie included a section on "ethnic groups" in his study of *Social Organization*, but he adds little in clarification.[14] Francis published a paper on this topic and actually confirmed the earlier position referred to. He states that "an ethnic group is not a race . . . an ethnic group is not a nation."[15] But in a footnote on the same page, Francis takes issue with Warner and Srole and explains, "We need not emphasize that in this context ethnic group is not limited to ethnic fragments and minorities within a larger culture. In our terminology not only the French-Canadians or the Pennsylvania Dutch would be ethnic groups but also the French of France or the Irish in Ireland."

But this was not to be the final clarification. In 1952 Paul Walter suggests that *ethnic group* may "refer to a distinct racial grouping, to one whose distinction is not racial but cultural, or it may apply where both racial and cultural differences coincide."[16] The encyclopedic inventory *Anthropology*

Today,[17] which was edited by Kroeber, does not include *ethnic* in its index, while its counterpart *Sociology Today*[18] indexes *ethnic relations* but cross-lists this with *intergroup relations*. In social psychology, the concept "ethnic" has been used to refer to both racial and sociocultural differences.[19] Recently Gamst and Norbeck returned to stir up the muddied water when they explain that "an ethnic group is a recognizable sociocultural unit based upon some form of national or tribal distinction, which lives among other people rather than in its own country."[20]

By extension, if *ethnic groups* refer to sociocultural groupings, they would be dynamic. Warner and Srole felt that "the future of American ethnic groups seems to be limited; it is likely that they will be quickly absorbed."[21] Robert Lowie, writing with a ten year advantage but also with an anthropological perspective, takes the opposite view. For him it is "clear that the old notion of the American melting pot is unsound, for . . . many immigrants do not want to be melted down, preferring somehow to preserve their identity."[22] This seems to have been confirmed by Glazer and Moynihan in *Beyond the Melting Pot*.[23]

The foregoing discussion has shown that at least five meanings have been assigned to the concept of ethnic or ethnic group. In the first sense, the term was equated with race. This is present in its early use by Huxley and Haddon as well as in point six of the statement by United Nations experts on race. The term is also used to refer to specific major races, as was done in point seven of the United Nations statement. The third reference is to a sociocultural group such as the French, either in France or in another country. But some writers have narrowed this down and set as prerequisite that *ethnic* really refers to a subgroup living among others in a foreign country. The fifth meaning uses *ethnic group* when a group of people contrast themselves or are contrasted by others, on the basis primarily of sharing certain cultural criteria such as language, beliefs and values, religion, or history. Such an ethnic group may have geographical contiguity and may include "racial characteristics," though neither of these is required.

II

It is to be expected that derivatives of the original root ἔθνος soon appeared. One of the earliest was *ethnocentrism*. Writing in 1906, William Graham Sumner devoted a subsection in the first chapter of *Folkways* to this subject.[24] He sees it as the technical term for the view according to which "one's own group is the center of everything, and all others are scaled and rated with reference to it." Stated differently, it is "a peculiar blend of inner solidarity with outward hostility."[25]

Studies of ethnocentrism have been done basically by social psychologists and political scientists. Among the former, Emory Bogardus developed the social distance scale to measure interethnic attitudes, and others have used this measure in various forms.[26] Political scientists have tended to use the concept in studies of nationalism.[27] More recently, wider applications of the concept occurred in situations where interethnic also involved phenotypic differences[28] as well as in situations that lacked the color contrast.[29] Brewer and Campbell link ethnocentrism to innergroup self-regard. "From the perspective of an individualistic, hedonistic psychology, the process of inculcating loyalty to a social group is one of convincing the individual to identify the satisfaction of his or her basic wants and ego needs with the survival of the group and his or her continued membership in it."[30] These references obviously have a sociocultural referent, but once again the British definition states that *ethnocentrism* refers to "one's own race or ethnic group."[31] We are here, it seems, confusing the issue of *racism* and *ethnocentrism*. We will return to this topic in the next section of this discussion.

Perhaps the most important derivative concept is *ethnicity*. One of the earliest uses of this concept is by David Riesman, who uses it in the same context as Warner and Srole use *ethnic*. Speaking about the old hierarchical class-based antagonisms, he suggests that a new form of antagonism is taking place in America, namely, between "the groups who by reason of rural or small town location, ethnicity, or other parochialism, feel threatened by the better educated, upper-middle-class people."[32] Instead of "class-consciousness," "group pride," or "racism," the concept of ethnicity has for the time being gained recognition as a neutral, unemotional referent to those characteristics and qualities that mark an ethnic group, irrespective of whether the group is defined basically on sociocultural or basically on phenotypic grounds.

Ethnicity does not appear in standard dictionaries before the 1960s. It appears for the first time in 1961 in *Webster's Third New International*, in 1972 in the *Supplement to the Oxford English Dictionary*, and in 1973 in the *American Heritage Dictionary of the English Language*. The Oxford dictionary defines it as "Restrict + obs. rare. . .2. Ethnic character or peculiarity," while the Heritage gives two meanings: "1. The condition of belonging to a particular ethnic group; 2. Ethnic pride." Given these somewhat vague and subjective meanings, Glazer and Moynihan suggest that "one senses a term still on the move."[33] The last five years have seen a number of studies on this subject,[34] though some authors think strictly of the cultural pluralism model.[35]

In the discussion that follows, we will turn our attention to Africa. How have the concepts ethnic and ethnicity been used in reference to African peoples or communities? What terms or referents did these concepts replace?

III

As far as we know, the first reference to the term *ethnic* in African material is that of Seligman in 1936. He was no doubt influenced by his British countrymen's use of the term[36] the previous year and used it in two forms: "ethnic reality" and "ethnic accuracy," and the summary at the head of the article mentions "ethnic groups."[37] At no place in this article does the author define, discuss, or clarify this concept or explain why he uses it at all, since he still speaks of "racial groups," "language groups," and "tribes" and since we are told that the "Nilotes show an aloofness and pride of race." Most of the major studies of that period continue to speak about tribes, language groups, and geographical divisions rather than employing the concept *ethnic*.[38]

A number of quite sophisticated analyses were being conducted during these years. Nadel had completed his research in the Korofan region of the Sudan. Although he uses the concept *tribe* in both his major studies, he explains that the concept "hinges on a *theory* of cultural identity, which ignores or dismisses as immaterial existing variations, and ignores or disregards uniformities beyond its self-chosen boundaries. The tribe exists, not in virtue of any objective unity or likeness, but in virtue of an ideological unity, and a likeness accepted as a dogma."[39] It would seem from this description that the boundaries of tribes, as the boundaries of ethnic groups, are subjectively drawn and exist where members want them to exist. This has much in common with the situation among the Nuer, whose political system includes all the people "who speak the same language and have, in other respects, the same culture, and *consider themselves to be distinct from like aggregates.*"[40] This subjective criterion also applies among the Alur, where people who are considered to be alien, i.e., by "social descent," form ethnic groups.[41] In all three of the ethnographic examples, "tribal" and "ethnic" boundaries do not necessarily coincide with any objective reality.

Meyer Fortes, on the other hand, was dealing with an ethnographic situation that lacked centralized political authority by which one group could be contrasted with another. Thus when he discusses the "Meaning of 'Tallensi,'" he prefers to speak of socio-geographical regions or areas.[42] This seems to be the closest approximation to ethnic group we have found. Though the authors, all writing in the British tradition, did not actually use the concept ethnic, they also did not give their referent the phenotypic implication as was done by Huxley and Haddon.

In 1959 Immanuel Wallerstein read a paper before the annual meetings of the American Sociological Society.[43] He stated that "membership in an ethnic group is a matter of social definition, an interplay of the self-definition of members and the definition of other groups. The ethnic group seems to need a minimum size to function effectively, and hence to achieve social definition. Now it may be that an individual who defined himself as being of a certain tribe in a rural area can find no others from his village in the city. He may simply redefine himself as a member of a new and larger group."[44] In the very first paragraph, Wallerstein speaks of "ethnicity (tribalism)," which seems to suggest that he sees them as interchangeable concepts. He goes on to suggest that ethic groups really only find expression in modern situations, for instance, from common occupation "rather than from a common language or traditional polity." He furthermore states that ethnic groups must be defined in terms of urban situations, for "by ethnicity we mean the feeling of loyalty to this new ethnic group of the towns." This reminds one of the statement by Clyde Mitchell that "tribalism as a significant factor in human relationships, in fact, arose in situations where people from widely different situations . . . were thrown together in social interaction. In other words, tribalism is a phenomenon arising out of culture contact."[45] Arguing along much the same lines, Mafeje has suggested that tribalism is more a product of the social scientist's mind than an empirical reality.[46]

According to both Wallerstein and Mitchell, one would expect tribalism and ethnicity to be products of the growth of modern towns and cities in Africa. I find this extremely hard to believe if by *ethnicity* one refers to that which characterizes an ethnic group and by *tribalism* one understands ethnic loyalty and identification.[47]

This discussion has reintroduced that old term which, for various reasons, has become anathema in anthropological discussions, namely, *tribal*. During the mid-1950s researchers were recording the importance of "urban tribalism"[48] and supertribalism.[49] Shortly after this, the concepts ethnic and ethnicity were introduced among others by Wallerstein, and for a decade there was a period of confusion when these terms were interchanged. Wallerstein tends to use *tribe* for the social group in the rural area, but when they become urban residents, he refers to them as "ethnic group." When he refers to a particular traditional social organization, he denotes this as "tribalism," but when he refers to a "persistence of loyalties and values, which stem from a particular form of social organization," this is denoted as "ethnicity." Although recognizing that "tribalism has become a dirty word in the new African states," Lloyd speaks of *tribal* society, *tribal* life, and *tribalism*, which is on the increase in independent African countries.[50]

A person who seemed, at that time, to be even less clear in his use of these terms is Abner Cohen. He states that he will use *tribalism* mainly as a native term, *ethnicity* as a sociological term. The ethnic group, then, is an informal interest group whose members are distinct from the members of other groups within the same society because they share kinship, religion, and language. *Ethnicity* "refers to strife between such ethnic groups, in the course of which people stress their identity and exclusiveness."[51] Yet he later speaks of "ethnic distinctiveness" and, four lines later, of "tribal distinctiveness."[52] We also learn that "tribal groupings" in central African urban centers may have been "detribalized" in their struggle against colonial authority and then "retribalized" in their struggle for power after independence."[53] Although Gluckman rather clearly designated a political frame of reference for his use of tribalism, Southall seemed to use it without any qualification.[54]

Though writing at about the same time as Gluckman and Southall, Paul Mercier attempted to come to grips with this terminological problem. Mercier pointed out that some people saw the ethnic group as a closed group, descended from a common ancestor (or at least having a common origin), sharing a homogeneous culture, and having the same language. The fact of political unity could be a further criterion. Mercier, however, agreed with Nadel that the ethnic group might not be the same reference group for all people and, we would suggest, might not be the same at all times. "An ethnic group is identical with the *theory* which its members have of it," Mercier suggests.[55] Following this, the author returns to the use of *tribal, tribalism, supertribalism,* and *tribal nationalism.*

However, much of the early discussion suggested a physical or racial frame of reference. This is also inherent in the criterion of a common ancestor or common origin. This may underlie ethnocentrism, but it is more clearly expressed in racism. Van den Berghe defines the latter as "any set of beliefs that organic, genetically transmitted differences (whether real or imagined) between human groups are intrinsically associated with the presence or absence of certain socially relevant abilities or character-istics"[56] and notes that these beliefs then form the basis of distinctions between individuals and groups supposed to belong to such socially defined races. Racism is commonly associated with whites, but it does not necessarily require the presence of whites. Van den Berghe discusses the example of Rwanda and Burundi, where the Tutsi aristocracy, the Hutu majority, and Pygmoid Twa represent an internal example of racism.[57] In discussing northern Nigeria, Smith explains that the "Fulani rulers of Zaria distinguish on racial grounds between themselves and their Hausa subjects."[58] Events during the late 1960s in Iboland might be a further example of this. In terms of the generalized use of the term *ethnic* and varying interpretations that have been given this concept, it seems that both Van den Berghe and Smith could have used *ethnic* or *ethnicity* in this context.

Race, or phenotype, seems a rather clear-cut criterion on the basis of which to base identity, but in many cases ethnicity is a changing concept. It will be recalled from our earlier discussion that Nadel saw the ethnic group as existing by virtue of an ideological identity, i.e., a subjective evaluation. Mercier also points out that the boundary a certain group claims at a certain time is arbitrary and can shift due to linguistic, social, or any other changes. "All the regions of Africa show examples that one is not a member of an ethnic group in an intangible way. Individuals, lineages, and class have become, or are in the process of becoming, members of a given ethnic group, either because their original group disappears completely or because they separate themselves from it."[59] But this process is even more clearly expressed in urban contexts, where a person may claim identity with any one of a number of ethnic divisions depending on the context and the need. Such a person will belong to a natal group, but in the absence of other members of that group, he may identify with a local or linguistic category, or even with a regional classification. In African cities, in the last analysis, the person would be black as compared to European businessmen or tourists, Lebanese traders, and Indian fellow citizens. This dynamics of ethnicity has been well documented for certain African cases, either as regards ethnic categories[60] or as situational ethnicity,[61] but in the past it was variously designated rural and urban tribalism, migrants versus townsmen, red versus school, or traditional versus modern (educated). Today it is often called cultural pluralism.

The most important point to keep in mind is the fact that ethnic boundaries, in spite of their importance and implications, are fluid and flexible. They have social, biological, and interactional implications and for that reason may be manipulated to serve given ends at a particular place and time. Barth has discussed this concept in some detail and emphasized the implications of boundaries in canalizing social life.[62] In recognizing boundaries, it is essential to keep in mind that they may be subjectively drawn and that they contain a fluid dimension. A group may thus use their group identity at different times to identify or contrast themselves with another social group. This is the theme of a number of chapters that follow.

In summary, we have noticed here two major perspectives, namely, the difference between the views and criteria employed by the scientist, who frequently was not dealing with an actual social situation, and the views and criteria employed by the people themselves and reported by the ethnographer. In the former case, the scientist could define the term *ethnic* and set limits to its use, either in physical or phenotypic terms or on the social and cultural level. But neither of these has cognitive value in an ethnographic sense. This forces us to return to the community level and to accept the fact that *ethnic* and its derivative *ethnicity* are a matter of social definition and that they are based on ideological criteria. Such a social definition will always involve a self-definition and a categorization by others with whom one is in contact. The ideological criteria will almost certainly include socio-behavioral, phenotypic, and spatial criteria. *Ethnicity* then would refer to the characteristics and attitudes of those who consider themselves and are considered by others to form a distinct ethnic group. Such a group may not satisfy empirical measures nor have any temporal permanence, but it does have sociological importance at the ethnographic level.

IV

In the chapters that follow, the authors have used the concepts ethnic and ethnicity in different ways and at different levels of analysis. A number of important themes, however, emerge from these studies.

Phenotypic characteristics, especially marked by pigmentation, are mentioned by a number of authors. The clearest example is Adams's discussion of the coloureds. She describes a situation in which a category of people are not only defined by what they are not, do not only contrast with the white and African fellow South Africans, but also recognize some internal differences based on skin color. This aspect is also mentioned by Pendleton with reference to Windhoek, by Coetzee in describing the emergence of Afrikaner ethnicity, and by du Toit. In the case of both Sandawe and Mbuti, the authors state that skin color and physical features are used as distinguishing characteristics.

The interplay between contrasting ethnic groups, in some cases referred to as "tribes" and marked by differences in language, geographical contiguity, and cultural aspects, is also mentioned a number of times. Guillotte uses the Magugu term *Kabila*, which refers to a group of people who reside in an ethnic homeland and usually have a distinctive language. In analyzing the position of the Mbugwe in Magugu, he contrasts them with nontribal people and refers to the Christian-Muslim dichotomy and the loyalties that supersede the tribe and that can be seen as citizenship in a modern nation-state. Newman and Turnbull also emphasize the ethnic homeland concept. In the case of the Sandawe, this land is tied to the traditions of the people, and the forest houses the spirits of the Mbuti and gives meaning to their life. The tribal-supratribal dichotomy is very clearly expressed in Gordon's discussion of the way in which the Ovambo, Kavango, and other African mine workers evolved the "brotherhood" that united them against white authority. In the case of urban residents, loyalty to ethnic group members, particularly to persons who speak the same language and come from the same region, is strong. This loyalty tends to decrease or become more diverse as persons become better educated and increase their economic positions. This is clearly illustrated in du Toit's discussion, but at the upper levels of education and income ethnic consciousness and competition among males seem to reappear, possibly due to the homeland policy of the South African government.

Language has been mentioned, but the Sandawe and the residents of Windhoek, as discussed by Pendleton, recognize linguistic and speech criteria more clearly. Landman specifically looks at the interplay between the language of the major ethnic group, the local lingua franca, and the official language in all modern African countries.

Several authors refer specifically to the emergence of ethnic group consciousness, and all of them see ethnicity emerging from situations of stress, situations in which identity was threatened. Coetzee discusses the ethnogenesis of the Afrikaner with time and space referents. During the period of emergence, there were threats to their identity both from indigenous pressures and from British administration. Pachai sees this same threat on an economic level as underlying hostile attitudes and discriminatory legislation against Indians in South Africa, and this in turn has emphasized their identity. To illustrate this point even more clearly, Gordon suggests that "ethnicity achieves its collective importance only through negation." The hostility and denial of any worth on the part of whites toward African mine workers in fact created and enhanced their ethnic consciousness and finally the brotherhood, which is ethnicity at a second level of analysis. Black brothers were still different from white mine officials. At a different level the same point is made by Mazrui and Skinner. The latter suggests that Africans traditionally ignored ethnicity, either incorporating groups that differed from them or moving away. The coexistence of groups that recognized and emphasized differences, almost as in a class structure, emerged after European administrations established colonial rule.

A number of students of the African urban scene have suggested that at a certain phase of urban commitment, ethnicity loses its significance and social class, labor union, or some other unifying factor gains significance. In a thought-provoking discussion, Little suggests that with the improvement of women's status in cities, women have come to form groups that ignore ethnicity. Reminiscent of the issue discussed in the previous paragraph, this new phase of women's cohesiveness is based on "female social solidarity initiated by militant leadership." It is equally a result of women being less traditionally oriented than men, for they have "little or nothing to gain from maintaining the status quo." Looking at a particular ethnographic situation, this point is confirmed by du Toit's material. Women in the better educated and higher income categories are less likely than males to emphasize ethnic choices. This pertains to friendship groups, mutual support associations, and even the selection of a marriage partner.

A last point not emphasized by many contributors but inherent in much of the material is the fluidity of ethnic identity. This may be expressed in the form of different types of identification at different times, it may find expression in situational ethnicity, or it may be as clear as it is among the Mbuti. Here, Turnbull states, individuals as well as groups have a plural identity, and "each moves from one frame of reference to another as suits the context of the moment." Whether we are discussing the Mbuti or the people of Africa in general, when the criterion of color is ignored, this is the ethnicity of modern Africa. It is an ethnicity that Nussey sees, perhaps optimistically, as the solution for the social ills of southern Africa. It does not ignore differences and does not aim for a melting-pot solution, but it allows plural identities depending on time, place, or context.

NOTES

1. *The Shorter Oxford English Dictionary on Historical Principles* (Oxford: The Clarendon Press, 1936), s.v. "ethnic."

2. T. K. Penniman, *A Hundred Years of Anthropology*, 2d ed. (London: Gerald Duckworth & Co., 1952), pp. 65-66.

3. Julian S. Huxley and A. C. Haddon, *We Europeans: A Survey of "Racial" Problems* (1935; reprinted., London: Penguin Books Ltd., 1939), pp. 91-92.

4. Ibid., p. 142. In an earlier study, A. C. Haddon, *The Races of Man and Their Distribution* (London: Milner & Co., 1909), had dealt with the topic of phenotype on a worldwide basis. In the 1929 edition of the latter study published by Cambridge University Press, there is still no mention of the concept "ethnic," even though he discusses culture and language as alternative criteria for the classification of mankind.

5. A. C. Haddon, review of *Environment, Race and Migration* by Griffith Taylor, in *Man*, vol. 38 (1938), p. 140. Emphasis added.

6. Gunnar Myrdal, *An American Dilemma, The Negro Problem and Modern Democracy* (New York: Harper & Brothers, 1944), p. 115.

7. Ibid., pp. 619-620.

8. W. Lloyd Warner and Leo Srole, *The Social System of American Ethnic Groups* (New Haven: Yale University Press, 1965), p. 28.

9. Ibid., pp. 285-286.

10. For instance, A. H. Gayton wrote about the "Areal Affiliations of California Folktales," *American Anthropologist* 37, no. 4 (1935):583, and refers to "the three ethnic areas of California" as conceived by Kroeber. The latter, however, used folktale or culture areas.

11. Franz Boas, ed., *General Anthropology* (New York: D. C. Heath & Co., 1938). This volume contains chapters by the prominent anthropologists of the day, including Lowie, Bunzel, Reichard, and Benedict.

12. This quotation is from the 1961 edition of *The Oxford English Dictionary* (Oxford: The Clarendon Press) which also lists *ethnical* as being "of or pertaining to race or races." A team of United Nations experts seems to lean in favor of the racial sense of the term *ethnic*. This is certainly the way in which it is used in Point Six of their statement as prepared by E. Beaglehole et al., "A Statement by Experts on Race Problems," *International Social Science Bulletin* 2 (1950):391-396. However, Point Seven of this same statement lists the major races of mankind as the Mongoloid, Negroid, and Caucasoid, while Point Eight refers to "subgroups or ethnic groups within these divisions," such as Armenoids, Dravidian, or Nordics. It is of interest that the UNESCO publication, *The Race Concept* (Paris, 1952), had among the coworkers Gunnar Myrdal and Julian S. Huxley.

13. *Webster's New World Dictionary of the American Language* (New York: The World Publishing Co., 1962), s. v. "ethnic."

14. In his study of *Social Organization* (New York: Rinehart & Co., 1948), Lowie includes a section on "ethnic groups" in his chapter discussing "Social Strata."

15. E. K. Francis, "The Nature of the Ethnic Group," *American Journal of*

Sociology 52 (1947):394-395, also discusses the European uses of the concepts *Volk, Volker, narod, groupe ethnique, Volksgruppe,* and related concepts. Much the same position is taken by Milton M. Gordon, *Assimilation in American Life* (New York: Oxford University Press, 1964), pp. 27-28, when he refers to criteria that create "a sense of peoplehood." In a recent book, *Interethnic Relations* (New York: Elsevier, 1976, p. 394), Francis clouds the position he took some years before. He explains now that an ethnic group is "a subgroup of a modern society formed by ethnics whose differential treatment in the distribution of social rewards is causing specific problems."

16. Paul A. F. Walter, *Race and Culture Relations* (New York: McGraw-Hill Book Co., 1952), p. 21. Much the same position is taken by Ralph Piddington, *An Introduction to Social Anthropology,* 2 vols. (Edinburgh: Oliver and Boyd, 1950-1957), 2:756.

17. A. L. Kroeber, ed., *Anthropology Today* (Chicago: The University of Chicago Press, 1952).

18. Robert K. Merton, Leonard Broom, and Leonard S. Cottrell, Jr., eds., *Sociology Today* (New York: Basic Books Inc., 1959).

19. See Gardner Lindzey, ed., *Handbook of Social Psychology* (Reading, Mass.: Addison-Wesley Publishing Co., 1954), p. 1022; and also the discussion in S. Stansfeld Sargent, *Social Psychology: An Integrative Interpretation* (New York: The Ronald Press Co., 1950), pp. 443-444.

20. Frederick C. Gamst and Edward Norbeck, eds., *Ideas of Culture, Sources and Uses* (New York: Holt, Rinehart & Winston, 1976), p. 240.

21. Warner and Srole, *Social System of American Ethnic Groups,* p. 295.

22. Lowie, *Social Organization,* p. 287.

23. Nathan Glazer and Daniel P. Moynihan, *Beyond the Melting Pot* (Cambridge, Mass.: Harvard University Press, 1963).

24. William Graham Sumner, *Folkways* (New York: New American Library, 1960).

25. R. A. Schermerhorn, *These Our People: Minorities in American Culture* (Boston: D. C. Heath & Co., 1949), p. 8.

26. For the initial work, see Emory S. Bogardus, "A Social Distance Scale," *Sociology and Social Research* 17 (January-February 1933); and idem, "Social Distance and Its Practical Implications," *Sociology and Social Research* 22 (May-June 1938). Later studies were done by G. Murphy and R. Likert, *Public Opinion and the Individual* (New York: Harper, 1938); and E. L. Hartley, *Problems in Prejudice* (New York: King's Crown Press, 1946). With reference to African material, see the discussion by du Toit in this volume.

27. P. C. Rosenblatt, "Origins and Effects of Group Ethnocentrism and Nationalism," *Journal of Conflict Resolution* 8 (1964). This line of inquiry has most recently been pursued by Marc Ross. However, he is more interested in the immediate small group (linguistic, political, or geographical) to which a person belongs. He differentiates between the person's own psychological and subjective view regarding his ethnic group membership in contrast to the social and objective perspective of others viewing him. See in this regard "Measuring Ethnicity in Nairobi," in *Survey Research in Africa,* ed. William M. O'Barr, David H.

Spain, and Mark A. Tessler (Evanston, Ill.: Northwestern University Press, 1973) and *Grass Roots in an African City* (Cambridge, Mass.: M.I.T. Press, 1975).

28. Henry Lever, *Ethnic Attitudes of Johannesburg Youth* (Johannesburg: Witwatersrand University Press, 1968).

29. Marilyn B. Brewer and Donald T. Campbell, *Ethnocentrism and Intergroup Attitudes* (New York: John Wiley and Sons, 1976).

30. Ibid., p. 73.

31. *A Supplement to the Oxford English Dictionary*, vol. 1 (Oxford: The Clarendon Press, 1972), s. v. "ethnocentrism."

32. David Riesman, "Some Observations on Intellectual Freedom," *The American Scholar* 23, no. 1 (1953), p. 15.

33. Nathan Glazer and Daniel P. Moynihan, eds., *Ethnicity: Theory and Experience* (Cambridge, Mass.: Harvard University Press, 1975), p. 1.

34. A number of studies dealing specifically with ethnicity, rather than ethnic groups, include: Andrew M. Greeley, *Ethnicity in the United States: A Preliminary Reconnaissance* (New York: John Wiley and Sons, 1974); John W. Bennet's edited volume of papers discussed by the American Ethnological Society and entitled *The New Ethnicity: Perspective from Ethnology* (New York: West Publishing Co., 1975); and Pierre van den Berghe's edited volume, which is entitled *Race and Ethnicity in Africa* (Nairobi: East African Publishing House, 1975).

35. Edward O. Laumann, *Bonds of Pluralism* (New York: John Wiley and Sons, 1973).

36. Huxley and Haddon, *We Europeans*.

37. C. G. Seligman, "Human Types in Tropical Africa," *Discovery*, June 1936, p. 167. In 1930 Seligman had published his *Races of Africa*. The third edition, which is completely revised, merely points out at the end of the introduction that political and geographical may "have no ethnic significance." The term is not clarified. See C. G. Seligman, *Races of Africa*, 3rd ed. (London: Oxford University Press, 1961), p. 14.

38. A random sampling includes A. R. Radcliffe-Brown and Daryll Forde, *African Systems of Kinship and Marriage* (London: Oxford University Press, 1950); M. Fortes and E. E. Evans-Pritchard, *African Political Systems* (London: Oxford University Press, 1940); Edwin W. Smith, ed., *African Ideas of God* (London: Edinburgh House Press, 1950); I. Schapera, ed., *The Bantu-Speaking Tribes of South Africa* (Cape Town: Maskew Miller Ltd., 1937). In the modern version of the last book, namely, W. D. Hammond-Tooke, ed., *The Bantu-Speaking Peoples of Southern Africa* (London: Routledge and Kegan Paul, 1974), the word *tribe* is dropped from the title but appears in the index as *tribe, tribalism, tribesman*. *Ethnic* and *ethnicity* are conspicuous by their absence.

39. S. F. Nadel, *A Black Byzantium* (London: Oxford University Press, 1942), particularly the second chapter; and idem, *The Nuba* (London: Oxford University Press, 1947), p. 13.

40. E. E. Evans-Pritchard, *The Nuer* (Oxford: The Clarendon Press, 1940), p. 5.

41. Aidan W. Southall, *Alur Society* (Cambridge: W. Heffer and Sons, 1953), p. 14.

42. Meyer Fortes, *The Dynamics of Clanship among the Tallensi* (London:

Oxford University Press, 1945), p. 231; see also discussions of this subject on pp. 14, 29.

43. Immanuel Wallerstein, "Ethnicity and National Integration in West Africa," *Cahiers d'Études Africaines* 3, no. 1 (1960).

44. Ibid., p. 131.

45. J. Clyde Mitchell, "Some Aspects of Tribal Social Distance," in *The Multitribal Society*, ed. A. A. Dubb (Lusaka: Rhodes-Livingstone Institute, 1962), p. 2.

46. Archie Mafeje, "The Ideology of Tribalism," *Journal of Modern African Studies* 9, no. 2 (August 1971). Many social scientists insist on the utility of this concept as a heuristic device, and others agree with Mafeje. See, in particular, the essays in Peter C. W. Gutkind, ed., *The Passing of Tribal Man in Africa* (Leiden: E. J. Brill, 1970).

47. This is how May Edel defines tribalism in her study, "African Tribalism," *Political Science Quarterly* 80, no. 3 (1965):371.

48. See J. Clyde Mitchell, *The Kalela Dance*, Rhodes-Livingstone paper 27 (Manchester: Manchester University Press, 1956); and A. L. Epstein, *Politics in an Urban African Community* (Manchester: Manchester University Press, 1958).

49. J. Rouch, "Migration on the Gold Coast," *Journal de la Société des Africanistes* 23 (1956).

50. P. C. Lloyd, ed., *The New Elites of Tropical Africa* (London: Oxford University Press, 1966), pp. 33, 58-59, 333.

51. Abner Cohen, *Custom and Politics in Urban Africa* (Los Angeles: University of California Press, 1969), p. 4.

52. Ibid., p. 119.

53. Ibid., p. 196.

54. Aidan Southall wrote a lengthy introduction to *Social Change in Modern Africa* (London: Oxford University Press, 1961), which is a collection of essays presented and discussed at the First International African Seminar, Makerere College, Kampala, in 1959. Max Gluckman's "Anthropological Problems Arising from the African Industrial Revolution" was the first chapter.

55. Paul Mercier, "On the Meaning of 'Tribalism' in Black Africa," in *Africa: Social Problems of Change and Conflict,* ed. Pierre L. van den Berghe (San Francisco: Chandler Publishing Co., 1965).

56. Pierre L. van den Berghe, *Race and Racism: A Comparative Perspective* (New York: John Wiley and Sons, 1967), p. 11.

57. Ibid., p. 12. See also J. J. Maquet, *The Premise of Inequality in Ruanda* (London: Oxford University Press, 1961).

58. M. G. Smith, *The Plural Society in the British West Indies* (Berkeley: University of California Press, 1965), p. 132.

59. Mercier, "On the Meaning of 'Tribalism,'" pp. 488-489.

60. The best-known studies are those of the Xhosa, particularly Philip Mayer, *Townsmen or Tribesmen* (Cape Town: Oxford University Press, 1961). See also Monica Wilson and Archie Mafeje, *Langa* (Cape Town: Oxford University Press, 1963); and Mia Brandel-Syrier, *Reeftown Elite* (New York: Africana Publishing Corporation, 1971).

61. Leo Kuper, *An African Bourgeoisie* (New Haven: Yale University Press, 1965).

62. Fredrik Barth, ed., *Ethnic Groups and Boundaries* (London: George Allen and Unwin, 1970).

PART 1
Tribe, Community, or Nation

Group identification and association may be based on a variety of criteria and find expression at a number of levels. When and in what contexts does a person activate his identification with his kingroup, his linguistic group, or the people among whom he resides? What factors influence him to identify on the basis of his membership in a world religion, a nation, a state, or a group physically different from neighbors?

These are the questions discussed in the following six essays. The subjects range from rural farmers in Tanzania to urbanites in Nigeria and from the descendants of the Ganda kingdom to hunters and gatherers. In each case, however, identification and association are fluid.

Citizens and Tribesmen: Variations in Ethnic Affiliation in a Multiethnic Farming Community in Northern Tanzania

Joseph V. Guillotte, III

This essay concerns the interaction of several groups of individuals in Magugu, a community in northern Tanzania in which scores of tribes are represented.[1] The term *citizen* could naturally be applied to most of these individuals, since they are technically citizens of the nation-state of Tanzania. Nevertheless, certain of them, the Mbugwe, usually choose to ignore this political identity and place priority upon the identity of tribe.[2] Many other residents have no alternatives but to stress the political identity of citizen.

Two of the crucial problems for the new states of East Africa are identity and unity. The concept *tribe* is a critical factor at the core of these contemporary problems. Sometimes it is a real division of people and refers to a definite spatial entity. As Gulliver has pointed out, the term means many things to many people. It is not the same thing to everyone or even the same thing to specific groups or persons at different times. Sometimes *tribe* is held to be nonexistent, and at other times it is used as a facile explanation for enormously complex social, political, and economic crises. "It is impossible to ignore the word and almost impossible not to use it."[3] Neither the term, nor the concept of ethnic exclusivity, nor priority of loyalty to ethnic group over state will go away just because it is legislated against or asserted not to exist. The problem of "tribe" is germane to any discussion of a community such as Magugu, which is inhabited by people who potentially may identify with scores of tribes. To most of them, tribal identity is not important in their everyday lives because they have been brought into associations of a nontribal nature. To the indigenous Mbugwe and a few other groups, however, the term may be applied in some contexts. In brief, ethnic affiliation is situational. As Southall has cautioned, there is a danger in stressing a unidimensional classification of sociocultural groups, because the units of classification and analysis will vary according to the phenomenon and the problem studied. "To hammer home the importance of interlocking, overlapping, multiple and alternative collective identities is one of the most important messages of social and cultural anthropology."[4]

With these caveats in mind, I shall describe the use of the term *tribe* in the analysis of the data from Magugu. The word most often used in Swahili, *kabila* (*makabila*), can be just as uncertain and ambiguous as the English term. The *Standard Swahili-English Dictionary* definition is: "tribe, clan—a smaller division than *Taifa* ("nation") and larger than *ufungu* ("relationship"), *jamaa* ("family")."[5] The most commonly used Swahili term that can be translated as *tribe* thus gives us nothing more definite than a category of human groups placed somewhere between kin-centered relationships and those of the nation-state. Another use of the word *kabila* does not apply to human groups at all but conveys the idea of type or class. For example, a Land Rover is a *kabila* of motor vehicle.

I have used the term *tribe* in this essay as the inhabitants of Magugu themselves use *kabila*, that is, to refer to a group of people who reside in an ethnic homeland and usually have a distinctive language. The population of Magugu, as I see it, consists of five salient groups. Three of these groups—the Mbugwe, the Chagga, and the Barabaig—are ethnically distinctive. Two of them are composed of representatives of scores of ethnic groups subsumed under Muslim and Christian religious categories. The multiethnic Muslims and Christians may well identify with a tribe when they reside in their homeland (if they have a homeland), just as the Mbugwe primarily identify with their ethnic group, often to the exclusion of other identifications. Yet "tribe" is only one of several categories with which the residents of Magugu might identify. As a rule, it is not a very important category of identification. In the presentation of some data such as that secured from censuses and other surveys, I identify *tribe* where it is appropriate, but in all cases, Mbugwe included, I use the term as a "preliminary label," as Southall has suggested.

In this study, I shall concentrate on the maintenance of boundaries between the Mbugwe and the Magugu Muslims. I hope that this will provide some insight into the multiple political identities of the population. The point that should be stressed here is that these boundaries are not unbreachable walls, but rather semipermeable membranes through which limited interaction may occur.

HISTORY

Magugu is a rural community in the Arusha Region of Northern Tanzania. It is a Branch territory of the Tanganyika African National Union—TANU—the solitary legitimate political party in Tanzania. Its population of about seven thousand represents eighty tribal groups within an eight-hundred-mile radius. The residents are for the most part dispersed over 150 square miles, with the exception of about a thousand who live in Kibaoni, a small town that straddles the Great North Road, ninety miles southwest of Arusha. Magugu lies four degrees south of the equator and thirty-five degrees east of Greenwich.

In 1943, sleeping sickness broke out among the labor force on European estates at Kiru, a few miles south of the present location of Kibaoni. The colonial administration ordered the immediate evacuation and quarantine of these African workers. After the quarantine period, the evacuees were offered two choices by the colonial administration: (1) they could clear bush and form a new community to hinder the northward spread of the tsetse fly, the vector of the disease or (2) they could be transported to their tribal homelands. About one thousand chose to stay and became the nucleus of the polyethnic community.

The settlement was planned in uninhabited land that was nominally the territory of the Mbugwe tribe, and the new community was placed under the authority of the Mbugwe paramount chief. The settlers elected a headman, and a Native Authority and Treasury were formed. For all practical purposes, however, Magugu was a separate polity largely ignored by both the colonial administration and the Mbugwe Paramount Chieftaincy. Local government was largely independent of the Mbugwe power structure, except for the referral of major criminal cases, which were heard by the Mbugwe judiciary. This relative political independence continued for the first ten years of the settlement's existence.[6]

From the founding of the settlement to the time of my field research in 1969-1970, mutual animosity had existed between the Mbugwe and the predominantly Muslim population of Magugu. This hard feeling was exacerbated in 1954, when the Mbugwe chief made an attempt to take over the heretofore semiindependent political structure of Magugu. At this time, the last paramount chief of the Mbugwe succeeded in having his brother appointed the headman of Magugu. He had the former Muslim headman arrested and incarcerated for theft. In late 1956, the first TANU organizer came to Magugu, and the recently formed independence party spread among the migrant population in spite of resistance from the Mbugwe chief and the colonial authorities.

After independence in 1961, the role of chief was abolished in Tanzania. Magugu then replaced the Mbugwe neighborhood as the center of government activity and services in the area. Kibaoni became the division headquarters as well as the site of the division magistrate's court, and the dispensary was expanded to the level of a twenty-bed hospital.

Factors of sociological isolation and opposition set the Mbugwe apart from the rest of the Magugu population. Foremost among these are language and social structure. The Mbugwe are a relatively recently arrived group of Bantu-speakers surrounded by Nilotic-speaking Masai to the north and east and southern Cushitic-speaking Iraqw and Gorowa to the west and south. Among predominantly patrilineal neighbors, they are distinct in possessing a matrilineal descent ideology. They apparently were never a unified political entity before the colonial period but rather a cluster of six chiefdoms and one republic. Their seventeen matrilineal clans were dispersed throughout these seven political divisions and were not territorial corporations in recent memory. The Mbugwe have been involved in population movements within their traditional boundaries and to and from the territories of neighboring tribes. They have always absorbed immigrants from other tribal groups and exhibit an awareness of their multiethnic origins. Their political units exercised autonomy in foreign relations with the Masai and Iraqw. They have carried on an exceptional amount of trade and intermarriage with the latter group and continue to do so at the present time. Thus, the Mbugwe, though standing out in Magugu as a distinct group, traditionally exhibited some of the characteristics of the multiethnic Magugu population of which many of them are a part. These are: multiethnic origins, relatively recent migration into the area, and separation into diverse groups and loyalties.

The Mbugwe were affected by the 1944 settlement scheme in several ways. They were involved in affairs outside of their traditional lands, even though this involvement was often unwilling and distasteful to them. In effect, they became more closely connected to an administrative, cultural, and communication network that encompassed all of East Africa. The building of Kibaoni and the settlement of other parts of Magugu brought large numbers of aliens into the territory, aliens with whom the Mbugwe were forced to interact. These aliens formed a cosmopolitan population. By 1955, Robert F. Gray noted that representatives of seventy-six tribes were listed on the tax register of Kibaoni. They were "drawn from all parts of Tanganyika and even further afield . . . from northern Rhodesia, Nyasaland, Ruanda-Urundi, Kenya and Mozambique."[7] Many of these settlers were, or eventually became, Muslims. For the first time, adherents of Islam were present in large numbers in the area.

Today the few remaining members of the Magugu population who took part in the original settlement scheme maintain that the settlers contributed to the well-being of the Mbugwe. As the Magugu settlers view the impact of their coming upon the Mbugwe, the Mbugwe were (and to a great extent remain) backward and uninformed, content to subsist on the brink of famine until the newcomers lifted them up. The settlers cite their "filling up of the country," which dispossessed tsetse, wild animals, and outlaws in the area. They see themselves as culture heroes who brought gifts of new domesticated plants to the retrograde aborigines. Like all culture heroes, they were unappreciated.

Informants from both the Magugu and Mbugwe sides of the confrontation were interviewed, and they essentially agreed on the events and attitudes outlined here. The tales of conflict reveal the antipathy that exists between the two groups. In essence, the oral history reveals the main events that bear on the residents of Magugu today:

1. A "tribe," the Mbugwe, occupied an area without much interference from outside elements. This group, although traditionally organized into several chiefdoms, was in 1944 linguistically and politically homogeneous.

2. A more amorphous entity developed with the inclusion of a large number of immigrants from other areas.

3. The Mbugwe leadership chose to ignore the newcomers for a decade but then began to exert supposedly legitimate authority over them.

4. The entrance of a national political party complicated the situation even more as far as the Mbugwe were concerned. At this point, the Magugu people, who were never overly fond of the Mbugwe, were being organized into a territory-wide African independence movement. Would-be leaders in Magugu saw the advantage of this. Consequently, they enrolled support for themselves as well as for TANU. The result benefited the national party as well as local aspirants to power. This situation was not unique to Magugu. All over the country, similar polarities between chiefs and people had been developing. Where they were most emphasized, TANU "set itself against the existing authority thus filling the vacuum which had been created by the alienation of the people from the colonial administration and the chiefs."[8] Nor was it unusual that the ethnic minorities were amenable to independence. TANU attracted ethnic minorities in the countryside and in towns. Where towns were dominated by one tribe, it was likely that the immigrants in those towns would take part in TANU activities.[9]

5. During the formative years of TANU organization, the lines were drawn that would separate Magugu from Mbugwe. With the achievement of independence, TANU organizational strategy became legitimized by government decree. Mbugwe and Magugu were separately administered, although both comprise Mbugwe Division. An old Mbugwe expressed the situation succinctly to me, "The power of the various tribes was increased with independence and the authority of the Mbugwe broken. They completely broke away Kibaoni from the authority of the Mbugwe."

The Mbugwe Resettlement Scheme reversed the process that had occurred in 1944. It moved unwilling Mbugwe immigrants to the domain of the heterogeneous settlers. This scheme had been planned for several years. According to Dr. Eckhardt, the regional health officer in 1964, it would take into consideration the improvement of health standards, clearing of tsetse fly bush, and the building of latrines. It would remove the constant threat of famine and sleeping sickness and improve land usage. Over a five-year period, some 2,500 Mbugwe families with 7,500 head of livestock were to be moved to the foot of the Sangaiwe Hills, the eastern boundary of Magugu. This settlement was to take place in five phases between 1964 and 1968. Eventually, the termination date of the scheme was accelerated to 1967.[10]

This resettlement plan uprooted hundreds of Mbugwe from their traditional homeland on the shores of Lake Manyara, ten miles north of Kibaoni, and placed them in an area they found undesirable. The following account by an Mbugwe expresses well the general sentiments about the transfer:

> In 1967, our people were told by the government not to go to Mbugwe again. They were intructed that Sangaiwe was a better place and the government was better able to help them there. There was a lot of resistance from the Mbugwe. They held a meeting at Mwada and wanted to obstruct the government. They would not agree to move. They did not understand that the government was finished discussing the affair. Many believed in the prophetical dreams of the Mbugwe diviners, who prophesied that there would not be any move. Another meeting was called by the government to make them understand. At that time they were told that the government was trying to help them and that they should obey the law. Some understood this, but others were contemptuous of the authority of the government. There were constant meetings between the Division Executive Officers [a government official] and the Mbugwe leaders. He tried to make them understand that the time for the move was near. The government got some rich men to move first and set an example. Still the Mbugwe were slow in obeying the orders. The people were moved in trucks by the police, and they stayed in camps in the bush until they were able to build houses. In general the Mbugwe felt that they were badly treated by the government, and the resentment still exists.

Not all the Mbugwe were sent to Sangaiwe. Some were relocated in a neighborhood called Taifa Njema just outside of Kibaoni and within the administrative boundaries of Magugu TANU Branch. Thus, the varied ethnic groups of Magugu were supplemented by the forced influx of another group. Some of the longest-settled people in the area were now under the aegis of the people who were formerly subordinate to the leaders.

This brief history indicates some of the events that have caused the Mbugwe and Magugu groups to view each other as political antagonists. Local and national political processes became more closely intertwined when TANU entered the region. Since that time, it has become increasingly difficult to examine either of the communities as isolated social and cultural entities. They have continued to interact with each other and with the colonial and subsequent independent government of Tanzania. The enthusiasm of the Magugu settlers for TANU was in contrast to the relative disinterest of the Mbugwe and is one example of the migrants' awareness of a political sphere larger than tribe. The events that took place in Magugu were to some extent mirrored throughout the continent wherever migrants settled down. That is, wherever migrants sought land rights in rural areas, the ground was laid for the political quarrels that came to the surface in the wake of nationalism in the late colonial era and after independence. The theme of "native" versus "stranger" is a recurrent one.[11]

A COMMUNITY OF STRANGERS: SPATIAL AND ECONOMIC
BOUNDARIES BETWEEN MAGUGU AND MBUGWE

People in Magugu often describe their town as a community of
strangers. This is statistically true. Most adults were born elsewhere.
However, there are indications that these strangers are beginning to put
down roots and become known to one another. Virtually the entire original
population came from the African labor force on the European estates in
the Kiru area. Apparently, most of them did not stay very long at Magugu.
When Wallace Dierickx, a geographer who was investigating land use,
interviewed 395 farming householders of Magugu in 1954, he discovered
that only twenty-eight of the original refugees were still there.[12] Only ten
years after the beginning of the community, voluntary migrants had
replaced most of the original evacuees.

Why did they come to Magugu? A survey of one hundred males of
Kibaoni in 1970 yielded the following reasons:

1. They were looking for work (forty).
2. They followed kinsmen to Magugu (twenty-seven).

(These two categories of answers are difficult to separate from each
other. The people who said they were looking for work often came to
Magugu specifically because they heard about the settlement from a friend
or relative who had already settled there. Those who said that they came to
join kinsmen very often migrated from their former location because it
offered little economic opportunity. Consequently, 67 percent of the survey
might fall into a broad category called socioeconomic with an emphasis on
economic motivation.)

3. Government transfers (eight). This is a special type of economic
reason. People such as medical personnel at the dispensary, employees of
the Tropical Pesticides Research Institute, and clerical workers in
government offices fall into this category.
4. Those who came to find work on European estates (eight). I have
separated this from the first category because that category is composed of
individual migrants who entered the region with the intention of being
independent farmers. This small number is not indicative of the position of
the European estates as an inducement to migration, nor does it indicate
the influence of these estates on the growth and development of Magugu.
The estates have always provided full and part-time employment for the
Magugu farmers and their wives and daughters.
5. Wanderers (five). Five men stated that they just happened to be in the
neighborhood, liked the area, and decided to stay since there was plenty of
land available for cultivation. They were vague about any specific
economic or social reasons bringing them to Magugu and indicated

an apparent aimlessness in their lives before settling down there.

6. Those born at Magugu (four). Only four of the sample were born at Magugu. These were young men who looked on Magugu as their home.

7. Kiru evacuees (three). Only three men were among the original evacuees from the sleeping sickness outbreak of 1943.

8. Those who came to visit kinsmen and stayed (three). Three men gave this as their reason for coming to Magugu. They might also be classified with the first and second categories, but they apparently had no intention of remaining at Magugu when they first arrived.

9. Mbugwe transferred by the government in 1967 (two). Two men were part of the Mbugwe Resettlement Scheme transfer that relocated the population on the Manyara shore.

10. One man said that he came to consult an *mganga*, or native curer, at Magugu. He liked the countryside and decided to stay.

11. Trouble in former location. One man stated that the reason he first came to Magugu was because of trouble in his former home at Kilimanjaro among the Chagga. He himself was not a Chagga, and he did not elaborate on the particular kind of trouble that he had. I believe that his reason might be related to the arrival of many other residents. This might be considered the "last straw," the factor that would cause a man to finally leave a residence. Gulliver noted that such a factor might not in itself be a strong enough reason to leave a place but might be the culmination of a long list of complaints about a location.[13] This was often brought up by other Magugu residents in conversation. In their former homes, they were unable to get along with tribal people who considered them strangers, so they finally came to a place where everybody was a stranger.

The Magugu population seems to be recruited from that small minority of people who leave their tribal homelands seeking work or adventure and ultimately never return. They are permanent migrants or at least long-term absentees from their ethnic homeland. In many cases, they never really have a tribal homeland but were born in multiethnic communities such as Magugu. After becoming dissatisfied with Magugu, they might leave, but from the pattern of their lives before arriving at Magugu, it seems likely that they will migrate to some other multiethnic community similar to Magugu. Their life histories breathe restlessness. This is surprising because Tanzania was and is a poor country and employment is at a premium. Nevertheless, time and time again one meets in Magugu people who had jobs elsewhere that generally paid more than could be earned in small-scale farming. They were dissatisfied with an assured income and chose to become independent farmers at Magugu. The criterion of rational "economic man" obviously is not always a viable one in assessing rationales for changing occupations. Apparently income alone was not enough of an attraction to keep many of the Magugu males from seeking a more independent existence in a community where plenty of land was and is available.

There is a difference between the Magugu settlers and the Mbugwe in the perception of economically exploitable space. The migrants see a vast area available for economic maneuvering—the nation of Tanzania as well as neighboring parts of East Africa. The Mbugwe, however, largely ignore these other areas and stick close to home, within the valley bounded by Lake Manyara to the north, the Dareda Ridge to the south, the Rift Escarpment and the Sangaiwe Hills to the west and east. Yet within this valley region, they exploit a wider area than do the settlers. Mbugwe disperse their farms to take advantage of different soil types and variable water-retention qualities so that in the event of a poor rainy season they will be assured of some harvest. They roam widely over the valley with their livestock and in 1970 were still hunting and gathering on a considerable scale.

Although they often come from afar, the Magugu settlers usually concentrate in specific neighborhoods. They cultivate wet rice, but they have no livestock to speak of. The settlers see the European estates nearby as a source of cash, but the Mbugwe almost never take advantage of wage-earning opportunities in these locations. The Mbugwe seldom cultivate rice. About half of the Kibaoni settlers grow this crop, and other migrants who live closer to water sources and boggy regions between the North Road and the Rift Wall grow rice almost to a man.

To the Mbugwe, all of Mbugwe Division is now Mbugwe land. Their relatives and friends are scattered through the Division area. The Magugu settlers, unlike the Mbugwe, usually do not have social ties throughout the area. The migrants see their farms and nearby homes as the basic social and economic units in their lives.

To the Mbugwe, lineage, clan, and tribe are the units of interaction. These units are dispersed over a wide area. The Mbugwe farmer who works in his distant field has the opportunity to visit relatives and friends during the heat of the day. As a rule, the Magugu settler does not have this opportunity. A farm distant from his dwelling would usually be among strangers, and he would be alone during the middle of the day. He would view this as unappealing, if not dangerous, especially in an Mbugwe neighborhood. No small part of the settlers' hesitancy to move far from home is due to the Mbugwe reputation for sorcery directed against humans and crops. Consequently, the settler stays close to home for sustenance and companionship.

The Magugu settlers often seize upon the subject of Mbugwe clannishness and insularity when denigrating them. This is pointed out as another example of Mbugwe barbarism and ignorance. Conversely, the settlers see themselves as having at least had some experience outside of the Magugu boundaries. Because of this and other factors, they see themselves as more sophisticated than the Mbugwe, who adhere to tribal symbols, locales, and associations.

Although the Mbugwe have more of an appreciation of the valley in which they and the settlers co-reside, the migrants have usually traveled from a greater distance to arrive at their present home. They have more of an appreciation of national boundaries than do the Mbugwe. The travel experiences of the two groups contrast. Magugu settlers have roamed widely over East Africa but stay close together at Magugu. Mbugwe wander all over their valley in tilling their dispersed farms and herding their livestock, but they seldom venture out of Divisional boundaries.

RELIGION AND SOCIAL ORGANIZATION

It is difficult to separate religion from social organization in Magugu, and there are several reasons why religious and social information should be combined and analyzed concurrently. Religion creates a boundary in the Mbugwe-Magugu neighborhood. It traces both variation and unity and forms psychological, sociological, and, in some cases, geographical delineations that set Magugu apart from the surrounding country. Religion is a mechanism of social interaction that serves to reduce ethnic variations and gives some degree of homogeneity to the diversified population of Magugu. Religion creates tensions that further define boundaries.

The Magugu settlers' concept of religion involves inclusion in a corporate group, an organized church, or mosque. This is more important in Magugu than any ideology or philosophy. A Muslim will define his religion as involving belief in God, loving one another, and praying and fasting during Ramadan. The Christians may give the essentially same definition, only excluding the Ramadan fast. Association with an ecclesiastical organization apparently takes precedence over such religious concepts as salvation or world rejection.

A sample of over half of households in Magugu reveals that this is a community in which the overwhelming majority of the population has at least nominal affiliation with a world religion. Muslims constituted 58 percent of the sample, Christians of all sects 31 percent, and pagans the remaining 11 percent. Almost all the pagans were Mbugwe. Only 1 percent were pagans from other tribal areas. One hypothesis to be derived from these proportions is that a multiethnic community such as Magugu is no place to be a pagan.

Magugu people, especially Muslims, are proud of their adherence to a world religion and often equate this affiliation with moral superiority over the Mbugwe. A constant theme in conversation with them about the relative merits of Muslim and Mbugwe is: *Sisi ni watu wa dini* ("We are religious people"). The implication is that the Magugu people belong to a world religion rather than a traditional or tribal religion. A correlate of this is usually some statement such as "the Mbugwe do not have any religion."

This is simply not true, even taking the settlers' criterion of world religious membership as the only canon of religious affiliation. A survey of religious affiliation in an Mbugwe area, the villages of Bondeni, Kerangi, and Makarinya, reveals that 406 Mbugwe out of a total of 503 claimed affiliation with a world religion. Twenty-six were Muslims, and the remaining 380 were Christians. Either the Magugu Muslims are unaware of the high incidence of Christianity among the Mbugwe, or they do not recognize any other world faith except Islam. This attitude of religious exclusion and ignorance is indicative of the antipathy and lack of communication between the two groups.

In Magugu, instead of blood ties, which create vertical, descent group boundaries, there exist horizontal ties, which cut across tribal and kin group boundaries. Through agreement among people of different traditions, an *umma*, or Islamic community, has been created under an all-embracing Islamic tradition. Such an environment, amenable to the creation of crosscutting ties, is a hallmark of the Muslim ethos. Arens noted that old Muslim residents of Mto wa Mbu, another multiethnic community on the northern shore of Lake Manyara, held a similar notion of the founding of their community. According to them, "there was no community in the sense of a group of related individuals until the establishment of Islam."[14]

Although the Magugu community may be looked upon as a microcosm of Islam, which embraces a multitude of peoples and cultures throughout the world, it is still a localized community whose members have worked out networks of interpersonal relationships among themselves. A newcomer must take these networks into consideration and fit himself or herself into them. The mere fact that a new arrival to Magugu is a Muslim does not guarantee him a hearty welcome; rather, his religion might be considered a prerequisite for minimal acceptance.

MUSLIM AND MBUGWE: MUTUAL ANTIPATHIES

A Muslim, a gray eminence, one of the pioneers of the community and a man of influence and wealth vehemently denigrated the Mbugwe in my presence one day. He saw them as worthless savages without any religion and given to the most barbarous practices. His stereotype is one held by most Muslims and a few Christians in Magugu. This antipathy is generally reciprocated by the Mbugwe, who see the Magugu settlers, particularly the Muslims, as being antithetical to their way of life. There is an attitudinal boundary between the Mbugwe and Magugu as well as a spatial one. This boundary is observable in cultural areas such as language, social organization, political organization and loyalties, and settlement patterns. The historical background of the founding and settling of the community—a process of forced accommodation of aliens—places this antipathy in perspective. Similar situations probably exist in other places where ethnically heterogeneous populations of Islamic newcomers have settled among or adjacent to a relatively homogeneous tribal society. Arens observed the same type of situation in Mto wa Mbu insofar as the Muslims of that community stood in relation to the indigenous Iraqw and Masai. The Muslims saw themselves as forward-looking "and in many ways primarily citizens of a new country as compared to the closeby Masai and Iraqw who are seen as relics of a colonial past."[15]

The paucity of the Mbugwe Muslims correlates with the Mbugwe reluctance to live in Muslim neighborhoods in Magugu. They avoid the more concentrated Islamic neighborhoods because they claim that they confront nothing but insults, rancor, and quarreling there. The specific reasons for Mbugwe aversion to Islam and Muslims are perhaps rationalizations of the major political events and processes that have occurred in the area since Magugu was founded. These are often minor points, but when taken together they assume major proportions. Islamic dietary restrictions are often seized upon as evidence of Muslim exoticism, impracticability, and snobbishness. The prohibition against eating animals not ritually slaughtered is impractical to the Mbugwe. Yet such dietary laws are important for the Muslims. Commensality is a sign of cohesion against pagans as well as exclusiveness. Like most prohibitions, they are frequently ignored, and Mbugwe see this as hypocrisy. Muslims drink beer, although some of them preach against it. Mbugwe find separation of men from women in religious ceremonies to be strange and at odds with their own somewhat feminist views. They consider Muslims to be liars and argumentative persons who call upon God to witness the most flagrant untruths and injustices. They see Muslims as torn by factionalism but presenting a united front to unbelievers. This is a shrewd observation of social reality. To the Mbugwe, this is another indication of Muslim attitudes of superiority and intolerance toward any other religion. In 1954, Dierickx observed this same Muslim hostility toward Mbugwe and Christian immigrants from other tribes. "To the suggestion that the Christians be given some sort of representation in the governmental councils of the area, the headman (a Moslem) has replied that if the Christians do not like it they can move."[16]

The Muslims' attitude toward the Mbugwe can be summed up in a bilingual pun. A short Mbugwe greeting is *Ta La La*. The Swahili *kulala* means to sleep or lie down. Hence the Muslim complaint spoken in Swahili, "All they say is *Ta La La, Ta La La*, and that is indeed true. *Watalala siku zote* ("they will sleep all the time"). This succinctly summarizes the Muslim appreciation of the Mbugwe ethos–sleeping, dancing, drinking beer, and singing. The Muslim appreciation of themselves in such comparisons is the opposite of this profligate existence. They see themselves as sober, hard-working, and vastly superior to the Mbugwe in morals and religion.

The stereotype of the Mbugwe as dissolute is also held by government officials. They consider the Mbugwe to be backward in their attitudes toward formal education and national politics. They are also critical of the patron-client relationships that exist between wealthy Mbugwe cattle-owners and people without cattle. The poor herd the cattle of the rich in return for the milk. This is, of course, antithetical to the government policy of socialism and is pointed out as another indication of the gullibility and stupidity of the Mbugwe.

The custom does insure a widespread dispersion of cattle throughout the country. Where these cattle are herded near Magugu farms, there arise claims that they have caused crop destruction. Perhaps a more subtle effect of herding is the presence of individual herdsmen through the environs of Magugu. Whenever a Magugu settler ventures along into the bush for firewood or wild plant foods, he will likely meet an Mbugwe or two with some cattle. These experiences probably reinforce the attitudes relating to the Mbugwe as "people of the bush" (*watu wa porini*), usually up to no good, who skulk around the fringe of more sophisticated and decent folks' homes and farms.

Mbugwe apparel is distinctive and an object of ridicule. Magugu settlers make fun of the Mbugwe's togalike cloths they drape over their heads in cold and dusty weather. The settlers say that the only way to tell an Mbugwe man from a woman is that the men carry spears or cattle sticks. This spear carrying is itself considered a patent expression of hostility toward non-Mbugwe. The government has also criticized Mbugwe traditional dress as being another sign of tribalism and has prohibited such attire for TANU meetings. The Muslim is still allowed to wear his *kanzu* and *kofia* to such meetings, and the Mbugwe believe this to be unfair.

These and other stereotypical antipathies and grievances help to maintain a boundary between the Mbugwe ethnic group and the multiethnic collectivity of Muslim settlers. These attitudes are "boundary-defining cultural differentiae."[17] Barth has suggested that the proper focus for analysis is not the ethnic group but the ethnic boundary. This certainly seems appropriate in the Magugu case. Values are the most crucial aspect of boundary maintenance because values refer to and reinforce ethnic identity. Values also provide a code of conduct to gauge appropriate ethnic behavior.[18] The delineation of boundaries sets apart the groups for analysis. These boundaries are not wholly territorial. Large numbers of Mbugwe live within the Magugu TANU Branch environs. To the Magugu settler, an Mbugwe living in Taifa Njema or Matufa is still an Mbugwe, and stereotypes that apply to larger concentrations of his fellow tribesmen living to the north also apply to him.

Muslims see boundary maintenances as a limitation of interaction created primarily by different religious values. The Mbugwe also see boundary maintenance as based on Muslim religious values, but because these values are external to the Mbugwe ethnic group, they are imperfectly perceived and evaluated.

Although interaction is limited, some does occur. Mbugwe become Muslims and even marry outside their ethnic group. Mbugwe and Muslims also interact in economic and political fields. Yet the boundaries are sufficiently effective to hinder the total integration of the Mbugwe into the Magugu community or the incorporation of larger numbers of Magugu settlers into the Mbugwe ethnic group.

Magugu Christians are affiliated with four sects: the Roman Catholic, Lutheran, Church Missionary Society, and Church of God. Except for the Roman Catholic priest who lives at Mbesi mission, a predominantly Mbugwe neighborhood, the clerics of these sects are Africans and reside in Magugu. Christian pastors and catechists make an attempt to proselytize, especially among pagans. Such missionizing is purportedly absent among Muslims. Christian pastors have had little success in converting Muslims and have generally given up on the idea. There seems to be little overt hostility between Muslim and Christian, although members of both groups will belittle the other in private conversation with their coreligionists.

Christian preachers have not had the success that Islam has had in breaking down tribal ties and in replacing diverse ethnic categories with one overriding religious identity. This appears, in part, due to the pattern of Christian missionaries staking out certain territories in tribal areas during the colonial period. Usually Christian missionaries would tolerate no competition from other Christian sects.[19]

There is some evidence that the Mbugwe look upon their Catholic and Church of God sects as variations of tribal religion. They have not converted to any other Christian sects in any noticeable numbers and in a sense look upon both of these as uniquely Mbugwe faiths. An individual can subscribe to most Mbugwe symbols, marry within the Mbugwe social network, and still affiliate with one of these world religions. When he travels, he can avail himself of the services of his world religion but still remain an Mbugwe.

Such a retention of ethnic identity after conversion to a world religion might occur in the Islamic context when there is a wholesale conversion of a whole ethnic group, such as the Yao. However, a Muslim who traveled away from his tribal home and entered a multiethnic community such as Magugu would probably find that strong ethnic identities had already broken down through the establishment of an Islamic *umma*. That Islam is more effective than Christianity in the merging of ethnic groups in Magugu is supported by the comparison of intertribal marriages among Christians and Muslims. Of a sample of 290 intertribal conjugal households, 233 (80.3 percent) were composed of Muslims. Only 33 (11.4 percent) of the intertribal marriages were between Christians. Almost all intertribal Christian marriages are within the same sect. Interreligious marriages are rarer still if one counts Muslim, Christian, and pagan as each being a single religion (14 households [4.8 percent]). Pagan intertribal marriages are even scarcer–10 households (3.5 percent).

Christianity was usually introduced at the tribal level in East Africa. Foreign beliefs concerning the supernatural were often introduced to a specific group that had definite cultural and often territorial boundaries. The missionary took cognizance of these boundaries and became a specialist in disseminating his creed within these boundaries. He made some attempt to learn the tribal language and customs and specialized in preaching to his ethnic group. To a certain extent, he also tried to isolate that group from other religions, whether they were other Christian sects or Islam. This process made the local group treat the new religion as they would any other foreign complex. The new Christian sect would be subject to manipulation by the tribally oriented converts in order that it would conform to their value system, just as they might manipulate any other borrowed technological item or organizational trait. It is true that this would generate culture change but probably a more localized type of change. This process of Christian incursion is also relevant to Christian sect endogamy in Magugu. People apparently find it easier to marry across religious lines (again, counting each world religion and pagans as a denomination) than across sect lines within the world (Christian) religion. There were thirty-two interdenominational marriages in the sample but only two marriages across sect lines. In each case, both partners in the conjugal household were Mbugwe, and in each case, a Lutheran husband was married to a Catholic wife.

Christians are hesitant to marry across tribal lines. One would expect that in Magugu, where sects are multiethnic, sect endogamy would encourage marriages across ethnic lines within the Christian sects. There has been little evidence for this so far in the brief history of the community. I noted only thirty-three intertribal Christian marriages within my sample, roughly corresponding to the sizes of the respective sects–twenty Catholic, nine Lutheran, three CMS, and two among the Church of God.

Islam in East Africa has generally spread at a nontribal, multiethnic level along trade routes and among migrants and traders. Where Islam entered at the tribal level, it did so with no intent of separating the traditional from the Islamic or group from group. Rather, it opened a doorway to the outside by providing a channel for the migrant to communicate with fellow Muslims in any part of the country. I believe that the comparative data on religion and intertribal marriage in Magugu demonstrate that Islam is more conducive to the formation of supratribal groups than is Christianity. The latter effects channels of religious and social interaction that tend to be more tribally centered and even fosters the inclination to ignore other Christian sects.[20] Islam is more absorptive of diverse groups and traditions; Christianity is absorbed by them. At least, this has been true in

Magugu. Yet in such a community the two religions might develop similar methods of recruitment and interaction and hence become equal forces for social interaction and integration.

As already noted, a multiethnic place such as Magugu is no place to be a pagan. A correlary of this is that the best place to be a pagan is among one's fellow tribesmen. An ethnic group such as the Mbugwe can tolerate religious diversity among its members, but in a multiethnic community such as Magugu, it is advantageous for one to affiliate with a world religion. Among the Mbugwe, kinsmen and other support groups such as neighbors and friends will consider other things besides religion when allying with an individual. In Magugu, where large numbers of such supporters are lacking, association with a world religion might result in social as well as spiritual security. Three hundred twenty-three of 433 pagans in the sample (table 2.1) are Mbugwe living in their own homeland. These data are even more geographically reinforced by a similar breakdown of a sample from several neighborhoods in the area where the largest concentration of Mbugwe live. Here, almost all of the pagans are Mbugwe. Although there were fourteen tribal groups represented in the sample, only 8 pagans from three other tribes were noted out of a total of 157 pagans.

The Mbugwe has several religious alternatives. He can be a pagan and follow his traditional beliefs with little tension or need to switch to another logical system of ordering the unknown; or he can convert to a world religion that he can manipulate as required. The fact that an Mbugwe is a pagan is no impediment to his participation in most aspects of Mbugwe society and culture. It is true that the Mbugwe do marry along religious lines within their ethnic group. This might be construed as a tendency toward religious isolation or the setting up of group boundaries such as those that exist in Magugu. I do not believe this to be the case. Although Mbugwe marry along sect lines, relatives and friends belong to different religions. Generally, the more elderly Mbugwe, those over forty years old, are more likely to be pagans. Thus, most Mbugwe under forty have parents, grandparents, aunts, and uncles who are pagans and with whom they have customary and affectionate relationships. At the moment, ethnicity seems to have priority over religion. One informant, a Catholic (very nominally) summed this up: "I have to live among my own people. I will be a Christian as well as I can and as long as it does not interfere with my life as an Mbugwe."

The Mbugwe create ties of marriage along religious lines, yet other ties of kinship, neighborhood, and friendship cut across these religious categories. They view their boundary as an ethnic one, and recruitment to their group is mainly a matter of birth. The Magugu Muslims also marry along religious lines, but their ties of kinship and affinity tend to cut across ethnic lines. The boundary of their community is one created by religious

TABLE 2.1

INCIDENCE OF INTERTRIBAL MARRIAGE AMONG SELECTED TRIBES

This table shows the total number of conjugal households in which members of the thirteen largest tribal groups are residing and the rates of intratribal and intertribal marriage. The latter category is broken down to indicate the percentage of both male and female outmarriage

		Percentage of Moslem in Total Sample	Intratribal Households	Intertribal Households			Total of Conjugal Households
				Total	Men	Women	
Large Tribal Groups	Mbugwe	(13%)	202 (73%)	75 (27%)	9 (3%)	66 (24%)	277
	Iramba	(79%)	66 (53%)	58 (47%)	40(32%)	18 (15%)	124
	Irangi	(94%)	27 (26%)	75 (74%)	21(21%)	54 (53%)	102
	Nyaturu	(88%)	47 (51%)	45 (49%)	19(21%)	26 (28%)	92
Small Tribal Groups	Nyamwezi	(85%)	3 (9%)	32 (91%)	22(63%)	10 (28%)	35
	Sandawe	(63%)	8 (25%)	24 (75%)	5 16%)	19 (59%)	32
	Gogo	(60%)	7 (28%)	18 (72%)	13(52%)	5 (20%)	25
	Iraqw	(50%)	4 (21%)	15 (79%)	4(21%)	11 (58%)	19
	Makua	(92%)	2 (14%)	12 (86%)	7 (50%)	5 (36%)	14
	Rundi	(55%)	2 (15%)	11 (85%)	9(70%)	2 (15%)	13
	Isanzu	(54%)	0 (0%)	11 (100%)	7(64%)	4 (36%)	11
	Kimbu	(99%)	2 (20%)	8 (80%)	3(30%)	5 (50%)	10
	Sukuma	(92%)	0 (0%)	9 (100%)	8(89%)	1 (11%)	9

priority. Recruitment to this group is a matter of conversion and migration as well as birth. Because the priorities of ethnicity and religion are reversed in the respective groups—Mbugwe and Magugu Muslims—the same structural principle of creating and maintenance of crosscutting ties results in different conceptions of boundaries.

SOCIAL ORGANIZATION

Some salient aspects of Magugu social organization are: a tendency toward affiliation with a conjugal household, the creation of crosscutting ties across tribal lines, the development of personal networks of affines, kinsmen, and friends, and the absence of large descent groups.

The 1967 census of Magugu listed 6,888 people in 2,105 houses with an average of 3.3 persons per household. Because this material was not wholly suitable for sociological analysis, an assistant and I took a sample of 1,186 houses containing 3,958 people (58 percent of the official population and residences). This was a house-by-house count of sections of Magugu along the Great North Road between the southern and northern boundaries of the community. This sample cuts across the more typical multiethnic neighborhoods such as Kibaoni as well as some predominantly Mbugwe neighborhoods. Married couples (63 percent of the sample) occupied 748 of the houses, in which resided 2,800 people (71 percent of the sample). In the composition of these conjugal households and in the composition of the community as a whole, there is noticeable multiethnicity. There were eighty ethnic groups represented in the sample.

The considerable incidence of outmarriage in Magugu can be considered a definite indication of social integration, a reduction of ethnic variance within the community, and a trend toward homogenization of the variegated population. There is nothing unusual about this in Black Africa. The process of incorporating new members into more or less homogeneous tribal groups has been mentioned throughout ethnographic literature.[21] Goody suggests that such a flexible marriage policy as exists in sub-Saharan Africa (that is, absence of rigid endogamy) leads to a mutual adjustment of domestic relations and results in a kind of "bilateral" system found in centralized states such as the Hausa, Nupe, Dagomba, and Lozi.[22] This is appropriate to the Magugu social organization and the configuration of the conjugal households. The lowest common denominators in Magugu are explicitly the religious environment and implicitly the proximity of diverse groups of people occupying an ecological niche. Religion, ethnicity, and neighborhood interact to bring about an organization that accommodates the variegated population. The process by which this is brought about is primarily outmarriage, or intertribal marriage. This is to be expected in such a community, especially where small ethnic groups are concerned, groups that in themselves might not be populous enough to provide a marriage pool for all their members. Larger groups whose members have the opportunity to marry among themselves would be expected to show a lower incidence of this type of marriage.

Looking at the thirteen largest ethnic groups in the sample, we can see that except for the Mbugwe, whose limited interaction with the rest of the community has already been discussed, the groups are predominantly Islamic (table 2.1). In smaller groups, intertribal marriage is the norm. The larger groups, no doubt due to sheer numbers as well as the ethnic boundary between the Mbugwe and the rest of the community, provide a larger pool for wives than do the smaller groups. There is a tendency to marry within the large groups since this is the line of least resistance. Irangi women marry out more than do women from other groups. They demonstrate how larger ethnic groups can provide wives for other ethnic groups as well as their own. In the nine smaller groups, marriages are overwhelmingly intertribal. A man or woman from a small ethnic population must either marry out or not marry at all.

There is a strong patrilineal bias in reckoning tribal affiliation. With this patrilineal principle observed, male outmarriage might be looked on as recruitment to an ethnic group and female outmarriage as a provision of recruits to another group. I do not feel that this is the case in Magugu. I suggest that women are links in personal networks of kinsmen and affines that actually break down ethnic boundaries. Tribal boundaries are submerged, in the individual's view, under a smaller, more immediate group of people related to him by marriage, by patrilateral and matrilateral kinsmen (who probably will be few in Magugu), neighbors, and friends. These networks represent a practical solution to the problem of support in a community where all consider themselves to be strangers.

The ethnic group, especially if it is a large one, can be important for introducing a new arrival into the community. When individuals or married couples arrive at Magugu, they will probably first settle near or with relatives from their tribal homeland. If they are unattached males who are unable to find a wife among their fellow tribesmen, they are forced to marry out of the ethnic group. They must generally look to their coreligionists to provide wives. This broad, nontribal category of individuals holds certain "common denominators" of religion, proximity, and the communication circuit of the Swahili language.

The high proportion of multiethnic households indicates the lack of emphasis placed on tribal alliances in Magugu. Tribal loyalties do exist, but as has been stressed, religion is a more important consideration than ethnicity in the community as a whole, especially among the Muslims. The settlers are very much aware of this priority and often refer to themselves as *makabila mbalimbali yameyochanganyika*—various tribes that have become mixed together and confused.

Tribal affiliation along patrilineal lines is the rule except where large numbers of matrilineal Mbugwe reside. In many cases this tribal affiliation is patently superficial and is just one of several means of identification and affiliation. An extreme case of superficial ethnic identification is noted in the household occupied by an Mbugwe woman, her Nyamwezi mother, one Irangi child, and one Sambaa child—four ethnic groups represented in one four-member household. The Nyamwezi woman married an Mbugwe man who died after the birth of her daughter. The Mbugwe ordinarily reckon descent matrilineally, yet the daughter used a patrilineal descent ideology to affiliate herself with a matrilineal tribe. The daughter in turn had children by men from different ethnic groups, and these children are identified with their respective fathers' ethnic groups.

The remaining 438 households (37 percent) in the sample were composed as follows:

single men—161 (13 1/2 percent)
single women—98 (7 1/2 percent)
woman with a child or children—61 (5 percent)
miscellaneous—122 (11 percent)

These are alternatives to the conjugal household. They represent individuals who have come to Magugu alone, former dependents or guests of a couple who have yet to begin forming their own conjugal arrangement, or the breakup of a conjugal household through divorce or death of one of the partners. The latter case is the norm with single women with children. Under the miscellaneous category, I have included such arrangements as a man supporting his mother; brothers with their mother; a man with his grandchildren; two brothers; single men with their children; two sisters; and a woman with older, dependent kinsmen. I believe that these households are significant in the social integration of Magugu because very often it is more important for these individuals to develop a personal network of supporters than for the more self-sufficient conjugal arrangement.

The Muslims refer to the Mbugwe as people without a religion. This statement is without substance and an indication of the boundary between the two groups. The Mbugwe also have a distorted view of the Magugu people. They refer to them as people without kinship. It is understandable why the Mbugwe feel this way about the settlers. The Mbugwe consider kinship to connect the individual to a large number of matrilineal clansmen. They accurately note the absence of clans in Magugu and consider the inhabitants to lack an essential level of social organization.[23] There is kinship in Magugu, but it is by no means the crucial factor in the order and defining of groups as in a traditionally oriented ethnic group such as the Mbugwe.

Magugu social organization is not a rigid structure that ties together several generations in vertical relationships along unilineal lines. Rather, it is a system of alternatives for the newcomer and resident to utilize as it suits him. These alternatives vary with the marital, religious, and tribal orientations of the migrant. If he is from a tribe whose members are represented by large numbers in Magugu, he might immediately avail himself of this support. Relatives can be important in introducing newcomers into the community. Religious affiliation of the newcomer is important, for it can open wider supratribal associations with other residents, particularly in the Islamic context. If the newcomer is from a tribal group not widely represented in Magugu, he is forced to place priority upon nontribal ties.

Except for the Chagga, Barabaig, and Mbugwe, groups in Magugu are nontribal associations formed on religious premises. Yet even the tribally oriented people interact with each other and sometimes cross their respective ethnic boundaries.

THE SWAHILI LANGUAGE: THE PREREQUISITE FOR ETHNIC BOUNDARY CROSSING

The Swahili language is a communication system that gives the migrant alternatives for maneuvering in Magugu. During the course of my field study, I made a language survey. The groups surveyed were approximately half of the population of Kibaoni and the majority of the population of Bondeni, an Mbugwe settlement a few miles north of Kibaoni. Although this is an Mbugwe neighborhood, there are fourteen other tribal groups represented in its population. Kibaoni is the more linguistically heterogeneous community, with forty-six African languages spoken. The total number of individuals in the Kibaoni survey is 597. There were 505 in the Bondeni survey.

In the whole survey of 1,102 people, only two did not speak Swahili. In Kibaoni, 25 percent of the chilren speak only Swahili (83 of 330). In Bondeni, 4 1/2 percent of the children speak only Swahili (13 of 295). It should be noted that this 25 percent monolingualism should be looked upon as a conservative estimate rather than an accurate figure. The data were collected from adults, usually the parents of the children. These parents were often carried away with enthusiasm when asked to assess the linguistic competence of their children.

There is a strong contrast in the mobility of the populations of Bondeni and Kibaoni. Only 12 percent of the Kibaoni adults were born there (30 of 267). This is in dramatic contrast with Bondeni, where 62 percent of the adults were locally born (131 of 210). This comparison reflects impressions gained from other data. Geographical mobility—the experience of moving across tribal and regional boundaries—is more congruent with developing a receptivity to national ideals than is remaining in the region of one's birth. At the very least, the physical horizons of the individual are widened, and he is taken out of a group that might be competitive with national interests. Such mobility also implies more recourse to a national language. To maneuver out of one's tribal area, one must often communicate with other groups through the use of Swahili. The settlers' travel experiences surpass the movement of the insular Mbugwe.

In Kibaoni there is a tendency for the adult settlers to retain the language of their tribal homelands and use Swahili for everyday communication. However, in Bondeni, there is also linguistic incorporation into the Mbugwe. The number who speak Kimbugwe in addition to the national tongue and their own tribal language indicates that because of the more homogeneous ethnic makeup of Bondeni, there is a tendency to use the local language. In Kibaoni, less than 14 percent speak Kimbugwe. In Bondeni, 55 percent, or 365 people speak the local vernacular. Of these Kimbugwe speakers, only 107 are adults who claim affiliation with the Mbugwe ethnic group. The 258 remaining speakers are adults from other tribal groups who have learned Kimbugwe since they moved to Bondeni (10), children whose parents are both Mbugwe (81), children with one Mbugwe parent (89), children of non-Mbugwe (47), and people who claim affiliation with other tribal groups but who were born in Bondeni (31). There is a similar linguistic incorporative process under way in both communities. In Kibaoni, there is a tendency for children to learn the national language to the exclusion of tribal vernaculars. Indeed, the local language and the national language are the same. Local and national communication circuits are congruent. In Bondeni, there is a tendency for non-Mbugwe children to learn the local vernacular in addition to the national language. In brief, in Kibaoni, vernaculars are lost rather than learned, but in Bondeni, Kimbugwe is adopted by immigrants and is also the vernacular of children whether or not their parents are Mbugwe.

The people of Kibaoni are very much aware of the role of Swahili in giving their community a degree of cohesion in spite of a multiethnic and multilingual population. They also see Swahili as being related to a political process different from the former, parochial Mbugwe hegemony, that is, a connection with citizenship rather than tribalism. The people of Kibaoni and the other multiethnic neighborhoods of Magugu have closer ties to the national government than do the people of Bondeni. In Kibaoni a significant number of young people have not acquired any other language but the national tongue. Of necessity, their interests are attached to pan-tribal or nontribal linguistic symbols, since they cannot communicate with any other linguistic communities except through the medium of the national language. They are forced to be part of the national communication circuit. Monolingualism in the national tongue denies them participation in activities that require a knowledge of a tribal language. Among the activities denied them is tribal politics. The national orientation also applies to Magugu aults. Although they still retain their tribal language, while they reside in Magugu they must use the national language most of the time, since their contacts are so often with people from other ethnic groups.

The incidence of Swahili varies between the two communities. In Kibaoni, the only way that individuals from two different language groups can communicate is through the national tongue. However, in Bondeni and in other Mbugwe neighborhoods, Kimbugwe is spoken whenever possible. Even Mbugwe drinking in Kibaoni beer hall will use their own language rather than Swahili. At times deliberately, at times unconsciously, they retreat behind an ethnolinguistic wall.

O'Barr noted that the Asu-speaking Pare consider their language to be a residual one "continuing to be used where Swahili has yet to make an incursion." The Pare say "that Swahili is an introduced language and serves in introduced situations, that Asu is the indigenous language and serves in traditional situations."[24] This is comparable to Mbugwe attitudes toward their own language. Swahili is associated with nation rather than tribe, with government and social service, schools and literacy, foreigners and commerce, a general enlargement of social and valuational scale rather than parochial interests. Swahili has the weight of government approval and encouragement and is the dominant language in the neighborhood. Nevertheless, this study indicates that in certain situations, as in Bondeni, where a few outsiders have settled in an ethnically homogeneous area, the local vernacular remains important in interethnic communications. In Magugu, Swahili has become the language for every aspect of the life of what I would call a "national" community to distinguish it from the tribal community of the Mbugwe, who prefer to use their own vernacular whenever possible. In any case, language is important to consider when discussing ethnicity, since it is often crucial to the assemblage or division of groups.

TRIBE AND NATION

In Magugu, just as people speak Swahili to form a circuit that overcomes the noise of diverse languages, so must they adhere to a political system that is nontribal. This system is of necessity a national one, and political identification concerns citizenship rather than tribe *while a person resides in Magugu.*

Cohen and Middleton have taken Fried's thesis[25] that ethnicity as a distinct group identification does not develop on its own but in reaction to a wider political membership and hypothesize that "the greater the pressure on a group from a surrounding hegemony (or hegemonies) then the greater is the possibility that such a group, no matter what its own multiethnic history, will form into a newly emergent ethnic unit in reaction to such pressure."[26]

The foregoing hypothesis is relevant to both Mbugwe and Magugu. The Mbugwe seem to coalesce their factions in reference to outsiders and present a united front. This is also true for the Muslim population of

Magugu in reference to outsiders. The Mbugwe and Magugu settlers have acted upon each other to develop more respective solidarity. A dimension of this solidarity is found in the political field. The Magugu settlers identify with the nation; the Mbugwe also identify with the nation but have the capacity to identify with the pre-nation or tribe. This generalization about the current (1969-1970) state of affairs implies that the Magugu community concerns itself with citizenship rather than tribalism. Mbugwe are more concerned with relationships within their ethnic boundaries rather than a whole network of national myths and power complexes.

The bonds that hold people together in a centralized state are differentiated and ranked, and this ranking places the authority system itself very high so that the legitimate set of offices in the society becomes more important than the sociocultural features.[27] In Magugu, the TANU structure became the closest thing there is to a social class, and the political offices involved are the most prestigious in the community.

Citizenship involves an awareness of the diverse levels of interaction that can take place within a territorial boundary. This awareness takes into consideration the cultural variety within the territorial boundaries and is a common denominator of interaction at the political level.[28] A citizen may ignore many of the requisites for tribal membership and still retain his identity with the national government. Most of the population of Magugu conceives of this type of political identity because, for the most part, tribal politics is not available nor appropriate in such a polyethnic locale. Citizenship is more highly valued there and is more often referred to than is allegiance to a particular tribal polity. I do not propose that once a person identifies with citizenship, this identification is irreversible or permanent; I merely propose that in Magugu, it is the appropriate one. There, the TANU authority supplants any other authority structure that might compete with it. This is not wholly the case for the Mbugwe. The Mbugwe are theoretically within the TANU organization, and their whole territory is divided into ten household cells and supervised by a TANU chairman. Nevertheless, they also possess a traditional structure that vies with TANU in some respects. In the field of traditional law, Mbugwe might resolve conflicts without ever consulting the governmental legal process, whereas among the Magugu settlers a conflict *must* be referred to the cell leader, who refers it up the chain of command until it reaches the Division court. The lack of any one corpus of tribal law forces the usage of the national legal apparatus for conflict resolution. Traditional leaders of the Mbugwe have had conflicts with the TANU officials, and these leaders still have sympathetic followers. Mbugwe loyalties to the government are not as a rule fervent. This is not merely my own opinion but also that of government officials in the area.

Magugu and similar multiethnic communities throughout Tanzania might serve as national way stations to people more closely tied to parochial interests and places. At these way stations, they are more intensely subjected to national values; there they find national services are more easily available. I do not infer that because the people of Magugu reside in a national community, they are ardent nationalists. They are really somewhat apathetic about politics. Very few of them attempt to manipulate the governmental structure for their individual aggrandizement. They even have to be prodded to exercise their prerogative of voting in national elections. In spite of this apathy, the structure and process of politics in the area does provide an overriding pattern and identity that develops and formulates a degree of cooperation.

TANU meetings are another aspect of social integration in Magugu. The TANU organization provides a chain of command for the relaying of messages of an integrative nature as well as physical amalgamation of the inhabitants into a unified whole. The meetings themselves are the most visible evidence of this political unity. When a citizen attends such a meeting, he is involved in the political field and the national consciousness. When complaints are made, government directives published, problems that relate to the whole community discussed—in brief, when matters of public and therefore political interest are discussed—then it can be said that the resident of Magugu is conducting himself as a citizen of a nation-state rather than a member of a tribe.

There is now a different type of community in the area, one that contrasts with the traditional organization of the Mbugwe. There are now two models of society in an area where formerly there was one. The national type of society that Magugu approaches in structure and ideology provides by its very presence an alternative style for those Mbugwe who desire or are forced to be incorporated into it. The presence of this national community also places some pressure upon the Mbugwe; their traditional organization and values are considered eccentric and undesirable in the national context. Magugu is also in part structured by responses to pressures from the Mbugwe society.

In sum, a multiethnic community created by fiat of the colonial government has developed in an environment of supratribal political identifications. Language, politics, marriage, and religion provide axes for social and cultural integration. Social change is occurring along these axes of integration and is indicated by change in statuses. These statuses tend to be directed toward achievement rather than the largely ascribed statuses that comprise the ethnic affiliation of the Mbugwe and other tribal groups. This change is moving toward the diminution of ethnic associations and the increase in emphasis on national ones. Barth has stated that ethnic statuses are superordinate to other statuses. They are similar to sex and rank in that ethnic status constrains the incumbent in all his activities, not only in defined social situations. "One might thus also say that it is an *imperative*, in that it cannot be disregarded and temporarily laid aside by other definitions of the situation."[29] I do not believe that this applies to many of the Magugu settlers. They *do* set ethnic identity aside while they are in Magugu. Perhaps it is wrong to call ethnicity a status; rather, it is a complex set of integrated roles. These roles are deemphasized when the individual leaves his ethnic field of interaction and adopts additional roles necessary for maneuvering outside the ethnic field.

NOTES

1. I carried out research in Magugu from September 1969 to August 1970. This was funded by a fellowship and research grant from the National Institutes of Mental Health (PHS Grant No. 1 F01 MH39425-01A1). I gratefully acknowledge this aid and that of the various agencies of the Tanzanian government that smoothed the way.

2. Much of my understanding of the Mbugwe derives from lectures and conversations with Robert F. Gray, who has been studying these people since 1950. I alone am responsible for any distortions that might arise in this essay.

3. Philip H. Gulliver, "Introduction," in *Tradition and Transition in East Africa*, ed. Philip H. Gulliver (Berkeley: University of California Press, 1969), pp. 5-7.

4. Aidan W. Southall, "The Illusion of Tribe," in *The Passing of Tribal Man in Africa*, ed. Peter C. W. Gutkind (Leiden: E. J. Brill, 1970), p. 44.

5. Frederick Johnson, *A Standard Swahili-English Dictionary* (Oxford: Oxford University Press, 1967), p. 164.

6. Charles Wallace Dierickx, "Magugu: Population and Land Use in a Resettlement Project in the Northern Province of Tanganyika" (Ph.D. diss., Northwestern University, 1955), pp. 1-34.

7. Robert F. Gray, "A Sleeping Sickness Settlement," (Newsletter for Institute of Current World Affairs, April 27, 1955; mimeographed), p. 3.

8. A. J. Temu, "The Rise and Triumph of Nationalism," in *A History of Tanzania*, ed. Isaria Kimambo and A. J. Temu (Nairobi: East African Publishing House, 1969), pp. 207-208.

9. Henry Bienen, *Tanzania: Party Transformation and Economic Development* (Princeton, N.J.: Princeton University Press, 1967), p. 42.

10. United Republic of Tanzania, Ministry of Agriculture, Monthly Reports, Minutes of the Coordinating Committee, Arusha Region, April 18, 1964. Accession Number 305, File A/MR/R/AR, National Archives, Dar es Salaam.

11. Immanuel Wallerstein, "Migration in West Africa: The Political Perspective," in *Urbanization and Migration in West Africa,* ed. Hilda Kuper (Berkeley: University of California Press, 1965), p. 156.

12. Dierickx, "Magugu," pp. 133-134.

13. Philip H. Gulliver, *Labour Migration in a Rural Economy* (Kampala: East African Institute of Social Research, 1955), p. 29.

14. William E. Arens, "Mto Wa Mbu: A Study of a Multi-tribal Community in Rural Tanzania" (Ph.D. diss., University of Virginia, 1970), p. 48.

15. Ibid., pp. 62-65.

16. Dierickx, "Magugu," p. 141.

17. Fredrik Barth, "Introduction," in *Ethnic Groups and Boundaries: The*

Social Organization of Culture Difference, ed. Fredrik Barth (Boston: Little, Brown and Company, 1969), p. 38.

18. Ibid., pp. 11-16.

19. For an example of some of the intrigue involved between rival Christian missionaries, see Betty A. Gray's account of Lutheran resistance to Catholic incursion among the Sonjo. Betty A. Gray, *Beyond the Serengeti Plains: Adventures of an Anthropologist's Wife in the East African Hinterland* (New York: Vantage Press, 1971), pp. 137-147.

20. J. Spencer Trimingham, *Islam in East Africa* (Oxford: Oxford University Press, 1964), pp. 54-56.

21. For an excellent introduction to this process of absorption in sub-Saharan Africa, see Ronald Cohen and John Middleton, eds., *From Tribe to Nation in Africa: Studies in Incorporation Processes* (Scranton: Chandler Publishing Company, 1970).

22. Jack Goody, "Marriage Policy and Incorporation in Northern Ghana," in ibid., p. 126.

23. I do not infer that Magugu settlers might not choose to affiliate with a clan under the proper circumstances. Several settlers stated in their life histories that they returned to their tribal homelands at some point in their lives to take part in clan ceremonies.

24. William M. O'Barr, "Multilingualism in a Rural Tanzanian Village," *Anthropological Linguistics* 13, no. 6 (June 1971): 295.

25. Morton H. Fried, *The Evolution of Political Society: An Essay in Political Anthropology* (New York: Random House, 1967), p. 170.

26. Cohen and Middleton, *From Tribe to Nation in Africa.*

27. Ibid., p. 17.

28. Ibid., pp. 15-18.

29. Barth, *Ethnic Groups and Boundaries,* p. 17.

SELECTED READING

Arens, William E. "Mto Wa Mbu: A Study of a Multi-Tribal Community in Rural Tanzania." Ph.D. dissertation, University of Virginia, 1970. An analysis of a multiethnic community about sixty miles from Magugu. Although Mto Wa Mbu is smaller than and quite different in composition from Magugu, it does present some structural similarities.

Cohen, Ronald, and Middleton, John, eds. *From Tribe to Nation in Africa: Studies in Incorporation Processes.* Scranton, Pa.: Chandler Publishing Co., 1970. The editors and contributors treat interethnic contact and absorption of foreigners into tribal groups. Especially valuable are the editors' Introduction and Elliott P. Skinner's article, "Processes of Political Incorporation among the Mossi."

Gray, Robert F. "Medical Research: Some Anthropological Aspects." In *The African World: A Survey of Social Research,* edited by Robert A. Lystad. New York: Praeger, 1965. Gray singles out Magugu in discussing the etiology of sleeping sickness and some of the methods of combating it.

Knappert, Jan. "The Function of Language in a Political Situation." *Linguistics*

39: 59-67. Stresses the political aspects of language choice with special attention to the linguistic problems of new nations.

Nyerere, Julius. *Ujamaa: Essays on Socialism.* Dar es Salaam: Oxford University Press, 1968. These essays by the president of Tanzania present the political ideology with which Magugu residents were being indoctrinated during the period of my field work.

Whiteley, W. H. *Swahili: The Rise of a National Language.* London: Methuen and Company, 1969. The history of Swahili and its development as the dominant tongue of Tanzania.

3
Ethnic Tensions and Political Stratification in Uganda

Ali A. Mazrui

Ethnocracy is basically a political system based on kinship, real or presumed. It can take the form of an ethnically exclusive state, in which citizenship is basically governed by biological descent. Or, it can take the form of an ethnic division of labor—with, say, Baganda for administrators and Nilotes for soldiers. Third, it can be based on quantified ethnic balance—for example, so many Nilotes in the civil service to balance so many Baganda in the same service.

When the British arrived in Uganda, they found a number of societies, some of which had state structures. Among the more developed, and certainly one that developed even further under colonial rule, was Buganda. Bunyoro was also a highly structured polity. The concept of citizenship in these societies was inseparable from the concept of kinship. All the Baganda were deemed to be descended from a single ancestor. The state rested on a principle of political consanguinity, a presumed descent from a shared forefather. The polity was ethnocratic.

In its most literal sense, consanguinity implies a blood tie, but in fact there were other ties connected either with marriage and adoption or with cultural assimilation.

New citizens of an African society did not become full citizens until they mixed their blood with the original members of that society or adopted more fully the language and culture of that society. Biological intermingling and cultural assimilation were the most effective ways by which foreigners could enter the mainstream of African citizenship.[1] President Amin's response to the cultural and sexual exclusiveness of the Asians of Uganda rested in part upon a primordial African conception of true citizenship. The Asians, by being distant culturally and by being reluctant to mingle their blood with black Ugandans, remained alien by this criterion. They could no more become Ugandans than an Acholi who refused to intermarry with the Baganda and resisted the adoption of the Ganda cultural ways could ever be a Muganda.

Among communities to the north, on both the Sudanese and Ugandan sides of the border, political ideas sometimes went to the extent of regarding all those who were not kinsmen as basically potential enemies. The distinction between a foreigner and an enemy could be very fine indeed. This did not mean that the groups automatically attacked strangers and foreigners who came in contact with them. But it did mean that they regarded them with the reserve and deep suspicion usually accorded to traditional enemies.

It is possible to be adopted as a kinsman fictionally or be given protection in terms of presumed kinship, but the stranger is then expected to behave as if he were a kinsman, permitting himself to be assimilated into the system of rights and duties of that society. As Evans-Pritchard wrote of the Nuer:

> If you wish to live among the Nuer you must do so on their terms, which means that you must treat them as a kind of kinsman and they will then treat you as a kind of kinsman. Rights, privileges and obligations are determined by kinship. Either a man is a kinsman, actually or by fiction, or he is a person to whom you have no reciprocal obligations and whom you treat as a potential enemy.[2]

East Africa as a whole betrays some ethnocratic tendencies that go back to these primordial conceptions of citizenship. What should be remembered is that ethnocracy in East Africa has both colonial and precolonial antecedents.

ETHNOCRACY: THE COLONIAL BACKGROUND

The colonial antecedents of ethnocratic tendencies include the consequences of racial and sometimes tribal stratification under British rule.

In the days of the legislative councils in East Africa, the most important cleavages in the electoral process were ultimately *racial* rather than "tribal." Kenya especially had a heavy European presence in the political and economic system and a significant Indo-Pakistani presence in the economy. Representation in the legislature was apportioned by race—European, Asian (Indian and "Muslim"), Arab, and African. Tanganyika and Zanzibar each also had an elaborate system of representation by races. Uganda combined the racial principle of representation with the beginnings of ethnic tensions among Africans themselves.

Since the entire colonization process was tied in with issues of race and ethnicity, the colonial legislative council was a school for ethnic politics from the start. As independence approached, the nature of ethnic politics began to shift away from "race" as the ultimate line of cleavage to "tribe" and region. (We use *ethnicity* in this chapter to refer to both "racial" and "tribal" identities.) Kenya and mainland Tanzania gradually ceased to be described as multiracial societies—even though the numerical inferiority of the few non-Africans was for a while disguised by their economic and political power. But with the expansion of the franchise and the coming of African majority rule, the political system became primarily multitribal rather than multiracial. The colonial legislative councils matured into the sovereign national assemblies. But the tradition of ethnic pluralism as a fundamental factor of electoral behavior now characterized the black majority itself and remained a major political variable in the life of independent East Africa. It had all the potential of ethnocratic evolution.

Both Kenya and Uganda have "heartland tribes." We define a "heartland tribe" as a community located relatively near the capital city, large numerically, politically active, and historically important. The heartland tribe of Kenya is the Kikuyu; the heartland tribe of Uganda is the Baganda, and the territory they occupy is Buganda.

The Baganda were privileged during the colonial period. Under the 1900 agreement, the British gave the Baganda considerable autonomy under the rule of their king, the Kabaka. The Baganda also responded well to the stimulus of acculturation and became among the best educated and most affluent of all East African communities. They increased their political preeminence over other communities. Their status sometimes resulted in an ethnic division of labor in Uganda—since they performed the more prestigious functions of the colonial polity. Later on, the legacy was to result in a system of quantified restoration of ethnic balance. Buganda was long the center of the society, while northern Uganda was clearly part of the periphery. Using Maoist terms, Buganda was the city writ large, while much of the rest of the country was functionally rural in this special sense of differential development, imbalance in the distribution of industries, and exploitative relationships. The Baganda were at the top of the emerging system of ethnic stratification. They were also at the center of an emerging ethnocracy.

For some time in its history, Buganda even assumed the role of a subimperial power, collaborating with the British in controlling and ruling significant parts of the rest of the country. Buganda provided many of the administrators for British rule assigned to different parts of the country. And through much of the colonial period the Baganda were clearly the heartland community of the country, displaying an impressive responsiveness to the stimulus of the new educational and cultural skills that came with the imperial power and European missionaries. The region became the best developed economically, the best educated, the best integrated through a network of communications, and the most influential politically. As Kampala evolved into the capital city, Buganda developed into the capital region.

As Buganda became more urbanized and consolidated its centrality in national affairs, it also became demilitarized. When the British arrived in that part of the world, the Buganda kingdom had been militarily one of the most powerful in East Africa as a whole. It had evolved impressive political and social institutions and had developed systems of collective organization that converted Buganda into an impressive military force.

The 1900 agreement between Buganda and Britain inaugurated a new era. The agreement itself put special limitations on numbers and types of arms that the king of the Baganda could acquire or keep for the protection of the palace. But beyond that, the process of demilitarizing the Baganda had got under way. The military profession, which had been one of honor and commitment, began to lose some of its luster. The 1900 agreement helped to shift the Baganda from a conception of national autonomy based on military might to a conception of their autonomy based on contractual rights. Over time they learned how to exploit effectively the terms and implications of the 1900 agreement. They learned how to use the courts with sophistication in pursuit of their rights against the British. Militarily they were of course no match for British military technology. The Baganda now realized that their ultimate weapons against the British were legal and political. The profession of arms was now left to "lesser" ethnic communities.

The British themselves also had a vested interest in the demilitarization of Buganda. British military recruitment turned to other areas, thus reinforcing the Baganda's own increasing inclination to look for alternative avenues of honor, income, and achievement.

But if Buganda was becoming a less promising area of recruitment for the King's African Rifles, where else should the British turn for those recruits? There were many alternative areas. The British could have turned to other Bantu areas of the country. To some extent, they did just that. But the Bantu areas were specially susceptible to the demonstration effect of Buganda's ways. Buganda's system of administration and cultural styles was to some extent emulated in the other Bantu areas of the country. Certainly the other areas of the kingdoms of Uganda, all basically Bantu, displayed a marked tendency to use Buganda as a reference point, if not as model. The demilitarization of Buganda was followed by a demilitarization of the other kingdoms.

Once again the British themselves had a vested interest in helping the demilitarization of the kingdoms. They assumed that those African societies that had been politically organized as states before they came were a greater military risk, once subjugated, than those African societies that were segmentary and politically acephalous. Buganda and Bunyoro, as kingdoms that had been particularly strong upon the arrival of the British, were regarded by the British for some time as potential military risks of a specially ominous kind.

The northern tribes of Uganda, since they were less centralized in their political organization, emerged as safer areas for military recruitment into the colonial armed forces. In reality the British had met upon arrival some resistance in parts of the north. And the record of northerners as fighters and warriors was already established. But northern political organization was such that the societies had collections of individual warriors, rather than units of organized armies. The two factors together made the north more attractive as a recruiting area for British colonial armed forces. The *individuals* so recruited were believed to have good martial qualities, but the societies from which they came were often not centralized enough to raise the threat of organized armies of resistance under the banner of a single political authority.

The relatively segmentary Nilotic and Sudanic communities of the north were already becoming politically peripheral in Uganda. But their very status as a political periphery made them attractive for military recruitment into the imperial armed forces.

What happened once again was an interplay among political, cultural, and economic factors in converting the rural areas of northern Uganda into a preeminently suitable source of recruits into the Ugandan army. The foundations of a fundamentally different system of ethnic division of labor were being laid.

Just as the British had made assumptions about the martial prowess of the Gurkhas and Panjabis, so they made assumptions about the martial prowess of the Nilotic and Sudanic peoples of northern Uganda.

An additional cultural factor was the interplay between food culture and physical anthropology. Eastern and western Nilotes and Sudanic tribes produced a disproportionate number of men who were tall and slim. This particular physique was interpreted in the colonial period as additional evidence of military suitability. The "tall and lean" were regarded as "good drill material."

Food culture over generations could have influenced the emergence of lean physical specimens, especially among communities that were truly pastoral. Reliance on milk and meat as the staple food, with periods when almost nothing else was added to the diet, had its impact on physical anthropology. Millet among other Nilotes was seen by the communities themselves as a diet fundamentally more relevant to physical strength than the *matoke* ("plantain bananas") of some of the Bantu tribes.

But whatever the relevance of food culture to physique, there is little doubt that the recruitment officers of the imperial power in Uganda came to look at Nilotic and Sudanic communities as being physically better "drill material" than most of the people of the Bantu kingdoms. In Ankole the ruling elite was sufficiently pastoral in its origin and culture that specimens of similar physique were available. But since Ankole was a kingdom, since the new criteria of prestige in colonial Uganda moved away from military symbolism, and since in any case Buganda was an important model for the other kingdoms, Ankole's representation in the Ugandan armed forces was as modest as that of the other kingdoms.

An ethnic separation of powers seemed to be under way in Uganda. There was a disproportionate number of the Bantu in administration and the economy. But there was also developing a disproportionate Nilotic and Sudanic presence within the armed forces of the new Uganda.

Buganda itself remained the most privileged part of the Bantu areas. It was indeed a city writ large. The Nilotic and Sudanic areas were virtually the most peripheral in the new national entity. The soldiers came from a part of the country that was rural in location, function, and status. The stage was set for the beginnings of a military-agrarian complex.[3]

THE MILITARIZATION OF THE COUNTRYSIDE

Partly with the disadvantages of physical and social distance, and partly with the presumed advantages of rural culture for military performance, peasant warriors began to join the army in significant numbers. For some villages, the army was second only to agriculture as a major source of livelihood and income for the local community. For some peasants, a military career was their first introduction to Uganda as a national entity. What was once said of young Turkish farm lads was also true of raw recruits from rural Uganda. These recruits "from isolated villages now suddenly felt themselves to be part of the larger society. The connection between their private life and public role became vivid to them—and this sense of their new personality they diffused around them when they returned to their villages."[4]

In some important sense the country boys became conscious of their membership in the Ugandan nation. Hundreds acquired some technical training relevant to some aspect of their military functions. Some became literate. All had to learn or improve their Swahili as a medium of interaction with lads from other ethnic areas. Those areas had indeed become partially militarized when they became converted into major grounds for recruitment into the armed forces. But the recruits themselves also acquired partially "national" perspectives, though they still retained serious rural handicaps.

The theme of rural status retained a critical relevance. As in many developing societies, the reduced opportunities in the rural areas tended to inflate the value of a military career to many in those areas. In the words of Marion J. Levy:

> insofar as membership in the armed forces is generally open to members of the society, the vast majority of the members of a given society are likely to be individuals of a single class, and hence if the armed force organizations are large scale, most recruitment is likely to come from people of more or less common origins. This is especially true, of course, of relatively nonmodernized societies. In such societies armed force organizations are frequently elite

organizations whose members are likely to come from representatives of a single elite class. If they are not elite organizations and recruitment is open class, the vast majority of individuals concerned are likely to come from agrarian social backgrounds.[5]

In Uganda's experience, the military did not recruit from an elite class. On the contrary, it had considerable difficulty in recruiting from the new educated elite. The overwhelming majority of the soldiers were therefore drawn from what Levy called "people of more or less common origins . . . from agrarian social backgrounds."

But were these people in sympathy with their rural origins and peasant compatriots? Their prejudices and predispositions were certainly considerably influenced by their social backgrounds. What Robert Scalapino said of the military rulers of Korea in the early 1960s has also been true to some extent of the majority of the officers and men of the Ugandan army. These were young men

who come from rural backgrounds and who, in many cases, have known poverty at close range. It is natural for these men to have a rural orientation—to feel empathy with the farmer. Such men must always regard urbanism with a certain ambivalence.[6]

But there is one important difference between Uganda's conditions and those of Turkey or Korea. Uganda is a *polyethnic* society, deeply divided along these "primordial lines." Lugbara peasants in the armed forces may have a bond of affinity with Lugbara peasants in the villages, but there is no guarantee that they would have a bond of sympathy with Kakwa or Acholi peasants inside or outside the army. The bonds of shared social origins are sometimes in conflict with the tensions of differing ethnic origins. In such conditions, a military-agrarian complex is seldom neat or stable. Conflicting loyalties—partly ethnic, partly social, and partly occupational—would periodically shake what might otherwise have been a bond of empathy between soldiers and rural folk.

The concept of a military-agrarian complex implies a *class alliance*. The concept of ethnocracy implies *ethnic power*. Uganda's modern history has been a continuing interplay between the forces that seek class alliances and the forces that seek ethnic power. In a country where ethnic groups are themselves graded and stratified, the issue of where class ends and ethnicity begins is hard to disentangle. The ethnocratic heritage of the country, going back to the days of "Ganda-centrism" in the political process, has continued to condition the whole process of class formation.

TRANSNATIONAL ETHNICITY

Ethnicity in Uganda has been linked not only to class formation, but also to transterritoriality. These are linkages across national frontiers.

Domestic tensions in Zaire or the Sudan could spill over into Uganda and vice versa.

This issue profoundly touched the fortunes of Uganda. The ethnic overlap across the boundary between southern Sudan and Uganda especially made the boundary itself uncertain from the time it was drawn. One school of thought quite early toyed with the idea of integrating southern Sudan with Uganda. Another school believed in uniting the Nilotic people of Uganda with those of the Sudan and creating a state separate from both Uganda and the Sudan. Among the more articulate of Nilotic irredentists earlier in the century were spokesmen for Acholi in Uganda. The Nilotes in Uganda felt underprivileged in relation to the Baganda and other Bantu; and the Nilotes in the Sudan felt underprivileged in relation to the Arabic-speaking north. The beginnings of Pan-Nilotism were at hand, but the movement never gathered momentum.

The fortunes of the Nilotes within Uganda began to change as independence approached. The Baganda were still preeminent, but the sons of the north had voices that were beginning to be heard. From Lango came Apolo Milton Obote, who was later to control the fortunes of the country as a whole for a number of years. But Pan-Nilotism even within Uganda was itself fragile. It was engendered briefly by a shared opposition to the Baganda. From 1966 the armed forces, recruited mainly from the north, derived a sense of national purpose from the very policy of keeping Buganda under a state of emergency and keeping the Baganda strictly subject to northern restraints.

Pan-Nilotism within southern Sudan did not fare much better, as internecine squabbles among the separatist movements reduced southern resistance to a deeply fragmented if still living force.

But before Idi Amin's coup of January 25, 1971, in Uganda, an important phenomenon was taking place. The word *Nilotic* refers both to a particular family of tribes in black Africa and to the Nile as a whole. Ethnic Pan-Nilotism envisaged the unity of a particular group of tribes in black Africa, some of whom may no longer be along the Nile Valley. The Luo of Kenya as western Nilotes and the Masai as eastern Nilotes (formerly known as Nilo-Hamites) are no longer inhabitants of the banks of the Nile, but they belong to the family of tribes bearing that name.

There is, however, an older sense of the word *Nilotic*. The word is of Greek derivation, and, according to the Oxford dictionary, it means quite simply "of the Nile." We may therefore have here another sense of Pan-Nilotism—not an ethnic sense denoting a community of black tribes but an ideological sense denoting commitment to the old doctrine of the unity of the Nile Valley.

If the Acholi in Uganda had once taken the lead in recommending Pan-Nilotism in the sense of a separate state for that family of black tribes, the Langi later came to take the leadership in championing an ideological Pan-

Nilotism in the sense of a shared movement of sympathy along the Nile Valley from Lake Victoria to the Mediterranean. The Langi who seemed to be exploring the possibility of ideological Pan-Nilotism were none other than Milton Obote himself, as President of Uganda, and his cousin Akena Adoko, a fellow traveler on the seas of radicalism. A major factor in the history of ideological Pan-Nilotism in the independence period was the burst of socialism along the Nile Valley, starting with Egypt under Nasser and then the Sudan under Numeiry. But would the virus of radicalism affect the body politic of Uganda as well? A good deal depended upon one Nilotic figure from Lango, Apolo Milton Obote. In the course of 1969-70 the idea of an ideological unity of the Nile Valley, from the waters of Jinja to the shores of the Mediterranean, began to take shape. Nasser's precedent in inaugurating a local brand of socialism and establishing the era of nationalism and state control with the historic takeover of the Suez Canal in 1965 was now finding echoes of solidarity from the reverberating stadium walls of Lugogo and Nakivubo in Kampala.

May 1969 had moved the Sudan to the left. October 1969 gave Obote's Uganda the *Common Man's Charter*. For the first time since independence, socialist solidarity across the whole course of the Nile Valley was possible.

Then, on January 25, 1971, a military coup overthrew Obote in Uganda. The Kakwa, Amin's tribe, were themselves Nilotes. So were the Langi, who produced Milton Obote. Internal Pan-Nilotism in Uganda was dead. Revised ethnic alignments were now in the making.

In my estimation, four factors aggravated ethnic tensions following Amin's coup; ironically, one of those four factors was the political magnanimity that characterized the first few days and weeks of Amin's rule. The first factor that fostered potential tensions included quite simply those prior tensions within the armed forces, tensions that focused more often on the Acholi than on the Langi. In the armed forces the Acholi had been the largest single group, going upward to one-third of the soldiers. A tradition of solid military performance had created a certain self-confidence among the Acholi, a self-confidence that might on occasion have been mistaken for military arrogance. It is sometimes assumed that sharing life together in the barracks helps to give armed forces in Africa opportunities for ethnic intermingling. That may be taken for granted. But it is further assumed that ethnic intermingling in itself reduces ethnic tensions and animosity. This latter assumption is not always borne out by events. Ethnic intermingling first results in increasing tension before it finally reaches a plateau of normalization and ultimate ethnic integration. Social scientists often underestimate the tension-generating effects of premature integration. Intertribal animosities within the Uganda army were taking shape well before the coup of January 25, 1971.

The second factor that aggravated the potentialities of an ethnic eruption was Amin's great "blunder," namely, his account of why the coup took place. Until Amin had his press conference, much of Acholi could have been recruited to his side. As indicated, the eighteen points mentioned on the first day of the coup could almost have been of Acholi authorship. They played up the issue of the Lango Development Master Plan written in 1967, ostensibly urging that all key positions in Uganda's political, commercial, military, and industrial life were to be occupied and controlled by Akokoro County, Lango District. In the words of the soldiers' eighteen points, "the same master plan decided that nothing of importance must be done for other districts, especially Acholi District."[7]

But then General Amin gave his press conference to explain how the coup took place. He said that former President Obote had sent a directive to certain army officers in Kampala instructing them to arm Acholi and Lango tribesmen within the army and disarm and arrest other army units consisting of different tribesmen. Amin said that on the weekend in question he had been at Karuma Falls, and on returning home Sunday evening, he found a tank and a personnel carrier outside his residence. In the carrier, the General said, was an injured soldier.

The soldier told Amin that Lieutenant Colonel Akwanga, commanding officer of the Mechanized Battalion, had been ordering Lango and Acholi soldiers to go to the armory to obtain weapons and ammunition. The soldier, realizing that something was amiss, had alerted some of his own colleagues and tribesmen and attempted to obtain weapons for his own self-defense. In the attempt to get the weapons he had been wounded. The other soldiers had by then realized that Amin was not party to what was happening and had gone to warn him and help protect him. The general, after making sure that the wounded man received medical treatment, took control of the situation.

This was Amin's account of the coup. Should he have given such an account? Should he not instead have suppressed it or have fundamentally distorted it in an attempt to eliminate the heavy tribal dimension implicit in it? According to this account, Obote had intended to trust only the Langi and the Acholi and to arm them, possibly at the expense of other tribesmen in the army. The whole account of the coup rested on an ethnic dimension. If it was accurate, it should deliberately have been made less correct for the sake of national survival. If it was inaccurate from the start, its errors were disastrous. Amin later promised to have the whole thing written out in detail and published, but wiser counsel within the government of the Second Republic prevailed. But, tragically, the ethnic dimension had already set the tone. The Acholi, who had been available for possible mobilization to the side of the coup, were now made insecure. Many Acholi in the army might indeed have been wary from the start, but civilian Acholi had enough grievances against Obote's regime to have been potential allies of Amin and his Second Republic. The Langi who might have been ready to dissociate themselves from Obote with a little persuasion and patronage retreated after Amin's press conference into a new sense of defensive insecurity. Meanwhile, the non-Langi and non-Acholi tribesmen within the armed forces, who need never have known in detail of plans to disarm them at the expense of Acholi and Langi (had Amin acted to reduce ethnic tensions within the armed forces), now found another deep grievance against their comrades-in-arms from the two northern districts. After all, these two groups had retained Obote's trust and were on the verge of being armed in a posture of combat against the rest of the armed forces.

The third aggravating factor behind the later ethnic eruptions in Uganda was Obote's decision to fight back. If Obote had behaved like old Farouk of Egypt or General Abboud of the Sudan and retreated into oblivion after being overthrown, the air of political magnanimity in Uganda might have lasted a little longer. But Obote's first press conference was a fighting press conference. He not only claimed that he was still President of Uganda but also made charges against Amin from a neighboring capital and sought to rally pan-African forces against the new republic in Uganda. Obote's travels to different capitals in a bid for diplomatic and, conceivably, military support began to look like a militant endeavor to make a comeback. Nyerere's diplomatic support for Obote and the massive utilization of Tanzania's diplomatic and journalistic influence to discredit the new regime in Uganda also contributed greatly toward the climate of tension within Uganda itself. Violence began to be regionalized in the East African Community and beyond. The politics of the Organization of African Unity and the wrangle about credentials, culminating in Sam Odaka's (Obote's minister) trip to Addis Ababa as a rival delegation to Amin's, could not but crack the aura of magnanimity that had characterized the coup from the beginning. One after another of Amin's former gestures of conciliation toward Obote were withdrawn. Symbolic of the shift in direction was the decision to do what Amin had promised not to do—rewrite history. Obote's photographs were banned, the grand medallion outside parliament was brought down, and a reward of a million shillings was offered for the person of Obote delivered alive to the new Uganda government.

In the beginning, Amin was cool and collected, and Obote, Nyerere, and the Tanzanian press responded hysterically to the events of January 1971. Tanzania's calculated measures to snub the new regime of Uganda, humiliate it diplomatically, and finally refuse to recognize its appointees to the East African Community—all helped aggravate the atmosphere and sharpen the postures of combat. Uganda's tensions were being regionalized both along the Nile and in East Africa. Tanzania's hysteria ultimately generated counter-hysteria from Uganda. The shrill cries of war on the border; allegations of pro-Obote guerrilla intrusion from Tanzania into Uganda; withdrawing working rights to Amin's namesake, Idi Simba, governor of the East African Development Bank; refusing to ratify the Appropriations Bill of the East African Community; pirating the helicopters intended for Tanzania—all these amounted to a posture of counteraggression against Tanzania, a posture clearly generated by the sense of frustration and humiliation that Tanzania's initial hysteria had brought upon the infant Second Republic of Uganda. It was not realistic to expect a military regime to remain magnanimous and forgiving towards Obote and his supporters while Obote and Tanzania pursued a policy of militant denunciation. The bad atmosphere created by Tanzania's and Obote's reaction to the coup had ethnic repercussions. It was genuinely feared that Langi and Acholi would be recruited to fight for Obote. Many Langi and Acholi soldiers, uneasy and insecure within Amin's armed forces, found disappearance the better part of valor. The fact that they had gone itself reinforced suspicions they had done so in order to rally—and fight again—*against* Amin.

There is probably reason to believe that some of these men did organize themselves into units for possible sabotage. Attempts to penetrate ammunition depots reinforced the new regime's fear of subversion and guerrilla attack. It was a classic vicious circle—suspicions leading to the realization of those suspicions. It was a classic self-fulfilling prophecy, since the fear of an underground movement made the Second Republic aggressive toward potential Obote supporters within the tribal areas, and since these supporters in turn became insecure and sought protection in the shadows of concealed existence and fearfully disguised intrigue.

The whole enterprise culminated in the invasion of Uganda by Obote's supporters from Tanzania in September 1972, clearly with the connivance of the government of Tanzania. The invasion was President Nyerere's "Bay of Pigs." The invaders were decisively defeated. Relations between Uganda and Tanzania dropped to an all-time low. Bukoba was bombed by Uganda planes. The two countries were on the brink of an all-out war.

The Nile Valley was nearly involved when President Numeiry prevented a contingent of Libyan troops from flying over Khartoum to go and help Amin. Somalia meanwhile entered into determined attempts to prevent the regionalization of violence, especially between Tanzania and Uganda, from getting worse.

The fourth factor that aggravated ethnic tensions was the initial magnanimity of Amin's coup. That magnanimity had sought to narrow the focus of hostility within Uganda to two individuals, Akeno Adoko and Milton Obote. Unfortunately, both were Langi. In current African politics, a focus on unique personalities tends to expand and include their tribal origins. To blame a catastrophe on an individual who happend to be a Langi exposes not merely the individual himself but also his ethnic roots. The community is held accountable for the "villains" it produces.

We are already familiar with a situation where individuals sometimes suffer because they are deemed to belong to the wrong tribe. Under Obote, a Muganda might sometimes be discriminated against because he was a Muganda. In Kenya a Luo applying for a job in either the private or public sectors might be discriminated against because he was a Luo.

But we also have a reverse phenomenon at times. This is when a tribe suffers because of hostility to an individual. The classic illustration of this second phenomenon is the case of the Langi and Milton Obote. The villain of the piece initially was not the Lango District as such. The villains were Akena and Obote. But because they were both Langi, and once powerful and influential Langi at that, their fall from eminence had adverse consequences for their tribe. The relentless interaction between ethnic factors and personal factors had found yet another arena in the very magnanimity of the Ugandan coup.

Had the vengeance of the coup been directed at Obote's government as a whole, it would have been directed at people from almost every corner of Uganda. The ministers were multiethnic, drawn from West Nile, Lango, Acholi, Kigezi, Ankole, Buganda, Bunyoro, Toro, and elsewhere. If the blame for the first errors of the First Republic were well and truly laid on Obote's regime as a whole, and not simply on Obote himself, the tribal repercussions would have been virtually national. If not every tribe, certainly every district was compromised in having participated in the First Republic. If the revenge against the First Republic had been based on a denunciation of the regime as a whole, of its performance and its scale of values, it would probably have had less potential for degenerating into pure ethnicity than if the blame had been pinned on two Langi and on two Langi alone.

This phenomenon has deep roots in the tradition of collective responsibility among kinsmen in African societies. Collective responsibility has been known to take a variety of forms, ranging from finding a job for an unemployed kinsman in Kampala who has just arrived from the home village to joint participation in an interclan feud. Collective responsibility may be among fellows of the same clan, members of the same age group, inhabitants of the same village, or even speakers of the same language, depending upon circumstance and the groups concerned.

Murder in African society is sometimes a question of collective guilt and collective obligation. Murder across the tribal line certainly often involves intertribal accountability and in some cases could result in intertribal conflict.

Considerations such as these can convert animosity against one individual into animosity against his tribe as a whole. In Uganda General Idi Amin repeatedly sounded a warning after the coup, addressed to Obote, advising him that continuing militancy by Obote against the regime could have adverse consequences on his kinsmen and his tribe. The magnanimity of the Ugandan coup in the initial stages, to the extent that it basically tolerated all the former ministers of Obote's government with the exception of Basil Bataringaya, minister of internal affairs, was, while it lasted, a major gesture in humanitarianism. What we have sought to demonstrate is that even such gestures sometimes have certain costs. In Uganda the cost of Amin's magnanimity was ethnically focused political blame. When he blamed Obote and Akena alone, the Langi became highly vulnerable. By forgiving almost all the other ministers, the rest of the country was excused from responsibility for the excesses of the First Republic. The blame was narrowed, and the barometer of ethnic tensions rose.

With the eruption of these further tensions, violence once again played its role as a penetrative agency. The Nile Valley took one more step toward becoming a valley of political blood.

But transterritorial ethnicity in Uganda had a religious dimension as well. It is to this that we must now turn.

ETHNICITY AND RELIGION

Aidan Southall has rightly reminded us that Idi Amin Dada was not "a bizarre or maverick intrusion upon the Uganda political scene" but was part of a historical process involving ethnicity and religion across Uganda's northern neighbor, the Sudan.

The link between religion and politics was established from the earlier phases of interaction between Uganda and the Sudan. The rise of the Mahdi in the Sudan provided a resurgence at once nationalistic and sectarian. Representatives of the British and Egyptian governments— ostensibly exercising a shared authority over the Sudan—were progressively overwhelmed by this new *jihad*. Emin Pasha, the Austrian-born governor of Equatoria in southern Sudan, temporarily survived in lonely splendor as the last figure of imperial authority after the fall of Gordon in Khartoum in January 1885.

Emin Pasha recruited his military forces from several different ethnic groups. The upheaval in the Sudan had uprooted large numbers of people. Urbanization in southern Sudan resulted in some cases in Islamization and also in the acquisition of a variety of colloquial Arabic. Out of these factors the Nubi community came into being.

Barri A. Wanji defines "Nubianization" as the "converting of members of such tribes as Acholi, Kakwa, Lugbara, Nyoro, and some others . . . to adopt Nubi Islamic practices, their language 'Lunubi' (Kinubi), dress and other cultural systems of the Nubi, who originally came from the Sudan."[8]

Wanji oberved that the process of "Nubianization" still continued. Has it been *accelerated* by the rise of Amin, his use of Nubi soldiers in his army, and the prestige and rewards that follow from being a member of a newly privileged group?

The Nubi community within Uganda can in part be traced to those soldiers of fortune recruited by Emin Pasha and later selectively incorporated into Lugard's own forces. Southall was certainly exaggerating when he asserted:

> The core tradition of the Uganda Army is a Nubi tradition. Any members of those ethnic groups from which the Nubi originally derive, especially the Bari-Kakwa who were the most numerous and central, are in the most fundamental sense Nubi as soon as they join the army, especially if they are, or become, Muslim as well.[9]

The bulk of the Ugandan army before the coup did not consist of Nubi, nor has it become so since the coup. Southall is in danger of diluting the term *Nubi* to such an extent that it becomes the generic term for any Nilotic or Sudanic individual who joins the army.

But the essential point that emerged from Southall's paper is that the Nubi presence in the Ugandan army is not a sudden alien intrusion; rather, it is solidly part of the history of that army from the beginning. It is inextricably linked to the pattern of ethnic affiliations between the West Nile district of Uganda and the adjoining areas of the Sudan and Zaire. The Nubi factor in the domestic balance of power in Uganda simply reaffirms once again that the boundaries drawn by the colonial powers were arbitrary and can sometimes result in ethnic reinforcement without regard for national frontiers.

A related distinction is between heartland tribes and peripheral tribes. Where those who exercise power in a given country belong to a community big enough and central enough to qualify as a "heartland tribe," the colonial boundaries might hold up as the outer frontiers of domestic politics. Neither the Baganda in the colonial period nor the Kikuyu in Kenya since independence have needed any Bantu ethnic allies beyond their borders in their confrontations with Nilotic rivals. Even the Hausa and Yoruba in Nigeria, though not "central" in the same sense as the Kikuyu, have not relied on their ethnic compatriots across their borders, partly because they themselves were already large enough and central enough in national politics to be satisfied with the colonially defined limits of Nigeria.

But where power at the center is captured by a geographically peripheral group whose own domestic size is small but that also has compensatory ethnic compatriots across the border, the colonial boundaries between countries become once again politically permeable. In Uganda the permeability began under Obote, when power shifted from the more central Baganda to the more peripheral Nilotes. The Nubi factor in Uganda politics is a continuation of that process. Amin's use of Zaireans and Sudanese is not a simple case of using "black mercenaries." It is even more a case of using ethnic compatriots.

The Islamic factor arises because the concept of "Nubi" itself fuses three elements—ethnicity, religion, and occupational specialization. Amin is a Kakwa Muslim: "It is not really possible to be a Kakwa Muslim without being enmeshed in a network of kinship ties with the Nubi, in and out of the army."[10]

Southall might have added "in or out of Uganda" as well. As for the occupational specialization of the Nubi, this is indeed the army and related occupations such as night watchmen and security officers at railway stations. As Wanji put it:

> The Nubi regard the army as their own historical occupation and the record of such a role is the pride of every Nubi family. For almost every Nubi man, a brother, father or uncle must have served in the army. The army was part of their life. Many Nubi still consider the army profession the most prestigious in terms of achieving manhood.[11]

What has happened in Uganda is that several peripheral ethnic groups and a peripheral religion called "Islam" have moved to the center of power and influence through the intervening medium of membership in the armed forces. Was the stage set for a new national religion?

In fairness to Amin, it must be emphasized that he, like Obote, manifested quite early a commitment to religious pluralism. His initial moves were certainly in the direction of creating an ecumenical state in Uganda, one in which the government was neither religiously neutral nor religiously monopolistic, but capable of serving as a referee among contending denominations.[12]

The question that inevitably arose was whether Amin was in a position to transcend and overcome the tradition of sectarianism in Uganda's history. Amin was caught in the paradoxical predicament of trying to create an ecumenical state from a sectarian base of power. From a denominational point of view, Amin's constituency were the Muslims of Uganda, including the Nubi. These were a minority. Yet one of Amin's ambitions was to be remembered in Uganda's history as a man who drastically reduced religious disunity in Uganda. How could he fulfill his ambition in the face of the obstinate sectarian factionalism of Uganda's heritage?

Amin's record has fluctuated between the ideal of the ecumenical state and the vulgarity of denominational favoritism. The pursuit of ecumenicalism has ranged from government-sponsored interdenominational conferences to the intervention by the state to prevent further internal divisions among Protestants. The tragic lapses into denominational favoritism have included disporportionate rewards for Muslims in the economic and military sectors of national resources and positions.

If Amin fails in his ecumenical ambitions and emphasizes instead a commitment to Muslim solidarity, the Nubi factor in Uganda could become more significant than ever. These marginal strangers could help change the course of the country's history. In addition, even non-Muslim kinsmen could basically be relied upon to maintain their "brother" in power, provided his patronage extends to both ethnic kinsmen and coreligionists.

Partly because of this, Amin has managed to get the support of both the Nubi and the Anyanya. The Nubi are credited with militant attachment to Islam. The Anyanya, the heroes of the Sudanese civil war, are credited with the sustained crusade in southern Sudan *against* northern Islam. Although both the Nubi and the Anyanya are regarded as being of Sudanese extraction, and are in practice overlapping categories, their religious orientations are often divergent. Yet sections of both the Nubi and the Anyanya have supported Amin in power and appear likely to continue to do so provided their support is rewarded. If Amin and his immediate successors remain Muslim, the proportion of followers of Islam will probably increase in Uganda before long. After all, Uganda became preponderantly Christian by the time of independence partly because the imperial "oppressor" was Christian. Uganda could now be substantially Islamized, partly because the indigenous military "oppressor" is Muslim. History is all too familiar with the correlation between religious conversion and political power. Both Christianity and Islam have thrived on their own respective political domination. Under Muslim rule, Uganda could become half-Muslim in another two decades.

From the point of view of the interaction between Uganda and the Sudan, this outcome would have its ironies. After all, southern Sudan conducted a civil war for seventeen years, partly in pursuit of the right to be culturally and religiously different from the Muslim North. Much of the support and equipment for the southern Sudanese fighters in their "crusade" against the Muslim North came through Uganda. This ranged from the personal support of kinsmen in Uganda to the religious encouragement of Christian missionaries, from the provision of military hardware by the Israelis through Uganda to the diplomatic "blindness" of successive Uganda governments.

It might therefore be said that Uganda played an important part in *preventing* for a while the Islamization of southern Sudan. With the support that Amin has since been receiving from the Anyanya and the Nubi, on the other hand, it could conversely be said that southern Sudan is now playing a part in *promoting* the Islamization of Uganda. If Amin does indeed survive for a while longer, or if he is succeeded by a regime that continues to need the support of Muslim and Sudanese soldiers, the balance between Muslims and Christians in Uganda is almost bound to change significantly. The religion of those who are powerful can always count on some additional converts if the rewards are attractive enough. The line between faith and functionality can be psychologically thin. That is what cultural diffusion is all about.

If Islamization in Uganda were to proceed at a pace that could make

Muslims of nearly half the population of the country before the end of this century, that old link between the Sudan and Uganda, between Emin Pasha and Amin Dada, would find a new religious meaning. The valley of the White Nile as a whole could be Islamized from Lake Victoria to the Mediterranean. Southern Sudan—which resisted Islamization by northern Sudan—would be then have been converted to Islam by Uganda. History would once again have played its dialectical game of eternal contradictions. It would have allowed marginal strangers to change its own sacred course.[13]

On the other hand, if Muslim power in Uganda is overthrown "prematurely," there are alternative scenarios to be evaluated. This takes us back to the experience of the Holy Roman Empire. In the words of Toynbee:

> It was in Germany, in particular, that the classical formula, *cuius regio eius religio*, was invented and applied; and we may take it that in Central Europe, at least, the secular princes did successfully use their power to force down the throats of their subjects whichever of the competing varieties of Western Christianity the local potentate happened to favor. . . . The readiness of all the competing factions of Western Christianity in the age of the Wars of Religion to seek a short cut to victory by condoning, or even demanding, the imposition of their own doctrines upon the adherents of rival faiths by the application of political force was a spectacle which sapped the foundations of all belief in the souls for whose allegiance the warring churches were competing.[14]

Under this scenario Amin's attempt to Islamize Uganda would help to neutralize the preceding attempt by Europeans to Christianize Uganda. This in turn could result in religious disenchantment and increasing secularization.

As Toynbee put it:

> Louis XIV's methods of barbarism eradicated Protestantism from the spiritual soil of France only to clear the ground for an alternative crop of scepticism. The revocation of the Edict of Nantes was followed within nine years by the birth of Voltaire.[15]

But apart from the two alternatives of Islamization or secularization of Uganda as a result of the impact of Idi Amin Dada, there is a third scenario of vicious hostility against Islam and the resurgence of Christian fanaticism. Here again the lessons of Germany under the Holy Roman Empire are appropriate.

The Bohemians rebelled against the Peace of Augsburg of 1555 and its formula that the ruler determined the religion. They decided on the alternative formula that the religion determined the king. In 1618 insurrectionary Bohemian nobles chose a Calvinist as their king, against the claims of the constitutional Catholic heir apparent. This reverse formula, linking religion to choice of a leader, also had its hazards.

The Catholic forces routed the Bohemian rebels in 1620 at the Battle of the White Mountain, not far from Prague. The Habsburgs confiscated the extensive lands of the fleeing rebels which they distributed among their followers. . . . The collapse of Bohemia spelled doom for Austrian Protestantism. Austria was subjected to a rigorous counter-reformation as was Silesia; between 1622 and 1628, Protestants throughout the Habsburg provinces were shorn of all their rights.[16]

This third scenario is a warning for both Christians and Muslims in Uganda against the excesses of sectarianism. A pendulum of revenge might be set in motion, perhaps from generation to generation. The ghosts of the Mahdi and Gordon, Lugard and Emin Pasha, might preside upon an unfolding tragedy of sectarian reprisals. And Idi Amin and his links with the Nubi might go down in history as one more episode in a record of passion that linked the fortunes of Uganda in the twentieth century with the agonies of the Thirty Years' War in Europe more than three centuries earlier.

CONCLUSION

The month of February 1977 witnessed two highly publicized acts of brutality reportedly committed by Africans against clerics. First came the news that seven white Roman Catholic missionaries, including four nuns, had been gunned down in Zimbabwe/Rhodesia. The sole survivor, Father Dunston Myerscough (sixty-five years old), was convinced that the murderers were nationalist guerrillas. The second event, less than two weeks later, was the apparent murder of the Most Reverend Janani Luwum, Anglican archbiship of Uganda, while he was in custody under the charge of plotting to overthrow the government of Idi Amin Dada. Amin's government claimed that the archbishop and two of Amin's own cabinet ministers under a similar charge were killed in a car crash, but most of the world was understandably skeptical.

In the case of the murder of the seven missionaries in Zimbabwe, it was assumed that they died as casualties of the racial war rather than as martyrs in a religious crusade. But in the case of the Ugandan archbishop, the world jumped to the conclusion that he was martyr of his Christian faith. Was the world justified in assuming that Archbishop Luwum died for religious reasons?

When the news of the archbishop's death broke, it reminded me of a night in Kampala six years earlier when my wife and I gave refuge to girls who were running away from potential rape by Amin's soldiers. The girls were all either Langi or Acholi. The previous night some soldiers had broken into Mary Stuart Hall at Makerere University and demanded to be taken to Langi and Acholi girls. On that occasion they did take away two girls, one of whom was saved from a serious fate by the fact that she was in her monthly period. The next night the Acholi and Langi girls were of course terrified, and some of them came to our house for refuge. Vice-Chancellor Kalimuzo and I had urgent consultations about the other girls left in Mary Stuart Hall. President Amin agreed to send us his more reliable soldiers to patrol the campus and keep the military rapists at bay. The situation was indeed eased, but periodic terror continued to haunt the life of every Acholi and every Langi from then on.

When six years after that night Archbishop Luwum was killed, the question sprang to my mind: did Luwum die because he was Acholi or because he was Anglican? If those Roman Catholic missionaries were casualties of an unfolding racial war in southern Africa, why could not Janani Luwum have been a casualty of continuing *ethnic* strife in Uganda?

After all, Cabinet Minister Oryema, who was killed with the archbishop, was also an Acholi. Before long, further news seemed to validate ethnic factors rather than religious ones as dominant behind the new atrocities in Uganda. Leading Langi and Acholi, including some at Makerere University, were either rounded up, assaulted, or at least briefly harassed. Hundreds of refugees from Lango and Acholi were soon reported to be pouring into Tanzania and Kenya. As for Amin's own statements, they seemed to echo some of the accusations he leveled against the Acholi and the Langi way back in the first week his assumption of power in January 1971.

But the problems of Uganda are not only a mixture of ethnic and religious factors. They are also a mixture of domestic and external factors, of national and regional variables. This is where the analogy between Uganda and Lebanon becomes striking. For both countries, part of the problem concerns the issue of where the imperial powers that ruled them decided to draw the boundaries. Lebanon was carved out of Greater Syria partly because the French wanted to create a separate Christian enclave, a kind of "Christian Israel," even before the Jewish Israel came into being. But the carving out of a Christian enclave was somewhat messy—there were still far too many Muslims in Lebanon. Although the Muslims were at the time a minority, their birthrate was higher than that of the Christians. Since then the Muslims of Lebanon have caught up with the Christians and have begun to outnumber them. The boundaries the French had so carefully drawn for their Christian enclave now provided a setting for sectarian confrontation.

The boundaries the British drew in East Africa were similarly messy. The British split up Amin's tribe, the Kakwa, between Uganda and the Sudan and helped the Belgians annex a third portion of Kakwaland. The Ugandan army under Amin reflects the untidiness of the colonial boundaries. As we indicated, Amin recruits *ethnic* compatriots (fellow tribesmen) into his army, even if they are not national compatriots and are Sudanese or Zaireans instead.

Similarly, while the Lebanese crisis was deepened by the presence of Palestinians in Lebanon, so has the Ugandan crisis been deepened by the Nubi presence in Uganda. Lebanon has suffered because of two partitions: the partition of Greater Syria in order to create a Chrisitan enclave and the partition of Palestine in order to create a Jewish state. Uganda has suffered because of *ethnic* partitions rather than denominational fragmentations. But both countries are now landed with a legacy of hate and recrimination that imperialism and militarism together have bequeathed to their unhappy people. When tension is militarized and factionalism is armed partly as a result of insensitive imperial frontiers, at least one entity is allowed to extend its ominous boundaries—the graveyard.

The basic struggle in such a society is the struggle to moderate social cleavage and civilize pluralism itself. That is perhaps what the agony of Uganda in the 1970s has been all about.

NOTES

1. Ethnic politics in Uganda are discussed more extensively in Ali A. Mazrui, *Soldiers and Kinsmen in Uganda: The Making of a Military Ethnocracy* (Beverly Hills and London: Sage Publications, 1975); Nelson Kasfir, *The Shrinking Political Arena* (Berkeley and Los Angeles: University of California Press, 1976); and James Mittelman, *Ideology and Politics in Uganda* (Ithaca, N.Y.: Cornell University Press, 1975). For a Marxist interpretation, see Mahmood Mamdani, *Politics and Class Formation in Uganda* (New York: Monthly Review Press, 1976).

2. E. E. Evans-Pritchard, *The Nuer* (Oxford: Clarendon Press, 1940), p. 183.

3. For related background issues, see Nelson Kasfir, "Controlling Ethnicity in Uganda Politics," in Kasfir, *The Shrinking Political Arena*; William Gutteridge, *The Military in African Politics* (London: Methuen and Co., 1969); and Claude E. Welch, Jr., and Arthur K. Smith, *Military Role and Rule: Perspectives on Civil-Military Relations* (North Scituate, Mass.: Duxbury Press, 1974).

4. See Daniel Lerner and Richard D. Robinson, "Swords and Ploughshares: The Turkish Army as a Modernizing Force," *World Politics* 12 (October 1960): 26-29.

5. Marion J. Levy, Jr., "Armed Force Organizations," in *The Military and Modernization*, ed. by Henry Bienen (Chicago and New York: Aldine, Atherton, 1971), p. 63.

6. Robert A. Scalapino, "Which Route for Korea?" *Asian Survey* 2, no. 7 (September 1962), p. 11.

7. *Uganda Argus* (Kampala), January 26, 1971.

8. Barri A. Wanji, *The Nubi Community, an Islamic Social Structure in East Africa*, Sociology working paper no. 115, Department of Sociology, Makerere University, 1971, p. 4. Cited by Southall, "Amin's Military Coup in Uganda: Great Man or Historical Inevitability?" Paper presented both at the annual African Studies Convention of the U.S.A., held in Syracuse in the fall of 1973, and at the

Third International Congress of Africanists, in Addis Ababa in December 1973, p. 2. A version of this paper has since been published in *The Journal of Modern African Studies* 13, no. 1 (March 1975). The page references used here are from the original conference paper.

9. Southall, "Amin's Military Coup," p. 3.

10. Ibid.

11. Wanji, *The Nubi Community*, pp. 8-9.

12. See Ali A. Mazrui, "Piety and Puritanism under a Military Theocracy: Uganda Soldiers as Apostolic Successors," in *Political-Military Systems: Comparative Perspectives*, ed. Catherine M. Kelleher (Beverly Hills and London: Sage Publications, 1974), pp. 105-124.

13. See also Ali A. Mazrui, "Is the Nile Valley Emerging as a New Political System?" (Paper presented at the Annual Social Science Conference of the Universities of Eastern Africa (USSC) held at Makerere University, Kampala, in December 1971). See also Akiiki Mujaju, "The Religio-Regional Factor in Uganda Politics" (Paper presented at the Third International Congress of Africanists, Addis Ababa, December 1973).

14. Arnold J. Toynbee, *A Study of History*, vol. 1-4 abridged by D. C. Somervell (London: Oxford University Press, 1948), pp. 485-486.

15. Ibid., p. 486.

16. See John L. Stipp, C. Warren Hollister, Alan W. Dirrim, and Harold L. Bauman, *The Rise and Development of Western Civilization* (New York and London: John Wiley & Sons, 1969), pp. 447-448.

4

Language Policies and Their Implications for Ethnic Relations in the Newly Sovereign States of Sub-Saharan Africa

Ruth H. Landman

The approximately thirty-five newly sovereign states in sub-Saharan Africa have been forging policies with respect to language use as part of their larger concern with governmental operations.[1] There is a striking disparity between the number of languages spoken by the inhabitants of these countries and the number selected as official and national languages. The former number well into the hundreds,[2] while the latter are a mere baker's dozen. Even within this small group, only two, English and French, are widely used. That is, in a potential field of choice of many indigenous languages, several African lingua francas (e.g., Swahili) and an earlier exogenous language of wide African distribution (Arabic), two languages that were instruments of recent colonial administration, have been chosen far more frequently.

It is the intention of this chapter to explore language policies of the governments of sub-Saharan states as these relate to questions of ethnicity maintenance. Decisions to select so few languages out of the total potential repertory have implications both in the short term and in the longer term for interethnic relations and for the distribution of power and access to power. Decisions made early in the post-independence period have had and continue to have far-reaching consequences for children of constituent ethnic communities as they enter the state school systems; language proficiencies determine access to the public job sector, both in the schools and in all areas of public administration. In turn, these choices suggest further consequences for the future: in what languages will laws and regulations be promulgated? Which languages will be aired on the (almost entirely governmentally operated) radio and television services? The data will be examined through a transactional perspective in which Paine's critique of Barth's relatively weak treatment of power is an important consideration.[3]

National language policies and their implementation in the independent states of sub-Saharan African have a sufficiently brief history and a sufficiently precise historical moment of inception to allow one to try to carry out a sociolinguistic inquiry in which the prior value orientations and the values gained and lost can be seen without too much historical overlay.

The data that will be discussed were collected in a rather unorthodox manner. I became interested in African language policy developments some years ago because I was struck by the relative lack of references to language questions and disputes in the literature on modernization in Africa, as compared to several other areas of the world where the process of nation building was also in full tide. My curiosity was therefore directed to Africa for the purpose of comparison. More recent reading on the subject confirms this earlier impression. For instance, the bibliography of Fishman's *Language and Nationalism* (1972) contains only about two dozen items on independent sub-Saharan Africa out of approximately 650 entries.[4]

Since I had no opportunity to carry out detailed field studies in Africa, I explored this apparent anomaly from a distance. In order to have as nearly complete coverage of all the independent states as possible, and in order to have as nearly comparable sources of information as possible, I eventually decided to interview an official representative of each country, namely, the information or cultural affairs attachés at the embassies of the various countries. Most were interviewed in Washington, D.C., but a few nations did not have diplomatic missions there, and insofar as this was possible, their representatives were interviewed in Paris and London. The respondents were limited to these single representatives–given the very small staff at some legations which would have made it impossible to use comparable cross-checks for all countries. A search through the literature made it clear that the information could not be compiled from library sources, although for some points handbooks were of considerable help.[5] Since all of the respondents served as information officers for their countries, they were able to supply me with the normative information that I sought, and were willing to respond to the second aspect of the questions, which dealt with their views about language policies and patterns in their home countries.

Interviews were conducted in two time periods about five years apart. This was done to permit examination of the retention or modification of initial decisions taken soon after independence had been gained. The first set of interviews was conducted over a period of approximately ten months in 1968 and early 1969; the second round was held in late 1973 and early 1974.[6] The attachés were almost without exception different individuals by that time. An examination of the accompanying tables shows that there are information gaps. It did not take me long to learn that when an anthropologist heeds Laura Nader's advice to "study up," she is likely to be met with potential respondents who "look down" and use their juniors to transmit a refusal to participate in her research efforts. I was unwilling to substitute alternate nationals where this was the case, since I wanted to hold to the comparability of the respondent pool. I therefore lack information for some countries. Furthermore, countries whose independent status postdates the time of the interviews are obviously not included, e.g., all three where the colonial language was Portuguese. From what has been said it is apparent that maximization strategies informed the researcher's decisions just as strongly as they did the nations under consideration.

Among the decisions that one country after the other had to make immediately upon assuming control over its affairs was the designation of an official medium of government communication. An inspection of table 4.1 shows that the choice almost always fell on a non-African language. Among the exceptions are three countries where one African language is very nearly universally known by the populace: Kirundi in Burundi, Sesotho in Lesotho, and Nyarwanda in Ruanda. Nevertheless, in all three cases, French or English also occur. Tanzania adopted Swahili and thereby became the only nation to select an African language exclusively. All others opted for an exogenous medium. The dual heading "national and/or official" language reflects the dichotomy that was adjudicated by the other three countries that selected an African language.

The official language was so designated because it was instituted for official purposes, including records, regulations and laws, competence required for civil service jobs, and the like. This decision was clearly pragmatic, because it allowed an immediate transition from the previous colonial to the now independent administration. However, the designation of a national, but nonofficial, language points up that other desiderata also entered into the decision. National languages are to be seen as symbolically significant and were so characterized by the diplomats interviewed. Parliamentary debates were visualized as taking place in the national as well as the official language. The nature of this consideration was brought out by the Ghanaian representative. He reported that although his country was relying very heavily on English for official communications and was continuing to teach it at all levels in the schools, no firm commitment had yet been made with regard to the selection of a national language. Five languages would have been in contention in Ghana at that time. Somalia was just readying itself for a language preference referendum, preceded by a period of intensive efforts to teach Somali as a spoken and a written language. Nigeria, as the only nation with a federal form of government, had no nationally designated official or national language. Guinea was hoping to proceed toward multiple African official languages in 1968-1969.

Over the succeeding five years, the picture changed slightly. Botswana added its widely shared African language, Setswana; the Central African Republic added a widely understood and spoken lingua franca, Sango; Ethiopia switched to Amharic and dropped English; Kenya adopted

TABLE 4.1

	1a. National and/or Official Language		1b. Major Vehicular languages or Lingua Francas	
	1968/69	1973/74	1968/69	1973/74
BOTSWANA	English	English, Setswana	Fanagalo	+English -Fanagalo
BURUNDI	Kirundi, French	no change	English, Swahili	none necessary
CAMEROON	English, French	no change	Pidgin (English)	no change
CENTRAL AFRICAN REPUBLIC	French	Sango and French	French, Sango	none
CHAD	French	no change	French, Trader's Arabic	+Sara -Arabic
CONGO, Brazzaville	French	no change	Lingala, Kikongo	no reply
DAHOMEY	French	no change	French	none
ETHIOPIA	English	Amharic	Amharic, English	+Tigrinya, Gallinya -English
GABON	French	no change	French	no change
GAMBIA	English	no change	English	no reply
GHANA	English	no change	Hausa, Twi, Akan	+English
GUINEA	French	no change	French	no change
IVORY COAST	French	no change	French	+English
KENYA	English	Swahili	Swahili	no change
LESOTHO	Sesotho, English	no change	English	not necessary
LIBERIA	English	no change	English	no change
MALAWI	English	no change	Chinjanja, English	no change

Swahili and dropped English; Nigeria further codified its federalism by continued reliance on English and by choosing Hausa as the language of the northern region; and Somalia's program of Somali education had resulted in its joint designation with languages of successive foreign occupations that had left their linguistic imprint on administration.

Despite these modifications, the situation seems not to have changed significantly over time. Most of the national/official language modifications strengthened a language or languages that do not single out any one ethnic group for special advantage on the national level. Those nations whose polity closely coincides with an ethnic-linguistic boundary did not need to concern themselves with this question. Nigeria had finessed the problem since Hausa speakers are concentrated in the northern region

TABLE 4.1 (continued)

	1a. National and/or Official Language 1968/69	1973/74	1b. Major Vehicular languages or Lingua Francas 1968/69	1973/74
MALI	French	no change	French	-French, +Bambara
NIGER	French	no change	Hausa, French	no reply
NIGERIA	none	English, Hausa in northern region	Hausa, English	-Hausa
RWANDA	Nyarwanda, French	no change	Swahili, English	-English, Swahili + French
SENEGAL	French	no change	French	no reply
SIERRA LEONE	English	no change	Creole	no change
SOMALI	in flux	Somali, English, Arabic, Italian	Swahili, Central Somali	-Swahili
SWAZILAND*		Swazi, English		none
TANZANIA	Swahili	no change	Swahili	+English
TOGO	French	no change	Mina, Ewe, French	-Mina
UGANDA	English	no change	East African English	none
UPPER VOLTA	French	no change	Dyula, French	-French
ZAIRE	French	no change	Swahili, Lingala	+French -Swahili, -Lingala
ZAMBIA	English	no change	English	no reply

*Not yet independent in 1968/69

and are the region's clearly dominant population element. In two countries, the added African language is ethnically neutral: Sango in the Central African Republic and Swahili in Kenya. Ghana, which was trying to come to ethnic language decisions in the 1960s, had continued to retain English and deferred any changeover. Guinea, which was trying to move to seven coordinately official African languages in 1968-1969, had nevertheless retained French as the sole language five years later.

One way of estimating the ripple effect of the official/national language choice is to learn what the requirements are for those who wish to enter the public employment sector. Even though almost all the countries have a highly centralized administrative structure, it was hypothesized that language requirements would differ for local as against national postings.

TABLE 4.2

Language Requirements for

	2a. National Government Posts		2b. Local-Level Government Posts	
	1968/69	1973/74	1968/69	1973/74
BOTSWANA	English (Setswana)*	English	Setswana (English)	English
BURUNDI	French (Kirundi)	depends on the job		depends on the job
CAMEROON	English and/ or French	no change	English: West French: East	no change
CENTRAL AFRICAN REPUBLIC	French	no change	French	French and Sango
CHAD	French	no change	French and local language	French
CONGO	French	no reply	French	no reply
DAHOMEY	French	no change	French and local language	French
ETHIOPIA	Amharic and English	Amharic (English)	Amharic and (English and local language)	Amharic
GABON	French	no change	Literacy in French, Fluency in local language	French
GAMBIA	English	no reply	English	no reply
GHANA	English	no change	English	English
GUINEA	French	any or several Guinean languages	French and local languages	any or all local languages
IVORY COAST	French	no change	French	no change
KENYA	English	English and Swahili	Swahili and local language	English
LESOTHO	English	English and (Sesotho)	(English)	English and (Sesotho)

* ( ) indicates that the language is highly
desirable, but not absolutely required.

Table 4.2 presents the information separately for the local and national levels. On the whole, a dual policy is indeed fairly common; one or another local language is either desirable or required at the local level in fourteen countries during 1968-1969, even where no local language is stipulated for national level jobs. In only one case, however, is the local language given as sufficient. In all others the civil servant must command the national/ official language as well, while the local language represents an added qualification. In other words, the key to government service is almost as consistently proficiency in a non-African language as table 4.1 would lead one to predict. Kenya previewed its switch to Swahili as a national/ official

TABLE 4.2 (continued)

	2a. National Government Posts		2b. Local-Level Government Posts	
	1968/69	1973/74	1968/69	1973/74
LIBERIA	English	no change	English and local language	English
MALAWI	English, Chinjanja	English	Chinjaja and local language (and English)	English
MALI	French	no change	French and local language	French
NIGER	French	no reply	French and local language	no reply
NIGERIA	English	no change	English	no change
RWANDA	French	no change	French and Kinyarwanda	French
SENEGAL	French	no reply	French	no reply
SIERRA LEONE	English	no change	local language	English
SOMALIA	English and/ or Italian	English, Italian, and Somali	English or Italian	Somali and English or Italian
SWAZILAND		English and Swazi		Swazi and English
TANZANIA	Swahili and (English)	no change	Swahili	Swahili and (English)
TOGO	French and Mina	French	French and Mina	French
UGANDA	English	no change	English and/ or local language	English
UPPER VOLTA	French	French and one local language	French	French and local language
ZAIRE	French	no change	French	no change
ZAMBIA	English	no reply	English	no reply

( ) indicates that the language is highly
   desirable, but not absolutely required.

language by requiring Swahili in 1968-1969 for local-level civil service positions but, interestingly, had adopted English as the language needed for such jobs at the second date despite the choice of its new national language.

At the time of the 1973-1974 interviews, several nations had added an African language to the national civil service qualifications, most of these being the same nations that had added an African national/official language as well. Upper Volta and Guinea, however, made this change without a concurrent modification of the official language. There is no

TABLE 4.3

Medium of Instruction - at Various Levels in the School Systems

	3a. Primary Grades 1968/69	1973/74	3b. Secondary Level 1968/69	1973/74
BOTSWANA	Setswana or English	no change	English	English and Setswana
BURUNDI	Kirundi – 1st year, French thereafter	increased use of Kirundi	French	no change
CAMEROON	English or French	+ local language	English or French	no change
CENTRAL AFRICAN REPUBLIC	French	+ Sango	French	no change
CHAD	French, some Arabic	–Arabic	French	no change
CONGO	French	no reply	French	no reply
DAHOMEY	French	no change	French	no change
ETHIOPIA	Amharic	no change	English	no change
GABON	French	no change	French	no change
GAMBIA	?	no reply	?	no reply
GHANA	English and local language	local language	English and local language	English and some local language
GUINEA	French and local language	local language	French and local language	no change
IVORY COAST	French	no change	French	no change
KENYA	English and Swahili	no change	English	no change
LESOTHO	Sesotho and English	increased use of English	English	no change

evidence that the use of local languages for local jobs increased during the five-year interval.

We can push the question further back in time by looking at the languages used for school instruction. These reveal that standardization of the school program around a European language intensifies as students move into secondary schools—in other words into the levels of schooling from which the power elite and the civil service will be drawn in modern times.

In 1968-1969 Guinea and Ghana were the only nations teaching secondary programs in local languages; all other systems conducted secondary school programs in French, English, or Arabic. In 1973-1974 only Botswana had changed this pattern and now accords Setswana joint

TABLE 4.3 (continued)

	3a. Primary Grades		3b. Secondary Level	
	1968/69	1973/74	1968/69	1973/74
LIBERIA	English in cities; local language elsewhere	English	English	no change
MALAWI	Chinjanja	no change	English	no change
MALI	French and Arabic	-Arabic	French and Arabic	-Arabic
NIGER	French	no reply	French	no reply
NIGERIA	Local language	no change	English	no change
RWANDA	Kinyarwanda	no change	French	no change
SENEGAL	French	no reply	French	no reply
SIERRA LEONE	English	no change	English	no change
SOMALIA	Arabic and English	-Arabic	Arabic and English	no change
SWAZILAND		English		English
TANZANIA	Swahili	no change	English	no change
TOGO	French	no change	French	no change
UGANDA	English and local language	no change	English	no change
UPPER VOLTA	Local language and French	-Local language	French	no change
ZAIRE	French	no change	French	no change
ZAMBIA	Local language	no reply	English	no reply

place with English. In all other countries, the facts remain as before: secondary school instruction is carried out in French, English, and Arabic alone.

At primary school, a child is more likely to be taught in his or her home language, and there seems to have been a good deal of change in the use of local languages at this level between the 1960s and 1970s, not all of it in the same direction.

At the time of the earlier interviews, fourteen countries were teaching youngsters in their home language in the lower grades. Sometimes this language would be truly local; in other instances it would be the widely shared national African language. In some countries the African language was used together with an exogenous one; in several exclusively. What has happened between 1968 and 1973 is a bit confusing. Burundi has intensified its use of Kirundi, the national language, while Cameroon has added a local language where none was used earlier. On the other hand, Lesotho has

TABLE 4.4

Languages of Radio Programs*

	1968/69	1973/74
BOTSWANA	Setswana, English	no change
BURUNDI	French (40%), Swahili, and Kirundi (60%)	no change
CAMEROON	French and English, + small amount of Dyula, Ewondo, Bamilike, Hausa, Fulani, and Bassa	English and French
CENTRAL AFRICAN REPUBLIC	French, Sango, Arabic, Lingala	French and Sango
CHAD	French (50%), Arabic (25%), Sonike (25%)	didn't know
CONGO	French, Lingala, Kituba, Portuguese, Kikongo	no reply
DAHOMEY	French, Fon, Yoruba, Dandi, Bariba, Mina, Fulani	+ English − Yoruba, Mina
ETHIOPIA	Amharic, Arabic, English, Somali	− Somali
GABON	French	+ Bateke, Pongo, Pongwe, Merye, Mina
GAMBIA	English, Wolof, Mandingo, Fulani, Dyula, Soninke	no reply
GHANA	English (25%), Twi, Fanti Ewe, Dagbani, Ga, Hausa, Nzema (total 75%)	Predominantly English, otherwise, no change
GUINEA	English, French, Creole, Arabic, Portuguese, Creole-Port., Wolof, Bambera, Mandyak, Malinke, Fulani, Susu, Kpelle	− Portuguese, Arabic
IVORY COAST	French (90%), Baoule, Ebre, Dyula, Attie, Senofo, Mossi, Dida	French and "dialects"

intensified the use of English as the medium of instruction, and Upper Volta has entirely eliminated the local language as a teaching vehicle. Somalia, Mali, and Chad have removed Arabic, and Liberia has adopted nationwide the "English only" pattern that was previously used exclusively in urban areas. Guinea and Ghana have gone over to local languages entirely.

Of all these changes, only the Guinean and Ghanaian seem to go against an apparent trend, which I would characterize as a trend toward concentration on a very few standardized languages. The deletion of Arabic leaves another standardized language in place; the intensification of Kirundi instruction is in harmony with the two-language pattern for Burundi, as is the intensification of Sesotho in Lesotho. Similarly, Liberia

TABLE 4.4 (continued)

	1968/69	1973/74
KENYA	English, Swahili, Hindustani, and 12 others (unnamed)	no change
LESOTHO	English, Sesotho	no change
LIBERIA	English and local languages (not named)	Kran, English, Kru, Bassa, Gio, Grebo, + 5 more "dialects"
MALAWI	English, Nyanja, Tumbuka	– Tumbuka
MALI	French, English, Bambara, Fulani, Sarakole, Wolof, Maure, Songhai, Tamacheque	French and Bambara, Fulani, Songhai, Sarakole
NIGER	French, Hausa, Djerma, Tamacheque, Fulani, Beri	no reply
NIGERIA	English, Yoruba, Hausa, Ibo, Fulani, Kanuri, Arabic, Efik, Ibibio, Edo, Uroboh	– Arabic
RWANDA	French and English (50%) Kinyarwandi (5%), Swahili (45%)	no change
SENEGAL	Wolof, French, Serere, Fulani, Portuguese	no reply
SIERRA LEONE	English, Mende, Temne, Krio	+ Creole, Limba, Sherbro, Fono, Fulani, Susu
SOMALIA	Somali, Arabic, Italian Swahili, English, Qoti, Maliaw	+ French, Danaki
SWAZILAND		English and Swazi
TANZANIA	Swahili, English, Comoro, Shona, Ndebele, Portuguese, Ovambo, Tlerero, Sene, Zulu	English and Swahili
TOGO	French, Ewe, Kabre, Kotokoli, Hausa, Moba, Bassari	French and 4 major African languages

and Upper Volta are moving further in the direction of concentrating their efforts entirely on a single, standardized language. In all three areas of activity that we have tabulated, this pattern is evident. The selection of a national/official language has been fortified through the further decisions that determine how children will be taught and through canalizing the instrumental efficacy of this language learning in the publicly controlled economic sector.

That leaves one arena of government language policy unexplored: broadcasting. Almost all radio transmissions are beamed from government corporations of the BBC type or from more directly government operated broadcasting facilities. (In addition, there are some transmitters that are run by the overseas divisions of the BBC and the French government's

TABLE 4.4 (continued)

	1968/69	1973/74
UGANDA	English, Luganda, Runyoro/Rutoro, Aleso, Luo, Karamajong, Hindi and 8 unnamed others	English, Swahili, Arabic - all others
UPPER VOLTA	French, Fulani, Mossi, Dyula, Gurunse, Gourmantsche, Lobi, Bobo, Senyufo	French, Mossi and "local dialects"
ZAIRE	Tshiluba, Kikongo, Portuguese, French, Lingala, Swahili	French and local stations broadcast in local dialects
ZAMBIA	English, Beriba, Nyanja, Tonga, Lozi, Lunda, Luvale, Kaonde	no reply

* by the Public or State Broadcast Systems

ORTF, and several religious organizations have broadcasting services based in Africa.) Table 4.4 lists only those transmissions that are handled by the systems of the African states themselves; this will give a clearer picture of the linguistic variety that is given this kind of official support. Even though it is immediately apparent that there are far more languages used in radio broadcasts than there have been until now, it is nevertheless appropriate to stress once more that even this much wider range does not begin to approach the several hundred languages that are spoken in the area. Unfortunately, it was impossible to obtain percentage figures for most of the countries, and so we cannot venture an estimate whether the pattern for the Ivory Coast is typical—that is, a heavy concentration on French broadcasts, with indigenous languages sharing 10 percent of total air time—or whether Chad, with its allotment of 50 percent to French, is nearer the norm.

As is evident, public policy decisions have tended to favor minority, exogenous languages and have given little support to the large variety of indigenous tongues.[7] I suggest that the relationships between this stress on minority language competence and majority language repertories are explicable on three major grounds: traditional patterns of language acquisition, a widespread instrumental attitude toward language, and the maximization strategies of the small groups who assumed power with the transition to independence. These will now be explored.

In many areas of the world, individuals are expected to learn only one language in the ordinary course of their lives. But many Africans have been enculturated in social systems where multiple language or dialect acquisition have been regarded as normal. In many areas of sub-Saharan Africa, the resulting repertories have included lingua francas, which have further expanded the communicative field in which individuals can perform.[8] The pervasive use of these vehicular languages in sub-Saharan Africa has been noted repeatedly.[9] Among various comments, Samarin's is particularly apposite. In discussing the adoption of neologisms and other unfamiliar terms by individual Congolese, he notes a positive inclination to adopt terms not yet locally provenant. Informants explained their usage as symbolic of cosmopolitanism. In my interviews, respondents were asked to list the languages that they knew, and the profiles presented typically included three to five languages. The constellations included local languages, one or more African lingua francas, and an exogenous language. In some interviews, when other clues suggested that respondents might also know Arabic and the information was not volunteered, it turned out that it was not mentioned because the individual regarded it as a "special purpose" language that was used mainly in a sacred context and therefore did not seem to qualify as a language to be included. In addition, in a good many conversations the respondents casually amended their inventory of linguistic range by the tag ending "and several dialects." Although the invariable inclusion of at least one and sometimes two European languages reflects the bias in the interviewing sample, the other parts of the repertories are regarded by the respondents as very usual for fellow countrymen (see table 4.1 and table 4.5).

This widespread multilingualism helps to explain the relative lack of challenge to the narrow range of languages that governments are promoting and requiring. When discussing the information on languages used in radio broadcasts, the attachés pointed out that listenership is by no means restricted to native speakers of the broadcast languages. They noted that audiences listen to transmissions in languages they have learned consecutively, and that they expect to listen and grasp the message of broadcasts in languages that are within a range of mutual intelligibility. In fact, they may well listen in order to broaden their competence. This willingness to acquire new languages has been reported by Tabouret-Keller for several Senegalese cities. She writes that in-migrating persons of any number of ethnic backgrounds learn Wolof eagerly because they regard Wolof acquisition as part of their desired progress toward becoming modern, urbanized Senegalese.[10]

The acceptance of English and French has similar connotations, since fluency in them has long been associated with urban life, elite status, and modernity. Because of this and the older standards of expected multilingual competence, parents would therefore not resist the schools' demand that their children use an unfamiliar code during lessons. Greenberg and Senghor,[11] as well as many of my respondents, pointed out how a widespread positive orientation toward French and English grew once the possibility of adding them to one's linguistic repertory became a reality. One may infer from these attitudes either that ethnicity

TABLE 4.5

Typical Linguistic Repertoires

	1968/69	1973/74
BOTSWANA	Setswana + English + Ikalanga OR Afrikaans + Setswana + English OR monolingual Bushmen	no reply
BURUNDI	Kirundi OR Kirundi + Swahili OR Kirundi + Swahili + French	Kirundi or Kinyarwandi + Swahili in cities
CAMEROON	two or three local languages + pidgin OR one local language + pidgin OR local language + English or French	English, French, pidgin English + one or more native languages
CENTRAL AFRICAN REPUBLIC	French + Sango + one or more local languages	French and Sango
CHAD	one or two local languages OR one or two local languages + French	Fa +/or French
CONGO	French + Kikongo + Lingala + one or more local languages	no reply
DAHOMEY	three or more normal: for Southerners French + Fon + Yoruba + Denda; for Northerners French + Baliba + Denda	French + Fon + Mina
ETHIOPIA	Amharic + local language OR English + Amharic + local language	Amharic + Italian + English OR Tigrinia + Amharic + Italian + English OR Gallinya + Guraginya
GABON	French + one, two, or three local languages	three or four typical; e.g., French + Fon + Mina
GAMBIA	three or four local lan- guages OR three or four local languages + English	no reply

maintenance is not keyed to a monolingual life experience or that other strategic goals outweigh ethnic and language identity questions.

Some of the evidence certainly favors stressing the latter possibility. In the discussions with respondents, instrumental reasons for shelving or essentially ignoring linguistic issues were repeatedly offered. Many of the attachés said that there had been a feeling of relief among independence leaders that they could concentrate on other, pressing problems and use the obviously practical solution of carrying on with the language of the colonial administration. In this respect it is noteworthy that the former Belgian areas never considered adopting Flemish, which offered no future advantages, while the retention of French offered a good many.

For instance, through retention of French and English many states could

TABLE 4.5 (continued)

	1968/69	1973/74
GHANA	e.g., Ga + Twi	more than one, e.g., Twi, Fanti, Ga + English and/or Hausa
GUINEA	local language + French OR several local languages + French OR several local languages	French, English, Spanish + several dialects OR French + 3 major local languages + several dialects
IVORY COAST	Bambara + French OR Bambara + French + Baoule OR Three local languages	unschooled persons: own dialect + a little French schooled persons: own language + French + English + Spanish
KENYA		own dialect + dialect of neighboring groups + Swahili; often + English
LESOTHO	Sesotho OR Sesotho + English	Sesotho
LIBERIA	local language and several related languages OR as above + English	Kru + Bassa + English and several local dialects
MALAWI	Chinjanja + English OR Chinjanja + Swahili + English OR Chinjanja + Tumbuku	Chichewa schooled persons: Chichewa + English
MALI		French + Bambara OR another major language + some of its dialects
NIGER	Hausa + French OR Djerma + French	N/A
NIGERIA	one local language + English OR one local language + Hausa in the north OR one local language + Hausa + English in the north	Hausa, or Yoruba, or Ibo + its dialects OR other local languages plus dialects + one of the three major languages, + English in many cases

move immediately toward their goal of universal education. France and Britain were only too pleased to enter into agreements with their former colonies for the supply of textbooks and curriculum materials. The French-speaking countries were invited to join in a concordat in which both certification of school performance, the *Baccalaureat*, and entrance qualifications for Civil Service became interchangeable with the same qualifications in France. English examination systems were either retained or slightly modified. Entrance to British and French colleges and universities was facilitated for African school graduates with such proof of their prior training. At the same time, many of the new states were able to use French- and English-speaking teachers through the Peace Corps and other voluntary programs, and they could hire salaried teachers from the

TABLE 4.5   (continued)

	1968/69	1973/74
RWANDA	Kinyarwanda + Swahili (for commercial situations) + French (for official situations)	French + Kinyarwanda + a little English
SENEGAL	Wolof + Fulani + French	no reply
SIERRA LEONE	Creole + Mende OR Creole + Temne OR Creole + either Mende or Temne + English	English + dialect
SOMALIA	Somali + Arabic OR Somali + Arabic + Italian (older pattern) OR Somali + Arabic + English (newer pattern)	several mutually intelligible dialects schooled persons: same, + English or Italian or Somali
SWAZILAND		English + Swazi
TANZANIA	local language + Swahili OR local language + Swahili + English	English + Swahili + an African language OR Swahili + one or more local languages
TOGO	Mina + Ewe + French and/or English OR Yoruba + French and/or English OR Fon + French and/or English	French + one or more African languages, many of which are mutually intelligible
UGANDA	three or four local languages OR three or four local languages + English	one or more local languages + English
UPPER VOLTA	Mossi + French OR Mossi + Diyula + French	Mali/Diyula + French OR other dialects + French OR Moré + French
ZAIRE	local language + French OR local language + variants of same	Two dialects + French
ZAMBIA	three or four local languages OR English + several local languages	no reply

French- and English-speaking world. Supplied curricula and textbooks were essentially those used in two European countries; they had not been recast to reflect the history, literature, geography or surroundings of the various nations' children. UNESCO eventually undertook to provide materials that stressed African content; nevertheless, these too were written in French and English.

Both the British Broadcasting Corporation and the Organisation de Radio et de Television Français mounted programs that reinforced the use of French and English. They taped broadcasts and made them available to the various state broadcast systems. More permanent and pronounced was the effect of the training schools for broadcasters: these were established in London and Paris, and Africans came to the two cities to prepare for careers in their home countries. Thus, entry into the radio world presupposed competence in one of two European languages.

A further advantage that a good many of the attachés stressed as they talked about the instrumental desiderata of French and English fluency was the rapidity with which one after another newly independent country was able to take its place in various international organizations, and the speed with which they could staff embassies. This, they pointed out, had enabled their countries to work effectively for development loans, to make their views heard at the United Nations and its subsidiary bodies, and generally to take their place in the international community.

If the exception proves the rule, the case of Guinea ought to be instructive. Unlike the other "Francophone" countries, Guinea decided to break many of the connections that France offered and other nations codified. Although the reasons for this were political, the repercussions were also felt in the linguistic sphere. In Guinea, the decision was reached by 1969 to try to administer and educate the nation in seven of the major African languages spoken by its citizens. This meant attempting to develop teaching materials in seven languages at each level of primary and secondary school and to train teachers who were well qualified to teach in each of these languages for each of these levels; it also meant financing all of this. Guinea did not entirely drop French, so it essentially committed itself to an eightfold effort. Of the eight languages, only French had readily available outside financial backing. (Financial aid, though not teaching materials, was offered by the People's Republic of China.)

In his comparative study of ethnicity and nationalism, E. K. Francis describes a number of European cases where the modern nation-state was created in settings of similar ethnic diversity.[12] I find a particularly striking parallel in European decision-making strategies and the ones more recently used in the countries of Africa. The establishment of the infrastructures so essential to national existence then, just as now, demand the adoption of a medium of technical and political communication. If railroads are going to run, roads going to be built, postal and telephone and telegraph services, not to mention banking and business, exports and shipping, and all the rest of the nonlocal apparatus are to run effectively, this shared medium of communication is essential. Only when all is well in place can smallish ethnic areas assert their more parochial claims without fear of losing the advantages of modernity. No Scottish nationalist expects to bring a halt to the British postal service or to found a separate phone system in which the operators will speak only Gaelic. There was ample evidence in the interviews that this lesson of history was taken into account and that Guinea was regarded with some astonishment for considering such a divergent language policy.

Nevertheless, the wonder is not that these pragmatic choices were made, but that they were made so easily. Doubtlessly, prior value orientations that appreciated and approved multiple language competence and instrumental attitudes toward modernity and cosmopolitanism were brought to bear on the language decisons that had to accompany independence. This same perspective allowed the rapid spread of Islam at an earlier period and the acceptance of Christianity on a wide front in non-Islamic areas more recently. The adoption of Christianity, as I interpret the record, may have been urged by the missionary as a monotheistic, exclusive substitute for other beliefs. It was frequently accepted, as I read it, as another, but not an alternative, belief.[13] The Christians who promoted their beliefs were obviously also backed by powerful invaders, and their religion was accorded respect. One might say that a traditional pantheism was capable of expansion and incorporated another set of deities that were quite evidently linked to groups with power and wealth. Thus both Christianity and the European languages in which it was cloaked became symbols of modernity, power, and cosmopolitanism. Muslim areas were much less penetrated by the Christian missionaries, no doubt in part because they countered the new religion from the position of another monotheistic, exclusivist faith.

If this line of reasoning has some validity, one can better understand why some of the regions and countries where Islam has long been dominant have acted more like other regions of the world where linguistic nationalism has been a factor in language policy decisions. Northern Nigeria and Somalia would be two cases in point. In both there have been decisions to favor and deliberately designate languages associated with the Islamic groups as national or official or both.

Up to this point, two elements affecting language decisions have been stressed: attitudes favoring widespread multilingual competence and the generally instrumental view of language facility. The third reason for the present sociolinguistic situation still needs to be examined. This derives from the particular conditions that prevailed when major linguistic policies were adopted. The small group of persons who had taken the most active roles in independence movements naturally also staffed the first governments of their respective countries. These individuals were a numerical and sociocultural minority. They were the small groups who had obtained a Western-style education, were literate in an exogenous language, and had been using the exogenous language as their lingua franca in which to lay the groundwork for their political goals. English or French (and more rarely Swahili and Arabic) were thus the vehicles by which they were able to communicate across widely dispersed and disparate ethnic and linguistic boundaries, and these languages constituted one of the ties that bound them together. Many of them had studied outside their own countries, and their ability to work in French or English first brought them into contact with fellow Africans at schools, colleges, and universities. For some the experience led to careers in literature in the

European languages. The writings of Senghor, Kane, Achebe, Kenyatta, and others are testimony to the extent to which they found these languages suitable and expressive vehicles in which to cast their observations of their changing societies.[14] Mazrui quotes Senghor as writing that he preferred French as his medium of expression.[15]

This is a sentiment a good many of the diplomats echoed when I first talked with them in 1968-1969. In one question I asked them to name for me what they regarded as "major or important languages" in their countries. Quite a few of them did not spontaneously refer to a single African language when they responded. Further explorations then elicited such comments as these: "well, they [the African languages] are just dialects"; "they're not written." This dismissive kind of comment had disappeared by 1973-1974. Something had clearly happened in the interim that made the diplomats much more eager to share their knowledge of the African part of the national language scene. In 1973-1974 none ignored the African part of the repertory entirely.

The leaders of the newly independent nations were a linguistic minority, drawn from diverse ethnic and linguistic constituencies. Perhaps one can label them an ethnic minority? By codifying their expertise in an exogenous language into the apparatus of statehood, they protected their own very special positions for the foreseeable future. By making competence in their chosen medium of literacy the yardstick for government service and for teaching and broadcasting, they were protecting the power they had just won. At the same time, they were making it more difficult for other forms of leadership, based on more traditionally gerontocratic principles, to reassert themselves, since the older generation was much less likely to be literate or fluent in the new language of administration and education.

Although French and English had been the most frequent choices in the independent sub-Saharan countries, it has already been pointed out that several other languages have been selected as well. Both patterns share one characteristic: they accelerate and accentuate the drive toward uniformity and standardization. The apparatus of the state itself and its preparatory training ground, the state school system, are concentrating expenditures and efforts on a highly selected few of all the area's languages. To date, there has been little evidence of countervailing activity. The decisions reached soon after independence seem to have been generally acceptable, despite ample evidence that ethnicity can be a most explosive force in many African countries. In this sense, then, it would appear that ethnicity maintenance is not strongly linked to the assertion of the ethnic-language tie.

The increased consciousness of the language repertory of their home countries and the change from expressed deprecation of African languages are two small straws in the wind: the linkage of national identity with a foreign language may not perhaps be acceptable for the long run. Although it would be very difficult to reverse early decisions with respect to national/official languages, there are signs that some regrets are being felt. Thus far they are muted and are voiced on behalf of nationhood rather than ethnic groups within nations. In addition, sentiments associated with Swahili cross national boundaries and have been mobilized intermittently on behalf of *ohuru*—freedom from outside domination.[16]

In the light of the data discussed in this chapter, it is clear that a transactional perspective allows one to identify maximization strategies successfully employed. Those who held power at the birth of the modern African states developed language policies that aided the smooth establishment of administration, diplomacy, and national institutions for the enculturation of citizens into the new polities. At the same time they made decisions that very specifically supported their own political careers.

POSTSCRIPT

The materials that have been discussed in this chapter all deal with a relatively short time perspective. The first massive injection of European languages is no more than about a hundred years old. The transition to colonial administration and from that to independence has all occurred in that brief span. The post-independence experience has all come in the last twenty-five years. Before all the traditional language acquisition patterns are completely overlaid with newer ones, it would seem to be very useful to learn more about the nature of language learning in these countries. Are the old modes of successive, apparently easy paths to multilingualism in any way related to the fact that the languages so acquired were learned as spoken languages? Was there a difference in those areas where Koranic Arabic was traditionally taught as a sacred, but written, language? Has the introduction of schools, with their stress on literacy and literacy-based skills in one or two languages changed the patterns of successive acquisition of languages, either in the oral or in the written and oral forms? Has the intense concentration on the language used for school and public affairs made people feel that it is better to concentrate their efforts on mastering that designated language? Does that mean that other languages, which were once important for intergroup communication, are falling into comparative disuse? Can monolingual countries such as the United States learn something important from the African experience to help their school systems improve their very poor performance in teaching languages?

NOTES

1. The work reported in this essay grows out of a long interest in questions of acculturation and interethnic relations, although research among Mexican and German Jewish immigrants in the United States raised questions of decision making in a very different milieu of minority-majority relations. Some of the materials in the present paper have been previously reported at meetings of the

Society for Applied Anthropology (1975), the African Studies Association (1974), and in a paper accepted by *Cahiers D'Études Africaines.* I should like to thank all the diplomats who gave me the benefit of their observations.

2. Classifications and enumerations such as those found in Joseph Greenberg, *The Languages of Africa*, 2d ed. (Bloomington: University of Indiana, 1967); and in the series, *Handbooks of African Languages* (London: International African Institute).

3. Fredrik Barth, *Models of Social Organization*, Royal Anthropological Institute, Occasional Paper no. 23 (London, 1966); and Robert Paine, *Some Second Thoughts about Barth's Models*, Royal Anthropological Institute, Occasional Paper no. 32 (London, 1974).

4. Joshua Fishman, *Language and Nationalism* (Chicago: Newbury House, 1972).

5. For instance, the annual editions of the *World Radio and TV Handbook* (Denmark: Hvidours).

6. The author gratefully acknowledges the assistance of some of her students in language and culture classes in the administration of several of the questionnaires.

7. The following linguistic census data from Senegal are an example of this situation. In 1962, 78 percent of the population age ten and older and 98 percent of the females did not read, write, speak, or understand French. In the Dakar region, however, 57 percent of the males over the age of ten had some French competence. Dakar is the site of political decision making and the arena in which French was selected as the official language of Senegal and as the medium of school instruction. Reported in A. Thieret, *L'enseignement du Français en Afrique. Le Sénégal, Population, Langues, Programmes scolaires* (Dakar: Centre de Linguistique appliquée de Dakar, 1966).

8. I use *competence* in the sense employed by Dell Hymes in, for instance, "Sociolinguistics and the Ethnography of Speaking," in *Social Anthropology and Linguistics*, ed. Edwin Ardener (London: Association of Social Anthropologists Monograph 10, 1971).

9. William J. Samarin, "Lingua Francas, with Special Reference to Africa," in *Study of the Role of Second Languages in Asia, Africa and Latin America*, ed. Frank A. Rice (Washington: Center for Applied Linguistics, 1962); and "Self Annulling Prestige Factors among Speakers of a Creole Language," in *Sociolinguistics*, ed. William Bright (The Hague: Mouton, 1968). See also Joshua Fishman, Charles A. Ferguson, and Jyotirindra Das Gupta, eds., *Language Problems of Developing Nations* (Boston: Wiley, 1968); Léopold S. Senghor, *Négritude et Humanisme* (Paris: Seuil, 1964); and W. H. Whiteley, ed., *Language Use and Social Change* (London: Oxford University Press, 1971).

10. A. Tabouret-Keller, "Sociological Factors of Language Maintenance and Language Shift: A Methodological Note Based on European and African Examples," in Fishman, Ferguson, and Das Gupta, *Language Problems.*

11. Ali A. Mazrui quotes Senghor to this effect in "Some Sociopolitical Functions of English Literature in Africa," ibid. See also Joseph Greenberg, "Urbanism, Migration and Language," in *Urbanization and Migration in West Africa*, ed. Hilda Kuper (Los Angeles: University of California Press, 1965).

12. E. K. Francis, *Interethnic Relations: An Essay in Sociological Theory* (New York: Elsevier, 1976).

13. An example of this may be found in Robert A. Lystadt, *The Ashanti* (New Brunswick, N.J.: Rutgers University Press, 1958).

14. Léopold Senghor, Jomo Kenyatta, and less politically active writers such as Chinua Achebe or Cheikh Hamidou Kane are a few who exemplify this pattern.

15. Mazrui, "Sociopolitical Functions."

16. For instance, A. Oladele Awolubuluyi, "Towards a National Language," *Ibadan*, October 23, 1966, pp. 16-18; and A. Moumouni, "Le Probléme linguistique en Afrique Noire," *Partisans*, November 23, 1966, pp. 37-48. In the opposite vein, Milton Obote, "Language and National Identification," *East African Journal* 4 (April, 1967):3-6.

5

Society and Sociality:
An Expanding Universe

Colin M. Turnbull

In the Ituri Forest of northeastern Zaire, one of the major difficulties for the anthropologist lies in defining the social unit he is studying. Be it a people, a culture, an aspect of a culture, or a single social institution, the boundaries are seldom clear and almost always fluid. Overlap and interconnection are, of course, common enough, but here they are heightened by the demographic situation, in which the original inhabitants, the Mbuti hunters and gatherers, have within the past few hundred years been surrounded by immigrant farmers of Sudanic or Bantu-speaking origin from the north, south, east, and west.

Add to this a history of Arab slave trade (a major branch of which was driven from the east coast, right through the Ituri, and down the Congo River to the Atlantic), Stanley's still remembered three disastrous treks through the forest, a half century of colonial exploitation, and an ideological warfare between Protestant (mainly Baptist) and Catholic missions, and the magnitude of the opportunity for cultural confusion is evident.

Overlap, then, is inherent in the situation; it has been imposed by demographic movements and relatively recent consolidation of the mixed forest populations within arbitrary and frequently unsatisfactory geographical boundaries; and by the influx of Arab slavers and colonial power, each of which in its own way imposed an unwilling physical mobility upon a small portion of the population while encouraging another kind of structural mobility on the total population as a defense mechanism. Two dominant themes arise and govern both interpersonal and intergroup relations: the necessity for establishing an unequivocal identity by which the person (and/or group) is clearly distinct and definable in the midst of this plurality, and the necessity for establishing a mechanism for social as well as physical mobility in such an uncertain and ever-changing world. Psychological as well as physical survival are thus at stake, and both seem to have been constant foci of attention for all segments of the forest population ever since the first wave of immigrant farmers arrived.

At times the individual is driven into himself, to a point where he becomes temporarily (and sometimes permanently) isolated from any group membership, where nothing stands between himself and the totality of the world around him. When he emerges from such isolation (and this is the trick), it may well be as a member of a different group, one quite distinct from the one he left, as though he were reborn. Change in group allegiance is indeed institutionalized in this way, and the period of isolation usually passes under the guise of possession, witchcraft, or sorcery (by which I mean spiritual agency, unwitting human agency, and witting human agency, respectively). Thus not only a man's group membership, but also his status and role, may change. Each population has its own charter for the process, but among none that I know of is it so clearly expressed as among the Mbuti, who are of course in the center of the changing world rather than on the periphery and who therefore have a rather different perspective. I shall refer to this later.

Something of the same situation pertains at the group level: groups have at one end an isolationist sense of their own identity and define themselves in institutionalized ways as distinct from all other populations around them; but at the other end of the social horizon, they see themselves as part of the whole. It is as though fission and fusion were placed between two reflecting mirrors so that the process is multiplied endlessly, for just as far as one cares to or needs to look. Each image is a reflection of the other, differing mainly in size, yet no two images are the same. Historically, tribal populations entering the primary forest from open grasslands were compelled to split into smaller segments as they competed for the fringe lands at the edge; and as they were driven deeper into the forest, they split further, ultimately forming a string of relatively autonomous and isolated villages. We get, in this way, the Bela (sometimes Bera), who still retain some of their original cattle culture and who managed to win the land between the eastern edge of the forest and the western shores of Lake Albert, while their kin, the Bira, hived off into two major segments across the banks of the Ituri River, each of those further subdividing into patrilineal segments. The Bira stretch westward to where the Epulu River crosses the old Arab slave trail. There the Ndaka peoples begin, similarly subdividing into the Mbo, and each segmenting yet further into minimal lineages. This process has characterized all the forest population in the Ituri, forming a circle of historically overlapping populations around the central core of Mbuti, which population alone has not subdivided in quite the same way (though the origin of the division between the eastern archers and western net-hunters is still obscure to me and may be related to the overall process in some way).

The process is formalized not only through the marriage circles that weld certain groups of these autonomous villages together but also through the institution of *nkumbi* initiation.[1] This seems to have been at one time an exclusively Kumu institution (though other peoples here considered probably had comparable forms of initiation), but it has progressively opened itself to outside membership until today it has all but closed the circle around the Mbuti, who may be initiated but never initiate. The political function is clearly seen by the acceptance of the circumcision ritual by some of the formerly noncircumcising Sudanic peoples during the internecine and mercenary warfare of the 1960s. This closed the major gap in the northeasten segment of the *nkumbi* circle, consolidating the political unity of the overall forest population. The enormous political elasticity of the institution and its role in redefining boundaries was seen by Joseph Towles in the early 1970s, when throughout the southern sector of the forest, at least, Mobutu himself was incorporated, through mime, into the traditional and hitherto seemingly exclusive institution, in just the same way that traditionally more locally important individuals are incorporated as "fathers" of the *baganza* (children of the *nkumbi*).[2]

Individuals as well as groups thus have a plural identity, and each moves from one frame of reference to another as suits the context of the moment. In this situation, the term *ethnicity*, if it is applicable, implies something different from its normal connotation. Yet the term *is* applicable, and ethnocentricism is an important positive structural institution in the Ituri. The application of these terms requires, however, that we return to an earlier usage where the focus of attention was on difference of religious belief rather than on difference of physical appearance or type. It is this focus of attention that dominates the Ituri, even when the relationship being defined is between the full-statured immigrant farmers and the small-statured (pygmy) Mbuti hunters. The physical appearance of the two groups is further sharpened by significant differences in skin color, hair form, body proportions, and body odor as well as by marked differences in dress and ornamentation. Under these circumstances, the lack of concern with the physiological factor is all the more remarkable.

For the sake of precision, from here on I shall concentrate the discussion on the relationship between the central (net-hunting) Mbuti and the adjacent Bira and Ndaka villagers. As seen in figure 5.1, the western Bira, on the eastern banks of the Epulu River, intermarry with Ndaka on the west bank, just as the eastern Bira tend to intermarry with the Lese, and the western Ndaka with the Budo. For each "tribe," the extremities are not directly connected; the connection is indirect, through a series of overlapping marriage circles. These marriage circles form loose federations of villages that perceive themselves as having distinct identities through the intermarital connection. This is manifest at the political level through the informal ascendancy of a dominant male member of any one of the villages (sometimes a dominant female, of local lineage) within the federation. This ascendancy is temporary, contextual, nonhereditary, and charismatic.

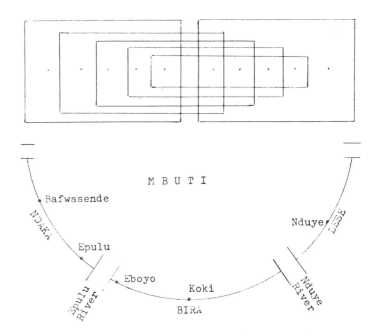

Figure 5.1 The first section presents an example of the kinds of overlappings created by village marriage circles. The second section represents the southern sector of the *nkumbi* federation.

It generally rests with someone who has hereditary ties with the area, however, and not with one of the "displaced persons" who have relocated themselves and redefined their allegiance and identity following the period of isolation referred to earlier. Such foreign (from our point of view) members often occupy other positions of importance, however, notably in the corresponding, but further-reaching, *nkumbi* initiation circles. These circles tend to reach to the center of the tribal area; thus even initiations performed on the eastern banks of the Epulu will draw participants from the central village of the western subsection of the Bira, providing a direct link with those Bira who associate with the Lese (due to the recent expansion of the *nkumbi* circle) on the one hand, and with the eastern subsection of the Bira on the other.

The units formed by marriage and by *nkumbi* are therefore different both in size and quality, the smaller unit providing a sense of social identity, the larger providing a sense of political identity. It is notable that at points, such as the Epulu River, where tribal boundaries meet, both marriage and initiation rituals are in the process of change. At the village of Epulu, on the Ndaka side of the river, the influence is predominantly Ndaka, but even a marriage between two Ndaka is attended by Bira friends and kin, and the ritual will incorporate some Bira elements. When the marriage is between

Bira and Ndaka, there will frequently be interminable arguments as to whether the ritual should be Bira *or* Ndaka. The fact that for both (patrilineal) peoples marriage is clearly patrilocal is not seen as being of any particular significance in this respect. Each side rather takes the point of view that only its own ritual is binding on its members, regardless of sex and locality. Numerous factors contribute to the solution of the problem, which most frequently, but by no means invariably, lies in favor of the groom. But the ritual is seldom clearly either Ndaka or Bira, and often ritual officiants from both sides participate, doubling each other's actions where the rituals coincide, each going their own way where there is no coincidence.

As a purely subjective statement, let me express a sentiment that, however, is shared by most Bira and Ndaka with whom I have discussed the matter and with whom I have attended marriage ceremonies. There is a distinct feeling of division and distinction that the joint participation in a combined ritual does not negate. Many of the ritual elements, including those of reversal and rebellion, emphasize the ultimate distinction between the sacred and the profane. Religious (spiritual, sentimental) affiliation is not changed by the political and economic aspects of the union, even if these are prominent (a rare occurrence here). The very problems involved in the performance of a dual ritual for which there is (as yet) no established model, the clumsiness of the ad hoc improvisations by both officiants and participants—all point to the vital and essentially religious and sentimental division between the two groups.

To a lesser extent, the same thing is true at any point where marriage circles overlap, even within the subsection of a tribe. Thus while at Eboyo, on the Bira side of the river, intertribal marriages are fraught with the same problems as they have to face at Epulu; further east—at Koki, say—there are undercurrents of similar tension at many marriages that take place there, since while this is a point of marriage overlap exlusively between Bira individuals, there are those on the one side that may have affinal or *nkumbi* links with the Ndaka and those on the other side that may have similar links with the Lese or eastern Bira. Significantly, if the link is not affinal but through the *nkumbi* alone, the tension is either less, or even totally absent or nonapparent.

When we consider the intergroup relationships as defined by joint participation in any *nkumbi* ritual, however, there is a difference. For one thing, there is much less ambivalence. An initiation at either Epulu or Eboyo will draw participants and officiants from both Bira and Ndaka villages, in each case from as far as the central village of the tribal area, Bafwasende for Epulu, Koki for Eboyo. It is not unknown for the chief of one tribal subsection to cross the river in order to participate in an *nkumbi* at either Epulu or Eboyo, though this generally tends to create otherwise absent tensions and will only be done if there is a significant number of boys from the other group being initiated. Thus when four of the nine *baganza* at an Epulu *nkumbi* were Bira, the Bira chief made a point of attending the opening festivities.

Perhaps the most significant difference between the *nkumbi* ritual and the marriage ritual, where both involve dual tribal affiliations, is that the dual aspect of the ritual itself is clearly formalized for the *nkumbi*, leading to a total lack of the sense of ambivalence that characterizes an intertribal marriage ritual. Distinction is still maintained; each officiant will be dressed and ornamented according to his own custom, so that while the similarity is dominant, the individual distinctions are nonetheless evident. This applies to both the doctors who oversee the entire *nkumbi* ritual and their acolytes, who perform the actual circumcisions. Normally a boy is circumcised by a circumciser of his own tribe, but this is not essential. However, it should be pointed out that when Mbuti/Villager relationships are considered, the system is such that an Mbuti can never be circumcised by an Mbuti, since they are never allowed to become officiants.

Perhaps reference should be made at this point to the distinction made in the title of this chapter between society and sociality. Whereas the Ituri may be seen as a series of overlapping, formal social units, each with its own definable boundaries, sociality is an informal continuum that, structurally, is just as significant, if not more so. Its manifestation may be determined by the societal context, explaining why interpersonal relationships are characterized by a constant flux between sociality and nonsociality or even unsociality; but what is evident is that whereas relations between two individuals (or two groups) may vary and fluctuate constantly at a political level, as defined by the bonds of *nkumbi* for instance, they remain much more constant at a social level as defined by kinship. It is not so clear-cut that we can say that whereas *nkumbi* determines effective political relations, kinship determines affective social relations; but there is a clear distinction between the *nature* of the affective, sentimental bonds created by each institution. Whereas the bonds created by common membership of the *nkumbi* are backed by supernatural sanctions, membership is far too diffuse and dispersed for effective economic or other sanctions that could directly affect daily life. Further, the *nkumbi* bond is invoked contextually; it is part of the process of constantly selecting at which point in the chain of mirrors one chooses to see oneself.

Bonds of kinship, however, while also backed by supernatural sanctions, are also backed by sanctions that are applicable in almost every day-to-day context, and the bonds are permanent and all-pervasive, affected much less by contextual considerations than are the bonds of *nkumbi*. It is perhaps a question of spiritual versus political identity, and once again we see the Mbuti/Villager relationship as being much more circumscribed than intravillage or intervillage relationships, for the prime structural factor relating the two groups is *nkumbi*, intermarriage being exceptional. Thus

Mbuti/Villager relationships are almost entirely contextual, political in nature, effective rather than affective (which is not to deny the existence of a great deal of a certain kind of affectivity), and rooted in a mutuality of needs rather than in any real sense of common identity. This means that Mbuti/Villager relationships are, like those of all *nkumbi* partners, simple and direct, unlike the multiplex (to borrow from Gluckman) relationships established by marriage. The net result is that there is less hostility in Mbuti/Villager relationships than there is in intertribal relationships, and less between tribes than between villages within any one tribe. Areas of hostility seem to coincide approximately with marriage circles, where man's most basic spiritual identity is in a sense questioned, challenged, threatened by ambivalence, and ultimately polluted and desecrated by the law of exogamy.

For the Ituri, at least, insofar as ethnicity refers to religious differences, it refers to differences of self-perception rather than to differences of any outwardly visible identity. In noting that Mbuti/Villager relationships are more stable than intervillage relations, we also note that their self-perceptions are more different. All villagers share the same sense of separateness from and hostility toward the forest, though to different degrees. Furthermore, they all make more of the distinction between man and other animals than do the Mbuti, who see the forest, its flora as well as its fauna, as a whole of which they are no more than an integral functioning part. The Bira in particular see themselves in their context of relatively recent history, as an immigrant people, once this and now that; but they have lost touch with their more distant past, with their source of origin. The Mbuti see themselves only as part of the timeless forest. They habitually only use the present tense, letting the context make it plain whether anything is behind or ahead in time, and similarly with space. The Bira are almost overly concerned with the exact location of everything and everyone in time and space. They have a clearly linear sense of time, though some occasionally speculate as though their concept were cyclic rather than linear. In either case, it is vastly different from that of the Mbuti, which is neither cyclic nor linear, but spherical.

The Mbuti see themselves as, each one separately and all together, in a time-space capsule. As the Mbuti moves, so does the capsule; he is always in the center of his time-space universe. Disorientation comes if he moves too quickly, particularly if he reaches the periphery before his world has had time to catch up with him. Such precipitate action (or thought) can cause sickness, unsocial action (another form of sickness), or almost any other kind of aberration. If such movement is too violent, say the Mbuti, the shell can be pierced. What happens then is that you emerge into another existence. It is likened to what may happen if you immerse yourself totally in water. What, after all, is the essence of that reflection that sends its foot up to meet yours as you take the first step; as your leg goes down into the water, the other comes up, into yours, and disappears it seems . . . or is it your own leg that has disappeared? And when you are finally almost totally immersed, and take a last look upward into the world you are leaving, what has happened to that other body that came up through your body, using it as a pathway to your world? And when, after total immersion you lose all contact with your world, and (on crossing to the other side of the river) you emerge on the other side, some other body comes from the void, passes through yours and into the water, step by step as you leave it. Perhaps this is what happens when you die; perhaps this is why sometimes you shoot and hit an antelope, but both the antelope and the arrow disappear, or the arrow is found with no trace of blood on it: it was really an antelope from the other world, temporarily made visible, perhaps because it had shaken a branch and sprinkled itself with dew.

There is no set form to this Mbuti concept, it is a subject of endless discussion and speculation, but this is the core, a clear focus on the here and now and an equally clear lack of concern with what is not here and now. This is a rather typical hunter-gatherer attitude, well suited to their circumstance; it allows for change, even for aberration (though that word now takes a different meaning), but without altering the shape of things, only the content. Anything and anyone that are "of the forest" partake of this form and are within the overall capsule; those that are of the other world, like the villagers, should stay there, be kept there; they have an existence appropriate to them but not to the forest, they are not of the here and now. This is the basis for the constant endeavor of the Mbuti to keep the villagers out of the forest, for if they penetrate, their clumsy, over-fast, aggressive, violent actions will continually pierce the shell, and they will pass through, allowing other elements to replace them.

Happily for the Mbuti, the villagers, with a few rare exceptions, want to stay as clear of the forest world as they can. Their supernatural world contains what for them are facts rather than speculation, an ordered hierarchy of spirits, ghosts, and ancestors; and the forest is the habitation of all that is harmful. They recognize its existence, and they need to relate to it and to its inhabitants (human and nonhuman); but whereas the various immigrant peoples relate to each other by marriage as well as through the *nkumbi*, all exclude the Mbuti and are equally excluded by the Mbuti from any kinship tie other than fictional. The *nkumbi* is the major structural link between the two worlds. It imposes a limited congruence of social action, but none of belief.

The rare exceptions to the rejection of intermarriage have never to my knowledge involved an Mbuti male and a village female; it is invariably a village male who will take an Mbuti bride. Sometimes it is said this is because she commands less bridewealth; but I have found that even where this is the case it is only part of the story, and the net result is much the same as where the intent of the marriage is, as it most frequently is, to acquire a closer relationship with and control over the forest world. The few intermarriages between Mbuti and villager that occur are nearly always clearly ritual in nature and intent. The offspring, although denied access to the forest as Mbuti, nonetheless are in every sense medial and play a vital role in one or other ritual capacity, frequently in connection with the *nkumbi*.

In the 1960s the Ituri was rent by warfare. Military activity was confined mostly to the road that bisects the region, and the villagers fled into the forest. Once there, tribal boundaries became blurred, but all villagers alike suddenly found themselves at the mercy of the Mbuti. Those whose social horizons were too fixed and exclusive and who tried to survive without Mbuti cooperation died of exposure, malnutrition, and from eating poisonous foods. It has been estimated that nearly half the village population died this way during the post-independence period of anarchy. Those who joined forces with the Mbuti survived: but they never became an integral part of the forest world. The Mbuti kept them as isolated as ever, continuing to render all the services they rendered formerly, thus keeping the villagers ignorant of forest lore and without the technological knowledge necessary for survival. It was as though the old villages, and the old system, had merely been relocated. Most important, the ideological distance between the two was maintained, if not heightened; each side maintained its integrity untouched, its identity unpolluted.

In 1970, when all but a few last Simba "rebels" had either surrendered or been captured or killed, the villagers began to return to the roadside. Their old village sites were for the most part completely overgrown, and in any case the initial flight had dispersed the population so that people did not always emerge at the same point at which they entered. As villages were reconstructed, then, new members were incorporated purely on grounds of their place of emergence from the profane forest. In the Epulu region, villagers from the north (Budo, Mangbetu, and Azande) were incorporated in this way, being adopted into Bira and Ndaka families. A few villagers, overanxious to return to the former status quo and thinking that in their years in the forest they had learned enough of forest lore to do without the always problematic help of the Mbuti, tried to forage for their own food on the outskirts of the new villages while waiting for the first fruits of their new plantations. There was a succession of deaths through the consumption of poisonous roots and mushrooms and through accidents, including snakebite. Even back in their old village habitat, the villagers found themselves still dependent on the Mbuti.

Two things were particularly significant in this situation. During those years in the forest, neither Mbuti nor villagers attempted to change the nature of their relationship by intermarriage, though if a function of marriage had been to break down barriers, to create greater interdependences, one might have expected such a course. Second, during those years there had been no *nkumbi* initiations; the entire institution was suspended. Various reasons were given for this, the major one being that the drumming, singing, and playing of the *makata* (a portable and highly sonorous xylophone), which must continue for three months or more, would have attracted mercenaries and Simba alike. However this may be, the suspension of the *nkumbi* at this time effectively increased the political distance between Mbuti and villager and removed a prime mechanism by which the villagers saw themselves as acquiring control over the Mbuti.

When the situation returned to normal and the villagers reemerged from the forest, in a state of some disarray, the prime need was for the rapid reestablishment of economic and political ties between villages and between tribes. Marriage was *not* used as a vehicle for such reintegration, the *nkumbi* was. There was an almost immediate and widespread surge of *nkumbi* activity. This was partly due to the fact that for two, sometimes three, successive three-year periods there had been no initiations due to the political turmoil. That at least accounted for the size of the initiations, but not for the increased intensity. Further, it was evident in the planning of the initiations by each village that the overall reestablishment of political alliances was a prime goal. The *nkumbi* was used consciously and effectively to reestablish old alliances, forge new ones, and formalize those that had come about during the interim as a fait accompli. Nowhere did I find marriage being used in this conscious way for these ends. A new factor was that whereas prior to the 1960s the presence of more than one *nkumbi* doctor, or at least the presence of doctors of foreign tribal affiliations, was confined mainly to villages near tribal boundaries, it was now rare to find any village celebrating the *nkumbi* without doctors and circumcisers from at least two, if not three or four, tribal affiliations.

Once again, despite the fact that their total dependence for survival upon the Mbuti had only come to an end a matter of a few months back, the villagers still excluded the Mbuti from all initiatory roles, placing them in a position that might be taken by some to be a position of subservience or inferiority (and is undoubtedly seen as such by some villagers), but that is best understood strictly as a medial position in which concepts of superiority and inferiority, superordination and subordination, have little relevance. The field work of Towles (unpublished) shows very clearly both how the *nkumbi* is used consciously for the extension of political horizons between villagers, on the one hand, and as a way of relating, on the other hand, all villagers to the forest by incorporating the Mbuti in a medial role. In fact it is through their common relationship to the forest through Mbuti mediation that the overall forest society reaches its maximum extension.

The *nkumbi* also creates effective economic bonds, both on an intertribal and intratribal level. The formal bond brother alliance of *kare* is probably every bit as effective as that of marriage in establishing day-to-day economic relationships and in providing a means for the informal resolution of disputes, regardless of tribal affiliation. So effective indeed is the *nkumbi*, in almost all aspects of life and social relationships, that one might wonder just what might be the functions of marriage, other than in providing for the orderly procreation of children, for the systematic inheritance of wealth and status, and so forth. Certainly the wider economic and political functions often associated with marriage are here usurped by the *nkumbi*.

There are several clues to the answer. One lies in the crucial difference between *nkumbi* and marriage rituals when these involve the linking of parties of more than one tribal membership. Whereas an intertribal marriage gives rise to an ad hoc dual ritual, each side consecrating its own member, an intertribal *nkumbi* ritual is a unified whole and functions as such; it is different from the ritual (now fast disappearing, it seems) when only one tribe was involved, but nonetheless it is a tightly knit, integrated whole. Another clue lies in the exclusion of the Mbuti from both marriage with the villagers and from initiatory roles in the *nkumbi*. Yet another is the overall conceptual difference, especially with regard to time and space, between Mbuti and villagers. The latter difference means that in the process of overlap, both through marriage and through participation in the *nkumbi*, by which the villagers form a circularly disposed federation, the villagers see themselves as in any one of a sequence of positions, in each of which their horizons are bigger or smaller, with the horizons of the *nkumbi* being different from those of kinship, which, to a large extent, they crosscut. Further, the villagers see their fluctuation from one position to another as being a linear process of addition and subtraction, a movement back and forth in time and space, of increase and decrease. The essence remains the same, only the position is changed, and with it, role and status. For the Mbuti, quite to the contrary, such changes, which also occur in their own world, are seen as changes in essence. It is the position, the status and role, that remain the same, and the essence that is changed. For an Mbuti, one does not move from one state to another, one becomes another. Here is the ultimate barrier against the true integration of the two worlds, the mainspring of what effectively is a system of indigenous and mutually acceptable apartheid, more effective and far-reaching than anything that could have been conceived by the conventional racist who would have brought his own concept of ethnicity to this situation.

Identity is the key to the situation, and it is here that, in this expanding universe of the forest, the institution of marriage finds its particular function. If *nkumbi* membership provides a series of political identities, kinship provides the stable unchanging core of spiritual identity. However much the *nkumbi* ritual models itself after, and mimes, the physiological phenomenon of birth, its membership is still to some extent an act of volition, intellectual rather than spiritual, and its ritual is seen as an artifice, however effective it might be in securing the proper sanctions for appropriate political behavior. Nothing can touch physiological birth for intensity and intimacy in the sense of belonging. It is to the locality and lineage of one's birth that one's deepest sense of identity lies. The birth rituals of the Bira and Mbuti alike, while by no means as public or as dramatic as rituals for other rites of passage (perhaps because the situation is less equivocal), are charged with a vital intensity of their own. But, alas, even for patrilineal peoples (the Mbuti, incidentally, unlike the villagers, are not patrilineal, but bilateral) man can not be born of man alone, a regrettable fact that the *nkumbi* attempts to negate in its ideological rebirth of man without any female assistance. And equally regrettable for such lineally oriented peoples in the forest, the rule of exogamy compels each marital partner to an act of defilation by which his or her whole inner integrity and identity are threatened by the pollution that is marriage—for marriage is indeed licensed pollution. The ritual accentuates the fact and, at the same time, emphasizes the *separate* identities of the participants and the groups to which they belong. As mentioned, this is particularly obvious when the marriage involves members of different tribes, but it is always present. It is also evident in the preoccupation villagers have with incest, an ideal forbidden precisely because it *is* an ideal, too perfect and too powerful for mortals. It is evident again in the innumerable sanctions against marital cohabitation in various crisis situations in which the utmost purity is called for. The necessary act of copulation between members of different lineages is pollution enough; there is no need for the marriage ritual to compound the defilation by suggesting that the two have or will in any other way become one.

Thus, among the villagers, marriage serves primarily as a divisive factor, by emphasizing separate identities on the basis of lineage. This is the absolute reduction of the social unit for the villagers; this is their unchanging core identity that will remain constant throughout all the lamentably necessary changes of position through which the individual will have to pass. It is absolutely essential to the villager that in his ever-changing world, fraught with ambivalence, unpredictability, and with a long and consistent history of duplicity and danger, he should have this unshakable sense of self; it is part of the nature of the fragmented, isolated forest village life that he should reduce his basic horizons and his true sociality to the level of the localized lineage. Marriage formalizes the situation for him, breaking his world down into its smallest component parts; and the *nkumbi* puts it together again, allowing for almost any contextual changes of position that can be imagined. For the villagers, ethnic horizons can almost be said to be horizons of lineage, and the lineage can be said to be the basic community of believers. It is the interlineage relationship that is most beset with potential hostility.

For the Mbuti, it is a different world, one in which identity is reduced still further from lineage (which is of mininal importance to them) to the individual self, and one in which the systematic overlappings provided for by both the marriage and *nkumbi* circles of the villagers are neither necessary nor desirable. Each Mbuti has his own private personal world, isolated within his own time-space capsule; but the major part of his life is spent in conjunction with others in the common fellowship of the band. This is his prime community of believers, membership in which comes from community of action, not from community of breeding patterns. The fact that the band changes composition every month is neither here nor there, for the band always is. So also the territory always is: it is only the content that changes, the essence. The ultimate horizon for the Mbuti is the forest, at the fringe of which lies the nonforest world of the villagers. Unlike the progressively expanding world of the villagers, the Mbuti do not see themselves as part of an overlapping series of progressively larger or smaller units. For them there is nothing between the band and the forest, just as there are no intermediate units between the individual and the band. There is no federation of bands—the Mbuti would tend to agree with Berkeley that *esse est percipi*, only that which you see has existence, or only that band of which you are a member at that moment has validity.

What the situation reveals about ethnicity is that, in this context at least, clear ethnic perceptions, that is to say, clear concepts of identity, are the first essential to the orderly and nonhostile, nonviolent relationship between adjacent groups. Opposition is not only possible without hostility, it is essential to the absence of hostility. The greater the ambivalence with respect to identity, the greater the likelihood of hostility.

The *nkumbi* merely reflects the necessity for political and economic interaction inherent in the demographic situation. It provides the structural form for such interaction. Marriage, on the other hand, provides the means by which distinct and separate identities are formulated and maintained. It is the medial role of the Mbuti that forms the third element of an essentially triadic relationship; a single, stable constant making possible the perpetual subtle shifts in societal position and sociality characteristic of village life in such a way that every villager maintains a firm sense of self regardless of where he may find himself at the moment. It is also the medial role of the Mbuti that gives the expanding universe its widest extension (creating a sense of a greater unity that made itself felt in opposition to slavers, to colonial powers) and that in the early 1970s was making itself felt just as strongly in opposition to the new Zairoise administration. It may well be the genius of the system that it will undergo yet another extension and incorporate the state within its horizons, and indeed the introduction by mime of President Mobutu into the *nkumbi* is an indication that this is already happening.

The conflux that we find in the Ituri of so many diverse peoples and cultures is uncommon in both its extent and its intensity. There are major differences of language, of subsistence economies, of belief; and apart from the visible and invisible cultural differences, there are major and highly visible differences in physical type. These diverse peoples have all had to compete for the same area, or a portion of it, and each having won its portion exploits it in its own way. The greatest opposition is, of course, between the Mbuti hunters, who survive by maintaining the forest intact, with no cultivation or domestication, and the village farmers, who can only survive by cutting the forest down. Yet once these populations had reached a more or less satisfactory division of the territory, the situation became and continued to be characterized by a remarkable lack of violence and hostility. Opposition was expressed in terms of difference rather than in hierarchical terms of superiority or inferiority. Marriage as an institution came to have the special function of openly expressing difference, of reinforcing group identities in terms of spiritual affiliation, and it is to the exogamous lineage that we can best apply the term *ethnic group*, because this is the basic community of believers (for the villagers). Marriage, through this expression, helps to expel hostility through expelling ambivalence, not by creating any mystical union.

The *nkumbi* initiation provides the mechanism for transcending the isolationist tendencies manifest in the marriage ritual and the sense of ethnicity as defined above. Just as the continuity and growth of the sacred ethnic group is only possible through the pollution of marriage, so is political survival only possible through the coalescence of diverse elements in the framework of the *nkumbi*.

Marriage and *nkumbi* are two complementary processes, then, dividing and uniting, creating a functioning whole through the unequivocal establishment of the sacred ethnic (spiritual) group on the one hand and the plural political entity on the other. The third factor, acting as an integrative force, is the medial role of the Mbuti. The fact that they happen to be the most visibly distinct of all the forest populations is an accident of history, but that should not prevent us from recognizing what an enormous asset this outward, visible difference is. To the same end, various groups of villagers practice physical mutilation and differentiate themselves through clothing and ornamentation. Ethnicity then seems to be a structural necessity such that it has to be heightened or shaped when it is weak or without visible form; it is a phenomenon that when divorced from hierarchical ranking can be a highly positive and integrative force.

NOTES

1. Colin Turnbull, "Demography of Small-Scale Societies," in *The Structure of Human Populations*, ed. G. A. Harrison and A. J. Boyce (London: Oxford University Press, 1972).

2. Joseph Towles, "Symbiosis and Opposition in Inter-Group Relations," unpublished notes.

6

Place and Ethnicity among the Sandawe of Tanzania

James L. Newman

Although definitional details vary somewhat, the central tendency of ethnicity is how people define themselves vis-à-vis others, how distinctions are made between "we" and "they."[1] Any number of criteria can be used, either singly or in combination, in arriving at such distinctions, but the important point is that they create a feeling of sameness among people, a feeling that subordinates other possible forms of stratification such as age, sex, and class.[2] Ethnic groups are "whole" societies from a biological perspective, in that they have the necessary age-sex structures for reproducing themselves; and, in a very real sense, ethnicity can be thought of as a "living" concept because it comes out of the world of interaction, competition, and conflict among people as they attempt to survive and prosper.

Because spatial proximity facilitates the forging of a common identity, it is not surprising that most ethnic groups have a territorial basis. It also is not surprising in light of the fact that throughout much of the world land is the critical resource and thus must be secured if continued existence is to be possible. Indeed, rights in land are often the sine qua non for group membership, and consequently people and place become virtually inseparable. That this situation prevails in Africa almost goes without saying; it has been documented by numerous anthropological mono-graphs, land tenure studies, and several well-known novels,[3] and highlighted by the many instances of opposition to imposed policies of land conservation and population relocation during both colonial and postcolonial times.

Although we recognize that this man-land bond exists, we nevertheless have little knowledge about the nature of the cement. That is, what features of a place seem to be most intimately linked to a people's ethnic identity? And, as is so important in present-day Africa, how are the links changing? These questions will be probed in the present chapter by focusing on the Sandawe of the Kondoa Area in central Tanzania. Before proceeding, however, it will be necessary first to set the environmental stage and then to establish the basic dimensions of Sandawe ethnicity.

THE ENVIRONMENTAL SETTING

The Sandawe number about 30,000 and reside primarily along the slopes of a granite-based hill mass known, appropriately enough, as the Sandawe Hills (figures 6.2 and 6.3). Local relief averages slightly over 300 meters, with the crests of the hills rising to 1300-1400 meters. For the most part, the Sandawe Hills carry a cover of open, deciduous *Brachystegia* woodland, the type normally referred to in eastern and central Africa by the term *miombo*.[4] As is invariably the case with this vegetation formation, the underlying soils are highly leached sands and loamy sands having a low natural agricultural potential, though they tend to be quite easily tilled. Along the valley bottoms, heavier clay-loam soils predominate with an associated low bushland and thicket vegetation, whereas the hill crests are often bare outcroppings of granite. To the northwest the hills lose relief, and eventually they merge into the broad, gently undulating plateaus that dominate so much of western Tanzania.

Northeast of the Sandawe Hills are the similarly composed, but slightly higher and more rugged, Songa Hills. However, they are very sparsely populated because of the heavy infestation of tsetse (*Glossina morsitans*) and the consequent threat of both human sleeping sickness (*Trypanosomiasis rhodesiense*) and *nagana* in livestock.

Separating these two hill masses is a lightly settled belt of flatlands covered predominantly by dense stands of thorny *Acacia-Commiphora* thicket, which in places is virtually impenetrable to anything but elephant herds. The soils are mainly brownish clay-loams that are richer organically than those in the hills but are much more difficult to cultivate. Where slight depressions occur on the flatlands, the thicket gives way to open savanna and occasionally grasslands that are supported by black, heavy clay soils. These are known in Tanzania as *mbuga* and are infamous because during the rains they become veritable quagmires; in the dry season, however, the surface bakes out to a cementlike consistency.

Moving northward the flatlands give way to a continuous but gentle rise in elevation that persists beyond the borders of Usandawe to the flanks of the inactive volcanic cone of Mount Henang.[5] This again is tsetse country, with little human population. The vegetation cover consists of a mosaic of *Combretum* savanna, *Acacia-Commiphora* thicket, and *mbuga*, depending on local edaphic conditions.

South of the Sandawe Hills lies another broad flatland that is part of the great central Tanzanian peneplain. The easternmost portion that juts into Usandawe is dominated by *Acacia-Commiphora* thickets, but the western portion is covered by the unique Itigi thicket, a dense web of nonthorn,

Figure 6.1 A View from Kwa Mtoro across the flatlands to the Songa Hills.

Figure 6.2 General topographic regions.

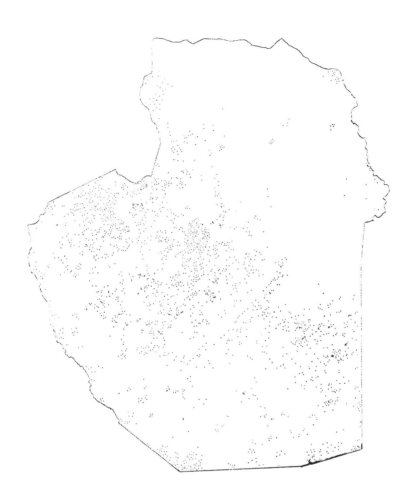

One dot equals ten persons

Miles

Figure 6.3 Population distribution.

deciduous shrubs overlying an ancient lake bed. So far there has been little human settlement here, and historically the Itigi thicket has proven to be a barrier to movement in this direction.

Most of the far west of Usandawe is bordered by the Mponde Valley, which lies beneath a fault escarpment of the Gregory Rift System. Here are found sparse grasslands where seasonal flooding is common and thickets where the channel banks are steep enough to circumscribe the water. Some tsetse exist in the thickets, and though they are not nearly as numerous as in the Songa Hills and the far north, they nevertheless have served to restrict the population to small pockets.

From a climatic standpoint, Usandawe is typically semiarid. There is a single rainy season from December through April or May, followed by an uninterrupted dry season. The two weather recording stations show annual precipitation averages of 640 millimeters and 605 millimeters, but with great variability from year to year. The lowest total on record is 280 millimeters and the highest over 1000 millimeters, so, in effect, there is really no such thing as an "average" year. Also, there tends to be great intrarainy season variation, with short periods of intense precipitation followed by long dry spells. All this makes for a very hazardous environment for crop production, and indeed drought and serious crop failure have been regular visitors to Usandawe as elsewhere in central Tanzania. In the present century they have come in 1911, 1918-1919, 1948-1950, 1953-1955, 1961-1962, and 1973-1974.[6]

Exacerbating the water problem is the fact that the streams flow only after heavy rains, and even the two principal rivers, the Mponde and Bubu, are effectively dried up by the middle of June. In addition, there are no lakes, and consequently irrigation is not a viable possibility for alleviating periods of drought. Water for human domestic uses and livestock comes predominantly from shallow wells dug in the dry stream beds and from numerous small springs. In the last several decades a few boreholes have been sunk, but these are used almost exclusively for watering livestock because the deeper groundwater tends to be highly saline.

An obviously important aspect of the Sandawe environmental setting, and certainly fundamental in attempting to understand their sense of ethnic identity, are the neighboring peoples. Much will be said about Sandawe and non-Sandawe interactions throughout the remainder of this chapter, but suffice it for now to indicate that to the west, south, and northeast reside the Bantu-speaking Turu, Gogo, and Rangi, respectively (figure 6.4). To the southeast are found the Burunge, apparently of Southern Cushitic linguistic affiliation; to the north, around Mount Henang, are the Barabaig or Mangatti, a branch of the widely dispersed Nilotic Tatog.

DIMENSIONS OF SANDAWE ETHNICITY

Following the normal pattern, Sandawe ethnic identity has been

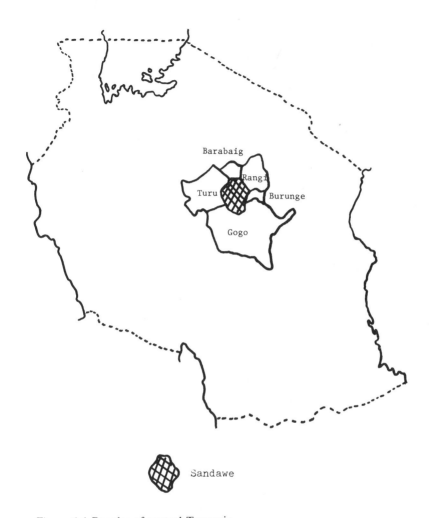

Figure 6.4 Peoples of central Tanzania.

established along several dimensions, with the most immediately apparent differentiating factor being speech. The Sandawe language is classified in the "Click" grouping and apparently is totally unrelated to any other language in eastern Africa.[7] Its closest affiliation, in fact, seems to be with some of the Nama Khoi Khoi dialects of southern Africa, although its precise lineage still awaits further analysis.[8] Some Turu speak Sandawe and vice versa, but generally there is not much crossing of language boundaries. As a matter of fact, virtually all non-Sandawe claim it is next to impossible to learn to speak Sandawe, mainly due to its tonal qualities.

For intergroup communications, Swahili is far and away the most frequently used medium, although many Sandawe only know its barest rudiments.

Almost all Sandawe will note that differences in physical appearance exist between them and their neighbors. They place special emphasis on their "lighter" skin color, for which there is a strong preference, particularly as an indicator of female beauty. Other, more general, remarks such as "we just look different" or "it is obvious we do not appear alike" are frequently expressed. As with most such "folk" assessments, there does seem to be at least a grain of truth in this contention: the Sandawe do show a relatively high frequency of Capoid (Bushmanoid) anatomical features, including yellowish-brown and wrinkled skin, shortness of stature, eye folds, peppercorn hair, and steatopygia,[9] whereas all the surrounding peoples are basically East African Negroids. Admittedly, the research relating to the anthropometric and genetic characteristics of the Sandawe and other central Tanzanians is far from adequate. Nonetheless, whether fact or fiction, there are indigenous perceptions of physical differences, and this is what matters in terms of defining Sandawe ethnicity.

The Sandawe are also recognized as the longest-term residents of the various inhabitants in central Tanzania.[10] This notion is embedded in their own oral myths, and they even go so far as to assert that there "always" have been Sandawe-like people in the area. The *always* is predicated upon a notion of derivation from a nomadic hunting people referred to as the *N/ini*, who play a central role in many Sandawe stories and legends. For example:

> Those people of who it is said they are called *N/ini*
> people, say, of old, they were very short
> and they were ruddyish.
> And their children were taken (in marriage) by Sandawe
> and at present they have disappeared for a long time;
> they have just become Sandawe.[11]

Similar legends describing such preexisting populations abound in central Tanzania, and, at least among the Turu, Gogo, and Barabaig, they are associated with the present-day Sandawe.[12] There is also a widespread belief that the "old Sandawe" are responsible for producing the many rock paintings in Kondoa Area, although this art form is not pursued actively today. In fact, many Sandawe attribute the paintings to the *Wareno* or Portuguese.[13]

Intimately associated with this conception of relative antiquity, and undoubtedly the single most important factor in delineating their ethnicity, is the Sandawe tradition of hunting and collecting. It seems virtually certain that not until 150-200 years ago did the Sandawe begin adopting technologies of crop and animal husbandry,[14] and even today most men still consider themselves as hunters first and foremost, no matter how much or how little hunting they really do. They pride themselves on their bowmanship and frequently can be seen with bows and arrows in hand as they tend their livestock or simply move from place to place. They are quick to point out the archery deficiencies of their neighbors, none of whom are very skilled and who rely more on spears for their weapons. Dogs are raised for tracking and killing small game, and there is spirited competition as to who has the best animals. Because of their economic function, the dogs are maintained in remarkably good condition, including being given regular meals. This is in marked contrast to the scruffy, half-starved beasts usually encountered in central Tanzania. Also in use are nets and a wide array of traps and snares, all of which make for a rather extensive repertoire of hunting gear. In addition, the Sandawe are proud of their abilities as honey collectors, and, in fact, honey ranks as their most highly prized food. Almost any other activity will be dropped in order to go out and search for honey. Although it is not true, as Bagshawe claimed,[15] that the Sandawe "will live and grow fat while other tribes are starving" because of their use of "wild" foods, there is little doubt that they are exceptionally skilled hunters and collectors, and the products obtained thereby are still an important part of their diet.[16] In contrast, the Rangi are almost exclusively cultivators; the Turu, Gogo, and Burunge mixed cultivators and livestock herders; and the Barabaig pastoralists.

These hunting and collecting facets of the Sandawe way of life are well recognized by their neighbors, but they are more often seen as negative, rather than positive, assets. It is not unusual to hear the Sandawe derided as "bush people," and the women in particular are stereotyped as lazy and very poor cultivators. Therefore, they are not highly sought after and even rejected as possible wives by many non-Sandawe, even in spite of their generally recognized physical attractiveness. All in all, the Sandawe do not quite measure up to proper "civilized" standards in central Tanzania, a notion that we will explore again shortly.

The Sandawe, then, can be said to be a recognizable ethnic group. Language, appearance, history, and subsistence have served to set them apart, both in their own and others' minds. Given this fact, we now can turn our attention to how Sandawe ethnicity is tied to place of habitation.

PLACE AND ETHNICITY

At first glance it would appear as if the Sandawe lacked any real notion of place identity. For one thing, they have been yielding territory to others in virtually every direction without significant conflict or even much apparent concern. The Turu are far and away the most numerous and widely distributed of the outsiders and now dominate in some of the far western portions of Usandawe (figure 6.4). Their encroachment has been going on for well over one hundred years and is made up of both colonizing efforts by whole Turu families and lineages and by intermarriage of single individuals with the Sandawe. Based on a survey of the parental affiliations of primary school pupils, nearly 70 percent of Sandawe to non-Sandawe marriages occur with the Turu, mostly in the direction of Turu male to Sandawe female.[17] On the whole, relations between the two peoples have been most amicable, and, as we shall see, strong acculturative forces emanating from the Turu have fostered a growing sense of kindredness.

The Gogo are the next most numerous of the outsiders, and they have been penetrating in significant numbers from the south for at least seventy-five years. Though contacts with the Gogo have been peaceful for the most part, they have not been nearly as close as those with the Turu. The Sandawe tend to be somewhat fearful of the Gogo, not because of their fighting prowess—Sandawe bows and poisoned arrows are more than a match for Gogo spears—but because of their reputed powers of witchcraft. A frequent Sandawe epithet for the Gogo is "dangerous," and they prefer to keep them at a distance.

Leaving aside the apparent presence of early iron traders, most Rangi in-migration seems to be of quite recent origin. Tsetse clearance schemes on the northeastern flanks of the Songa Hills during the 1950s and 1960s opened up new "living space" for the overcrowded Rangi, and since independence many of the government officers stationed in Usandawe have been Rangi. But in general, associations between the Sandawe and Rangi have not been very intimate, including very few cases of intermarriage.

Scattered throughout Usandawe are groups of Nyamwezi and Kimbu, both Bantu-speaking peoples from western Tanzania. The Nyamwezi are mostly descendants of nineteenth-century ivory hunters, and in fact the administrative center of Usandawe, Kwa Mtoro, is named after a Nyamwezi.[18] The Kimbu have come more recently, primarily as agricultural colonists during periods of flight from serious famine in their home area. The Nyamwezi and Kimbu have intermarried to some extent with the Sandawe, but both groups have tended to maintain their separate identities.

A goodly number of Baraguyu Masai have taken up residence in Usandawe, mainly on the flatlands surrounding the Sandawe Hills, in their seemingly never-ending search for more grazing lands.[19] Although a wife is occasionally obtained from the Sandawe, and from time to time a few head of livestock are either traded or stolen, there is very little contact between the two peoples. For their part, the Sandawe consider the Masai totally alien and avoid them if possible.

Barabaig migration from the north has been sealed off by tsetse, but there is no doubt about an extensive presence in the past, especially in the nineteenth century. The principal rainmaking clan of the Sandawe, the Alagwa, who were appointed chiefs during the colonial era, are of known Barabaig origin, and Sandawe legends are replete with references to Barabaig movements and activities.[20] Many battles seem to have been fought with the Barabaig, particularly after the Alagwa brought cattle to the Sandawe, and even today the occasional Barabaig warrior who does appear provokes anxiety and fear.

Finally, the Burunge and Sandawe seem to have had little enduring contact. Very few Burunge are settled in Usandawe, and, for a change, what little crossover there is consists of mainly Sandawe encroachment on Burunge territory. The few Sandawe-Burunge interactions that have occurred are not remembered as being overly cordial. No major battles have been waged, but, at least according to Sandawe views, quarrels and fights are common when Burunge and Sandawe meet, an opinion that is borne out by the number of conflicts that erupt between Burunge and Sandawe school children.[21] However, the exact reasons for this animosity have not been determined.

The picture, then, is one of quite permeable Sandawe borders. As we have seen, this is not because the Sandawe have been a weak and defenseless people, but for some reason they have not developed a protective attitude toward a given territory. Perhaps this has been due to limited demographic pressure in the past or perhaps because Sandawe political organization has never really transcended the lineage level. Whatever the reasons may be, there is no exclusive Sandawe domain that is perceived as inviolable.

At a larger scale of inquiry, there are further indications of a lack of strong territorial identity by the Sandawe. Some of the most important places for them are sacrificial sites for the clan ancestors, where offerings of livestock are made usually for help with drought or sickness. These sites are almost invariably associated with particular hills; however, they in no way affect the location of people.[22] Indeed, it is quite common to find individuals who have no idea where their clan hills are located. Clan lands as such are only very vaguely defined, and members of any given Sandawe clan can usually be found scattered widely across the territory.

Grave sites also are important to the Sandawe, and there is a frequently expressed desire to remain in the vicinity of ancestral graves, particularly those of parents. Nevertheless, this is a desire that is more often violated than realized, for the Sandawe are highly mobile people. Most of the movement today is associated with the shifting or swidden form of agriculture. A parcel of land can normally be cropped for from three to six years before fallowing is necessary, and the pattern is for the new land to be opened up immediately adjacent to the land being abandoned—in order not to move the whole homestead. Eventually, though, homes deteriorate to such an extent—a house lasts anywhere from six to twelve years depending on the wood used in construction—because of weather and termites that they must be vacated. New ones will then be built as close to the new fields as possible. Almost never will a family return to land it has cultivated previously, so the pattern of settlement movement resembles a slow, spatial drift across the landscape more than it does a cycle.

Quite often, however, individuals and families will pack up and move considerable distances, up to fifteen miles or more, in an attempt to find better agricultural land, sweeter water, friendlier neighbors, or sometimes (it seems) just to change surroundings.[23] Sandawe mobility, some of which undoubtedly is also a legacy from their nomadic hunting and collecting past, has been a persistent thorn in the side of administrators, a typical reaction being "how can you deal with people who are in this place one year and somewhere else the next."

Given this movement, it should come as no surprise that the Sandawe do not possess a system of land inheritance. There is no buildup over time of either a material or spiritual investment in a piece of property, nothing to foster the attitude that this is "my land" and this is "your land" forever. As a matter of fact, Sandawe often will laugh when asked about land inheritance and reply to the effect that "why would anyone want old, used-up land when there is plenty of fresh land in the bush for the taking." Among the Sandawe, as long as it is not being actively cultivated, land is open to anyone, even, as we have seen, to non-Sandawe. The only apparent constraint is that a person get along well with his neighbors, a feature of life that is extremely important because of the cooperative effort needed during the critical weeding-cultivation phase of the agricultural cycle. Local work groups of a half a dozen to a dozen families normally assemble and do each other's fields in turn, with a payment of "beer" or meat or both being provided by the resident. Close cooperation also is required in net-hunting. Here men guard a series of tight meshed nets with spears while the women and children drive small game toward them for the kill. All in all, friendship rather than kinship is normally the determinant of Sandawe local groupings.

The Sandawe, then, seem to lack both a recognition of a collective territory and an individual attachment to specific sites. Soja has argued that this is characteristic of hunting and collecting societies in general,[24] and it is probably fairly typical of swidden cultivators as well. This does not mean, however, that that place is conceptually unimportant in such circumstances. Certainly this is far from true for the Sandawe, who as we shall see, manifest a more generalized place identity, one bound up with what is considered to be the "good life" and definitely integral to their ethnic identity.

Basically this notion of the "good life" involves "being in the bush," which loosely translates as living in dispersed homesteads separated by extensive stands of uncleared land. The dominant homestead unit consists of a man and his wife or wives plus his sons and their wives, although strictly speaking the Sandawe would have to be classified as neolocal since there are many variations on this theme. As previously indicated, hunting and collecting are still important to Sandawe subsistence, and a very strong preference is expressed for having wild fruits, nuts and vegetables, honey, small game animals, building materials, firewood, and water close at hand. Considerable dismay is often expressed at how some of the surrounding peoples, particularly the Turu and Rangi, can live in almost totally cleared environments of nothing but fields and houses. Most of the important amenities of life just seem to be lacking in such places, and they certainly appear "far from home."

There is also a desire to separate people in order to minimize the number of close personal contacts, which most Sandawe view as inimicable to the "good life." As an informant put it, "people should be far enough apart so that private conversations cannot be overheard." When people live too close together, the assumption is that they will meddle in one another's affairs, with the consequence being quarrels, fights, and even killings.[25] The real fear, however, centers around the potential for witchcraft activity. Witches are thought to require detailed personal information about their intended victims in order to be most effective, and this is considered easier to obtain if people are concentrated spatially. Actions, interactions, and conversations can be observed more closely. Thus, the Sandawe try to disperse, staying close only to trusted friends and relatives, a process that Sahlins suggests is normal in societies based, as are the Sandawe, on a "domestic mode of production."[26] In these instances, distance substitutes for reciprocity and laws.

"Being in the bush" is qualified by "being in the bush in the hills." The Sandawe openly extol the virtues of life in the hills and express a strong distaste for the surrounding flatlands. Undoubtedly some connection is made between the hills and clan shrines, and although this relationship seldom is articulated as such, it probably can be inferred from remarks such as "the hills are the homes of our fathers" and "living in the hills is our way." But most often the reasons for preferring the hills are phrased in terms of the things of the "good life" being more abundant than in the flatlands. Small game such as dik-dik, duiker, and hyrax is found in greater numbers, the water is fresh rather than salty and can be gotten from numerous small springs and wells rather than from just a few boreholes, honey bees are far more abundant, and there is much better timber for building houses, making charcoal, and fashioning utensils.

A good illustration of this differential environmental evaluation is provided by the case of the Songa Hills. Sometime in the remote past the Sandawe leapfrogged across the intervening flatlands and settled these hills, although not nearly to the same extent as the Sandawe Hills. Then in the early 1920s tsetse began spreading through the Songa Hills, thoroughly covering them by the mid-1930s. Sporadic cases of human sleeping sickness began to be reported, and a full-fledged epidemic broke out in 1946-1948. Some families chose to abandon the area, but most decided to stay put in spite of the threat. This prompted the British colonial authorities to intervene, and the remaining people were rounded up and resettled on the new clearings in the flatlands. These clearings were designed to serve the dual purpose of removing the human sleeping sickness reservoir from the Songa Hills and halting the westward advance of the tsetse toward the Sandawe Hills. Close supervision of the new settlements, making sure the people stayed and the bush was kept cut, was maintained through the 1950s. But then it was relaxed, and as a result a drift back to the still tsetse-infested Songa Hills began. Even those people who have remained on the flatlands admit that they regularly visit the hills to hunt and gather honey, and they are prone to reminisce about how much better life was and could be again if they were allowed to return. Only the fear of being jailed or fined seems to keep them where they are.

Place and ethnic identity, therefore, are closely linked together among the Sandawe. Maintaining their image as hunters and collectors and holding on to those things thought to be essential to the "good life" are dependent upon preserving a particular kind of natural and human environment. However, several forces now are operative that augur for substantial changes in these environments and thus by extension changes in the character of Sandawe ethnicity. It is with these changes and their impacts that we will bring the present chapter to a close.

FORCES OF CHANGE

One of the forces of change among the Sandawe has actually been operative for quite some time, namely, acculturation. As already indicated, the Sandawe have been and still are in an active process of switching over from a hunting and collecting way of life to one based on crop and animal husbandry. The Turu definitely seem to have been the primary source of the inputs, with secondary contributions, particularly of livestock-rearing practices, coming from the Barabaig and Gogo.[27]

In the wake of this change in subsistence technologies have come other important modifications and additions to Sandawe culture. Polygamy has become more widespread; various new initiation rites, including circumcision and clitoridectomy, adopted; house types and clothing styles altered; and so on. However, perhaps most significant of all has been the spread of new attitudes about what is right and proper behavior. This has given rise to a dichotomy among the Sandawe called *Bisa* and the Sandawe called *Tehla*, one that Ten Raa claims is based on a "couth-uncouth" social distinction.[28] The Bisa are considered to be closer linguistically and culturally to the aboriginal Sandawe than are the Tehla; they are therefore uncouth. Put into American parlance, the Bisa are felt to be "hicks" or "backwoodsmen." Of particular Tehla disgust are the Bisa's ribald speech habits with frequent references to genitals and sexual activity, their casual attitudes about nudity even in mixed company, their failure to recognize the "right hand clean–left hand unclean" differentiation, and their rather catholic eating practices, including such items as snakes, monkeys, baboons, lions, leopards, and hyenas.[29] "Proper" (which is the translation of *Tehla*) Sandawe simply do not behave in such ways.

These are essentially Turu notions, and in fact virtually all Tehla are found in the central and northwestern sections of the Sandawe Hills, where the Turu presence is most prominent. This seems to be a classic case of contact diffusion, but the pattern also represents the process of marriage between Turu men and Sandawe women. The men bring with them their cultural heritage of ideas, and some of these are passed on to their children. Yet, the core of Sandawe culture remains, including the language, because it is the women who have the most control over the children during their critical formative years. Out of this kind of circumstance are born many Tehla Sandawe.

Another potent force for change is demography. Through in-migration and natural increase, population growth in Usandawe presently is over 2 percent per annum. This means greater pressure on land resources, which translates into more cropland and grazing areas and less "bush" for wild animals and vegetable products. Only a very few entirely nomadic hunters still remain, and many men admit that they hunt far less often than formerly because the game animals are scarcer near their homes, most having been hunted out or retreated to the tsetse-infested margins of Usandawe. It is not too uncommon to find teenagers and young adults who cannot identify many of the game animals that once would have been so much a part of their lives. With time and increasing population pressure, permanent field agriculture replaces shifting cultivation. This change already has happened around the mission stations of Ovada and Kurio and the trade center of Kwa Mtoro. Here population densities reach around forty people per square kilometer, and as a consequence fields are being permanently demarcated, land is passed on to sons, and disputes over rights in land and trespass are beginning to occur. Again, this is mainly Tehla country, and, excepting their language, these Sandawe are coming more and more to resemble the Turu.

Government policies during the colonial period accentuated the trend toward greater sedentarization by imposing taxation and cash cropping, but by and large there was little other direct pressure on the Sandawe. Their territory lay pretty much beyond the pale of effective administrative intervention. Such is not the case today, however, with the imposition of *ujamaa* villages. This is Tanzania's attempt to reorganize the countryside into nucleated population clusters in order to increase agricultural productivity, facilitate the bringing of basic health and educational services, and quicken the spread of a Tanzanian socialist ideology. In the words of President Julius Nyerere:

> In a socialist Tanzania . . . our agricultural organization would be that of co-operative living and working for the good of all. This means our farming would be done by groups of people who live in community and work as a community. They would live together in a village; they would farm together; market together; and undertake the provision of local services and small local requirements as a community.[30]

In other words, the traditional Sandawe values associated with dispersal and mobility are now at odds with official Tanzanian policy.

It appears, then, that Sandawe ethnic identity is destined for profound alteration. Although they have maintained themselves in the face of acculturation in the past, the cards are stacked against them now. Demographic and political forces are radically reshaping the environment so critical to their way of life, and they are neither numerous enough nor sufficiently well organized to form an effective ethnic political force. Even their language seems threatened with eventual extinction as Swahili continues its spread through both official and unofficial channels. The most likely prospect is that they will be absorbed, and probably without overt resistance, into an enduring Turu-like culture or, if present plans succeed, into a new, hybrid Tanzanian identity.

NOTES

1. T. Shibutani and K. M. Kwann, *Ethnic Stratification: A Comparative Approach* (New York: The Macmillan Company, 1965); also Fredrik Barth, ed., *Ethnic Groups and Boundaries* (Boston: Little, Brown and Company, 1969).

2. S. Lieberson, *Social Stratification: Theory and Research* (Indianapolis, Ind.: The Bobbs-Merrill Co., 1970).

3. See especially James Ngugi, *The River Between* and Camara Laye, *A Dream of Africa*.

4. *Miombo* is a Swahili word that seems to refer to all the various *Brachystegia* tree species. It does not refer to the total vegetation complex but has been widely used by English-speakers in eastern and southern Africa to do so.

5. The U-prefix in Swahili means *land-of*, and most Sandawe use *Usandawe* because they have no comparable term in their own language.

6. Eric Ten Raa, "Bush Foraging and Agricultural Development: A History of Sandawe Famines," *Tanzania Notes and Records*, no. 69 (November 1968), pp. 33-40; also J. L. Newman, *The Ecological Basis for Subsistence Change among the Sandawe of Tanzania* (Washington, D.C.: National Academy of Sciences, 1970).

7. J. Greenberg, *The Languages of Africa* (Bloomington: University of Indiana Press, 1966).

8. Eric Ten Raa, "The Moon as a Symbol of Life and Fertility in Sandawe Thought," *Africa* 39, no. 1 (January 1969): 24-53.

9. J. C. Trevor, "The Physical Characters of the Sandawe," *Journal of the Royal Anthropological Institute*, 77 (1947): 61-78.

10. There is some evidence of an earlier population. Hard-clay habitation sites and first millenium A.D. pottery remains have been found and cannot be accounted for by the Sandawe. See J. E. G. Sutton, "Archaeological Sites in Usandawe," *Azania* 3 (1968): 157-173.

11. Eric Ten Raa, "Sandawe Prehistory and the Vernacular Tradition," *Azania* 4 (1969): 98.

12. Eric Ten Raa, oral communication, 1974.

13. As far as I could determine, the origin of this myth seems to reside with the Germans. Apparently, when they entered central Tanzania and found the paintings but no practicing artists, they assumed that the only likely candidates would be the Europeans who preceded them to East Africa, namely, the Portuguese.

14. Newman, *Subsistence Change among the Sandawe.*

15. F. J. Bagshawe, "Peoples of the Happy Valley (East Africa), Part III: The Sandawi," *Journal of the African Society* 24 (1924-25): 224.

16. J. L. Newman, "Dimensions of Sandawe Diet," *Ecology of Food and Nutrition* 4 (1975): 33-39.

17. Newman, *Subsistence Change among the Sandawe,* p. 50.

18. Mtoro is not remembered with affection by the Sandawe because he helped lead a German punitive expedition into the territory.

19. The Baraguyu are Masai who engage in some cultivation and can be found widely scattered throughout Tanzania.

20. Ten Raa, "Sandawe Prehistory and the Vernacular Tradition."

21. Farkwa Mission, oral communication, 1966.

22. Eric Ten Raa, "Dead Art and Living Society: A Study of Rockpaintings in a Social Context," *Mankind* 8, no. 1 (June 1971): 42-58.

23. J. L. Newman, "A Sandawe Settlement Geography," *The East African Geographical Review*, no. 7 (1969), pp. 15-24.

24. E. W. Soja, "The Political Organization of Space," Association of American Geographers, Commission of College Geography, Resource Paper no. 9 (1971).

25. Although reliable comparative statistics are not available, the Sandawe are reputed to possess one of the highest murder rates in Tanzania.

26. M. Sahlins, *Stone Age Economics* (Chicago: Aldine, 1972).

27. Ten Raa, "Sandawe Prehistory and the Vernacular Tradition"; also Newman, *Subsistence Change among the Sandawe.*

28. Eric Ten Raa, "The Couth and the Uncouth: Ethnic, Social, and Linguistic Divisions among the Sandawe of Central Tanzania," *Anthropos* 65, nos. 1-2 (1970): 127-153.

29. Ibid.

30. J. K. Nyerere, *Socialism and Rural Development* (Dar es Salaam: Government Printer, 1967), p. 16.

PART 2
Urban and
Situational Ethnicity

Two opposing views have marked the study of the African in the city. Some have held that when the researcher studies the man in the city he must be seen as a miner, a factory worker, or a clerk, irrespective of his ethnic background. Others have maintained that irrespective of the current role a person was first of all a product of his past and had to be seen in terms of this cultural continuity.

The essays in this section suggest that the truth might lie in a combination of these approaches. This would allow a person to hold seemingly contradictory views and to behave in seemingly antithetical ways depending on the context in which the action takes place. Such a situational approach is inherent in these essays.

A second major theme results from the fluidity of ethnic boundaries and ways in which new criteria for association may emerge, urban versus nonurban, female versus male, black versus nonblack, worker versus overseer.

7

Urban Ethnicity in Windhoek

Wade C. Pendleton

Urban ethnicity is remarkably salient in Windhoek and shows no signs of diminishing in importance in the future.[1] Ethnicity there is the product of historical events, political and administrative factors, socioeconomic conditions, racial stratification, and a general lack of crosscutting mechanisms that would diminish its importance. Urban ethnicity in Windhoek has much in common with ethnicity elsewhere in Africa and other parts of the world. It is certainly not a unique occurrence, but particular local characteristics have given it some special features.

This analysis and description makes use of concepts that combine several theoretical approaches. Mitchell and others have described similar phenomena in central Africa as situational ethnicity or urban tribalism.[2] Barth added to the conceptual tools by emphasizing interactional criteria and boundary maintenance.[3] Mitchell in his discussion of the situational approach provided a model for the conceptualization of urban social structure in terms of external determinants.[4] These concepts have been used in this chapter in two ways: (1) to describe the features of urban ethnicity as they are demonstrated by the behavior and interaction of Windhoek's people, and (2) to describe and analyze how features of Windhoek's social structure create boundaries that emphasize ethnic group identification.

Cohen has defined ethnic groups, "as a collectivity of people who (a) share some patterns of normative behavior and (b) form part of larger populations, interacting with people from other collectivities within the framework of a social system."[5] He defines ethnicity as the extent of conformity to expected ethnic group patterns in interaction with members of other ethnic groups and coethnics. His formulation of ethnicity and ethnic group is compatible with the previously described ideas and is used in this chapter.

BACKGROUND DESCRIPTION OF WINDHOEK

Windhoek is the capital of Namibia and the most important urban center in the country.[6] The modern beginning of Windhoek was in October 1890, when German occupation troops arrived in the area and built a fort. Prior to that, the Windhoek area had been periodically occupied by Herero and Oorlam groups. The Herero still claimed the area as theirs, but it was unoccupied when the German troops established themselves there. The town grew into the administrative and political capital of the territory during the period of German colonialism (1890-1915). When South Africa took over control of the country under a mandate from the League of Nations in 1920, Windhoek remained the capital and continues so today. The town grew from a population of 3,324 in 1903 to 61,369 in 1970. The composition of the Windhoek population is as follows: Africans 20,917 (39 percent)—Damara 41 percent, Herero 25 percent, town-Ovambo 17 percent, Nama 8 percent, and other African ethnic groups 9 percent; whites 25,417 (48 percent)—Afrikaner 60 percent, German 34 percent, English 5 percent and other white ethnic groups 1 percent; and coloured people 6,947 (13 percent).

Urban migration to Windhoek is the result of a complex of factors. Initial African migration was unavoidable. After the German army had defeated the Herero and Nama and imposed severe restrictions on owning cattle and land, they were forced to seek work in the towns and on white-owned farms. As a cash economy became established in the country and land for African pastoralism became more scarce, others were forced to migrate in order to make a living. The Damara did not suffer such severe restrictions but appear to have sought wage employment as an alternative to the increasingly difficult problem of pursuing a hunting and gathering way of life given the radically altered situation in the country. More recent African migration was a result of many factors similar to those elsewhere in Africa: the need for a job since wage employment is virtually nonexistent in the African reserves, the attractions of the city, better facilities, and educational opportunities. The breakdown of tribal institutions and solidarity also played an important part.

Afrikaners were attracted in the 1920s to South West Africa by government offers of large farms that could be financed with low interest loans, but not many migrated to Windhoek. After the National Party won the 1948 election in South Africa, Afrikaners did migrate to Windhoek. The government opened up many jobs in the administration and expanded public services and the railways. Most of these new jobs were filled with Afrikaners recruited in South Africa. German migration continued after the First World War and especially after the Second World War. Many Jews came to the territory during the 1930s. White migration to Windhoek continues from other parts of Africa where white settlers are no longer welcome or where they have been unable to adapt to the changed circumstances of African independence. Many coloured people have migrated to Windhoek from the Cape Province of South Africa seeking a better life. The growth of Windhoek as an administrative, business, and light industry center has attracted people because it offers the largest and most diverse job market in the country.

In 1975 the Windhoek urban area is composed of three geographically separate township areas: (1) white residential areas and the areas of business and light industry; (2) a coloured residential area west of Windhoek called Khomasdal; and (3) an African residential area called Katutura located northwest of Windhoek.[7] Only whites are allowed to live in Windhoek except for those African and coloured people who live on their employers' residences where they work as domestic servants. Only coloured people live in Khomasdal, and Katutura is exclusively for Africans. Within Katutura the township is divided into sections for each of the major ethnic groups living there: the Damara, Nama, Herero, and Ovambo. An area called the "mixed section" accommodates people of other ethnic groups. Within Windhoek, Afrikaners tend to be clustered in certain areas where they are provided with inexpensive housing as a consequence of their administrative or municipal employment.

The separation of racial and ethnic groups into different residential areas was established during the German period. It has been strengthened in recent years by the establishment of separate townships for African and coloured people in the 1960s. This and other measures reflect the extension of apartheid policies from South Africa to Namibia and Windhoek. Africans must possess service contracts, urban residence permits, and housing cards to reside in the area legally. If they want to travel from Windhoek, they must obtain travel permits.[8] The entire urban area is administered by whites, and the municipal government is exclusively controlled by elected and appointed white officials. The only voice in muncipal affairs that African and coloured people have is through the advisory board system, which can make recommendations to the Windhoek town council, but they have no real economic or political power of their own. The Katutura Board has members appointed from the major ethnic groups in the township.

Economically, the society is racially stratified, with Africans primarily limited to unskilled jobs, and coloured people to unskilled and semiskilled jobs. Whites control the skilled, executive, and managerial positions. Nearly all African and coloured people work for whites. The average annual salaries of white, coloured, and African men are $4,565, $1,424, and $592, respectively. Thus, although Windhoek has a large and diverse job market, Africans and coloured people are limited to only certain types of work, which consequently limits their income. Most employers, including the government and the municipality, maintain differential pay scales for whites, coloureds, and Africans who perform the same jobs. Women are also usually paid less than men.

Windhoek is a cosmopolitan town with many restaurants, theaters, an art gallery, shops with the latest European goods and fashions, and even a Scientific Society. It is an exciting place, and its importance to the country far exceeds its relatively small size by international standards. But more than half of its population is prevented from fully taking advantage of its attractions due to racial restrictions and limitations imposed because of socioeconomic considerations. In 1975 many of the racial restrictions that had operated in Windhoek for more than fifty years were relaxed, but it is too soon to know what impact this will have on society. At present Africans do not have the economic power to take advantage of these new opportunities.

THE ETHNIC GROUPS OF WINDHOEK

The radical changes that have been taking place in South West Africa since the German colonial era began have transformed most of the tribal groups of the country into ethnic groups. Tribal groups are no longer autonomous or semiautonomous societies pursuing subsistence economies; they have been integrated into a larger social system. This is reflected in the geographical distribution of the African population. Based on the 1970 population census of South West Africa, 40.6 percent of the southern section African population lived in urban areas, 43.8 percent worked on white-owned farms, and only 15.6 percent resided in African reserves.[9] Significant percentages of each African ethnic group are found in the urban areas, on farms, and in the reserves.

Traditionally the Damara were hunters and gatherers, and politically uncentralized. The Herero were cattle pastoralists with a double descent system and chiefs. The Nama were patrilineal pastoralists, had chiefs, and were the northwestern extension of the Khoikhoin of southern Africa. Ovambo is a collective term for the seven related tribal groups that live in the north-central part of the country. They were, and still largely are, settled cultivators who keep some cattle. They have a matrilineal kinship system. They were politically centralized with many of the characteristics of divine kingship. The Ovambo social system has changed less than the other groups, but the institution of divine kingship has become obsolete since the government started appointing chiefs in the 1920s. The influence of heavy migrant labor out of Ovamboland has had far-reaching effects on the area, which is not surprising since as much as 37 percent of the adult male population is away from Ovamboland at any given time.[10]

The patterns of normative behavior found in Windhoek today, which are identified as typical of the major ethnic groups of the country, have little to do with the normative patterns that distinguished tribal groups from each other in the past. The general importance of ethnic groups in Windhoek is not reflected in different subsistence economies or traditional political organization. They are demonstrated by widely shared stereotypes of what members of different groups look like, how they are supposed to behave, and common stereotypes in value orientations. This importance is reflected in the use of ethnic categories for recruitment to voluntary associations, conjugal relationships, and friendships. In fact, ethnicity influences practically every aspect of social life. The cues for ethnic group identification are many and include such things as physical appearance, dress, language usage, gestures, and role behavior. It is a combination of cues that signals identification. The primary ethnic categories are Damara, Herero, Nama, Ovambo, coloured, Afrikaner, English, and German.

A detailed description of all important cues is beyond the scope of this chapter. A few of the important ones are described below. Skin color usually distinguishes Africans, coloured people, and whites—except for the Nama, who have a yellow skin color. Nama women often have steatopygia, which is a very distinctive characteristic. Herero and Ovambo men are well known for dressing well. The long Victorian-style dresses that African women wear identify their ethnic group. Each group has its own colors and patterns of material from which their dresses are made. Germans often wear European-style clothes, which are purchased from shops specializing in European clothes. Afrikaners wear clothes made in South Africa. Coloureds dress similarly to Afrikaners except for limitations imposed by socioeconomic considerations. Children are often distinctively dressed, which reflects their ethnic group. Some people, because of their knowledge of the customs of other groups and their physical appearance, can "pass" as a member of a different group. Some Africans, especially Nama, can "pass" for coloured, and some coloureds can "pass" for Afrikaners. Whites do not try to pass for coloureds or Africans: in such a racially stratified society there would be no reason to do so.

Languages and speech varieties are associated with each ethnic group.[11] Aside from the obvious correlations (Germans who speak German and English who speak English), the Damara and Nama speak a non-Bantu click language called Nama, the Herero and Ovambo speak different but related Bantu languages, and coloured people and Afrikaners speak Afrikaans. Different varieties of Afrikaans are spoken by coloured and Afrikaners, and the German spoken differs in some aspects of grammar and vocabulary from standard German. Communication between members of these different ethnic groups, if they do not speak the same language, is usually in Afrikaans. Afrikaans is a lingua franca of the Windhoek area and in South West Africa; it is taught in the schools, and most whites speak it to Africans. Older Africans usually have a speaking knowledge of German, but it is seldom used between Africans. A few Africans speak English, and its use carries some prestige, especially among the Herero, Ovambo, and Damara. Loan words from Afrikaans, German, English, and African languages are frequently used by members of all groups.

Stereotyped value orientations and attitudes, frequently reflected in reference group terms, are associated with the various racial and ethnic categories. Not all members of each group hold the following attitudes or beliefs, but they are widespread and representative. Whites think Africans are unreliable, irresponsible, and will steal when given the chance. Whites think Africans are less "developed," and they think it will take at least a generation (some think a century or more) for Africans to reach the stage of white "civilization."

Whites think that Africans are less intelligent as a result of genetic inferiority. Coloureds are thought to be somewhere between the stage of "development" of Africans and whites. These attitudes are frequently expressed as rationalizations for the white dominance of the society. Whites who contradict these attitudes are often labeled communists or liberals. One white who criticized these attitudes was accused of having African or coloured blood; presumably the accuser believed that liberal attitudes were acquired genetically, which would be consistent with a racial ideology. These views are held by Germans, Afrikaners, and English and are reflected in the frequently used derogatory term for Africans, *Kaffir*, which is often preceded by a descriptive adjective for emphasis.

Other less value-loaded reference terms for Africans were *native* (English), *Eingeborener* (German), *Inboorling* (Afrikaans), the official government term *Bantu*, and the frequently used *nonwhite* and *non-European* referring collectively to Africans and coloureds. One neutral but paternalistic term for African and coloured men is *Outa*. The various African ethnic groups are most commonly referred to as Herero, Nama, and Ovambo in all three white languages. The Damara are also called *Klipkaffir* (mountain Kaffir) in Afrikaans and *Bergdama* (mountain people) in German.

Germans call Afrikaners *Buren*, and English refer to them as *Boer*; both terms are derived from the Dutch word *Boor* for farmer or peasant. Both English and German people think Afrikaners are unsophisticated and less culturally "developed" than they are, and Afrikaners think Germans are snobbish. Afrikaners call English *rooinek* ("red neck"). Coloureds are called *Farbige* (German) and *Kleurlinge* (Afrikaans), which literally means color.

Africans have more value-oriented terms for coloured people. Nama call them /Hei-//goan ("mule"), Damara say /Hai-/hūn ("pale whites"), and Herero use the term *Ovikonde* ("useless refuse"). African informants said coloured people think they are better than Africans and that this attitude was getting stronger. Informants told how Africans and coloureds used to visit in the Old Location, but now, since coloureds live in Khomasdal, things have changed. Africans are rarely seen in Khomasdal, and those coloured people who come to Katutura usually only do so to go to the cinema. When they do, they buy the more expensive tickets and sit in the balcony, thereby separating themselves from Africans who occupy the main floor. Some coloureds have told African friends not to greet them if they are in a group with other coloureds. African informants said coloureds want to be treated like whites, and resentment was expressed about this. One informant said coloureds "treat you worse than whites." A few Africans expressed the desire to be classified coloured because they thought they would be treated better by whites, have a better life, and not have to carry passes.

Many terms and attitudes are expressed by Africans of one ethnic group about other ethnic groups. Terms such as *German, Herero,* or *Nama* exist in most languages in slightly modified forms to agree with the grammar and phonology of the speaker's language, e.g., a Nama term for the Herero is *Hereron,* but more descriptive terms are often used. The Herero are identified by all other ethnic groups as being very proud and arrogant. This attitude is reflected in the Herero term *ovatwa,* which is used for non-Herero people, and means, in a derogatory sense, a person of low status; the Ovambo are the only people who are not *ovatwa.* Herero call Damara *ovazorotwa* ("black *twa,*" or "black slaves"). This term and others indicate the low respect that Herero have for the Damara. Herero terms for the Nama and Ovambo are more neutral, and whites are called *Overumbu* ("yellow things"). The pride and arrogance of Herero women were remarked on by many informants, who said it could be seen in the way they walked, always slow, never running, even if it meant missing the bus, and the reluctance of Herero women to speak other languages.

The Nama also have an abusive term for the Damara calling them *Xhou-Daman* ("human excrement people"), although Damara have been heard to use this term jokingly for themselves. The Ovambo are called /*Nawén* ("swingers"), which has a derogatory connotation. Nama and Damara call Afrikaners /*Khorán* ("rough people") and Germans !*Om-Kxoin* ("bull without horns"). Nama call all whites collectively /*hūn,* a contemptuous term for the way whites smell. The recent placing of the Nama under the Department of Coloured Affairs may eventually have far-reaching effects on the image of the Nama held by other African groups and by the self-image of the Nama. This would especially be true if the Nama are moved to Khomasdal.

The Damara have a reputation for being friendly, ready to laugh, and easy to make friends with. Damara informants mentioned having a feeling of inferiority because the Herero and Nama look down on them as their former slaves. Some Damara also resent the fact that the Nama say they lost lost their own language and adopted Nama, but a definite joking relationship exists between the Nama and Damara. Members of other ethnic groups say the Damara are dirty and their women "loose." It does not help the reputation of Damara women that many younger women wear mini-skirts, pants suits, wigs, and lipstick.

Subethnic distinctions exist within some groups. Germans distinguish recently arrived immigrants from those who were born in the country or have lived there for a long period of time. Different varieties of German are spoken by these two subgroups, and their life-style differs. South West Germans make fun of recently arrived Germans and call them *Schneebantu* ("snow African"). Two different categories of Ovambos are distinguished: town-Ovambos and migrant contract-Ovambos. Town-Ovambos are considered hard to make friends with by members of other ethnic groups, but lifelong friends once a friendship has been established. A certain amount of community spirit exists among the Ovambo, as indicated by the fact that they are the only ethnic group where some members come together to hear reports from their Advisory Board representatives. Town-Ovambos generally have a good reputation, and they are clearly distinguished by themselves and others from migrant contract-Ovambos, who are considered troublemakers. Contract-Ovambos are only in Windhoek temporarily on twelve-to-eighteen-month contracts, and they must live in a barracklike compound in Katutura. They go around Katutura in homeboy groups and have a bad reputation for starting fights.

THE CORRELATES WITH ETHNICITY

The structure of the Windhoek social system establishes many boundaries that strengthen ethnic group identification and solidarity. Patterns of interaction and customs also have a similar effect. Cumulatively, both dimensions of the social system help to create a situation where ethnic group identification and categories are unavoidable. Described below are some of the important areas that correlate with ethnicity.

Public Life

Ethnic group identification plays a very important role in public life. It enables people in Windhoek, Katutura, and Khomasdal to identify people easily, to place them in categories, and to modify their behavior accordingly. Such categorical interaction, where the interaction is patterned on the basis of the most basic identity of a person such as sex, age, and ethnic group, is of three types. One type of interaction is between strangers, where ethnic and racial categories allow a person to identify the people who are, for example passing by on the street. There may be no direct interaction, but the knowledge of who the people are allows one to impose some order on situations made up of many heterogeneous strangers.

The interaction of strangers or near-strangers in public places such as the beer hall or on a bus is a second dimension of ethnic categorical interaction. In these situations the categories to which people belong may determine whether they will talk to each other, ask directions, sit together, discuss the news, or gossip. In a shop the ethnic group of the customer versus that of the clerk may determine how the customer will be treated and what language will be spoken.

The ethnic categorical relationship also applies to people who are not strangers. They have perhaps seen each other many times in the same shop, or they may belong to the same club or work together. However, due to the strength of ethnic stereotypes, although the people actually "know" each other, they may maintain social distance unless the barriers that separate them can be broken down and some common ground for interaction can be found. The strength of ethnic and racial boundaries in public life makes it difficult for people who belong to different ethnic and racial groups to get to know each other in anything other than stereotyped relationships.

Many everyday situations require ethnic and racial group identification. The most common ones are interaction situations in shops, stores, and offices. In some shops on Kaiser Strasse, the main business street in Windhoek, when a white customer enters, the salesman decides which ethnic group he belongs to, addresses him in "his" language (Afrikaans, English, or German), and shows him the sort of merchandise purchased by members of his ethnic group. In these shops it is not unusual to see a salesman serve one customer in Afrikaans and then switch to German for the next one. Salesmen sometimes compete with each other to see who makes the correct categorization. If the salesman is uncertain about the categorization of his customer, he simply smiles and clears his throat while he waits for the customer to say something first. Africans are spoken to in Afrikaans. These shops are considered to be "European" shops, and although Africans will be served, they are expected to switch their language to Afrikaans if they want to be served.

Different varieties of Afrikaans are used when speaking to a white or an African customer. To an Afrikaner the salesman will say, "*Wat kan ek vir u doen, Meneer?*" ("What can I do for you, Sir?" using the polite formal form of *you*), but to an African or coloured person the salesman will say, "*Wat wil jy hê?*" (What do you want?" using the personal, informal form of *you*) or simply "*Ja?*" ("Yes?"). By having asked the queston in this way, it is made clear to the African customer, white salesman, and any other people present that the African is being treated differentially. African and coloured customers usually wait until white customers have been served first, except at the large supermarkets, where everybody stands in line.

On the side streets off Kaiser Strasse in Windhoek and in Katutura, there are shops that cater exclusively to Africans. When an African customer enters one of these shops, attention is paid to his ethnic group. He will often be addressed in his own language by an African salesman. The range of goods carried in these shops is especially oriented for African customers, and they cater to ethnic group preferences. These preferences are more marked for women than men. For example, a Herero woman who wants to buy material to make a new dress will be shown the style and type of material that is the current Herero fashion. A Damara woman would choose different styles and colors of material for her dress, and she would be shown the material by a salesman who speaks her language and knows the Damara fashion. Other ethnic group differences in material culture are evident in such items as cosmetics, jewelry, and scarves.

Some shops, restaurants, and hotels are primarily oriented for a particular ethnic group often indicated by name, e.g., Ostora Ovambo is a general dealers shop especially oriented for Ovambo customers. It is well known in Windhoek and Katutura which shops are oriented especially for one ethnic group, and members of other groups would not normally go to these shops. They would not be served in their ethnic language, and the service or goods they wanted could be better obtained at a shop more oriented for their ethnic group. The social situation in an Ostora Ovambo emphasizes particular language usage connected with values, attitudes, and role behavior associated with a particular ethnic group.

Other categorical situations are common elsewhere in Windhoek. When an African goes to the housing office in Katutura or applies for a travel pass, he must tell the clerk his ethnic group. The ethnic group identity of Africans is stated on all the many documents they must carry. Thus, in formal dealings with the authorities and in making requests for permits, people must identify to which ethnic group they belong.

Separate entrances are maintained for whites and nonwhites (African and coloured people) in public buildings such as the post office and in many private businesses such as banks. This physical separation of racial groups helps to maintain boundaries between groups. In 1975 these measures were dropped, but the impact of this change has not yet been observed. The mass media including newspapers, magazines, and radio broadcasts also cater to ethnic groups by publications and broadcasts in ethnic group languages.

In general, social distance is maintained between racial and ethnic groups by the observance of numerous customs. Members of different racial and ethnic groups do not usually eat or drink together. In Katutura the *shebeens* (private houses where beer is brewed and sold by women) are located in the different ethnic group sections where people drink with friends. Men are reluctant to drink in *sheebens* in other ethnic group sections because they may become involved in fights and are reluctant to walk home through other sections of the township, where they may be assaulted and robbed. Whites and nonwhites do not shake hands and do not ride next to each other in automobiles. Whites call Africans by the first name, and Africans call whites Mister, *Meneer,* or *Baas*; these verbal markers structure the subordinate and superordinate interaction rituals between whites and Africans.

Conjugal Unions and Consanguineal Households

Within the Windhoek social system, ethnic endogamy in conjugal partner choice is very high. For the African population, two types of conjugal union may be distinguished—living together and legal marriage. The average ethnic endogamy of each was 82 percent and 86 percent, respectively. For whites, the ethnic endogamy of legal unions was calculated to be 82 percent. The ethnic endogamy of the coloured population was virtually 100 percent. Endogamy rates for the African population would probably be even higher were it not for demographic factors.

The Damara, Herero, and Nama each have more than 10 percent more women than men over fourteen years of age. Thus, some women are forced to seek conjugal partners outside their own ethnic group. The most frequent patterns of exogamous partner choice are Herero women with Ovambo men, Nama women with Damara men, and Damara women have many unions with men in all African ethnic groups.

Coloured exogamous unions are rare. Out of a coloured population of almost 7,000, only 49 had established conjugal union households in Katutura, but their pattern reflects the social system very clearly. Of the conjugal unions between coloureds and Africans, 75 percent were between coloured women and African men. Coloured women can not seek upward mobility by marrying white men; it is illegal. Thus some marry African men, who achieve upward mobility thereby. The coloured woman is assured of a husband who will probably treat her better than coloured man would. All exogamous coloured conjugal unions are found in Katutura.

Conjugal union households are the ideal in nearly all societies, and they are the ideal in Katutura. But a majority of the households in Katutura are not based on conjugal unions. Consanguineal households made up 54 percent (1,864) of all the households in Katutura, and this statistic reflects a very extreme situation. Two types of consanguineal households make up more than 90 percent of the total: those headed by women (matrifocal households) 60 percent, and those headed by men 32 percent. The high percentage of such households is a consequence of socioeconomic conditions, migrant labor, and sex ratio imbalance.

The low income of African men prevents many from accepting the responsibilities of a conjugal union and child support; they cannot afford to do so. Most African men are unable to earn an adequate, reliable wage income; 86 percent of the African men in Katutura earned sixty-nine dollars a month or less. Migrant labor places some men in the urban area temporarily, especially contract-Ovambos, and they are both unwilling and unable to establish conjugal households with Katutura women. The Katutura population has 6 percent more women than men, which is also a factor in the high number of matrifocal households.

Consanguineal households are an ethnic boundary maintenance mechanism. The people who are forced to live in this manner are primarily at the bottom of the socioeconomic stratification of the society. They have less means, fewer opportunities, and are more influenced by the boundaries that separate ethnic groups.

Friendship

The social networks of people in Windhoek reflect the pattern that most people restrict friendships to members of their own racial and ethnic groups. Social relationships with people in other ethnic groups tend to be instrumental. They are also normally single-stranded, being used for only one purpose. Social relationships among people of the same ethnic and racial group are frequently multi-stranded and have affect as well as instrumental characteristics.

Most friendships are ethnically bound, but by looking at those that are heterogeneous it is possible to see some interesting features. If a network is ethnically or racially heterogeneous, it is usually the result of some special circumstances: an atypical occupation or job, high school or university attendance, or travel. For example, attending high school or university resulted in meeting and associating with students from other groups. In recent years the social system has been changing, reducing such opportunities. High schools are now ethnically bound, and those who had an opportunity to make friends with members of other ethnic groups in the past no longer have such opportunities. When the Old Location still existed before 1968, many people lived next to members of other ethnic groups and were friends. With the relocation of people to Katutura, this is less likely because of the stricter enforcement of ethnic housing regulations and the more formal layout of the township.

Both kinship and friendship social relationships are of vital importance for the African population. Because of their precarious economic situation, people are more dependent on their friends for help and assistance than in western urban situations. There are few social services and institutions to help people in Katutura; most people must provide for themselves. This is reflected in the multi-stranded quality of social relationships and in the high density of networks.[12] The high density of networks is particularly characteristic of those people in a network who are coethnics. In general, high network density is usually associated with rural social relationships, but among the Windhoek African population high network density seems to reflect the need people have for help and assistance from a group of people of the same ethnic group who all know each other.

Voluntary Associations

Among the Windhoek white population, there are more than sixty voluntary associations. They range from the primarily English Shakespeare Society to the Afrikaner Railway Club. Most of these organizations are ethnic bounded: 37 percent German, 24 percent Afrikaner, 18 percent English, 12 percent German-English and 2 percent German-Afrikaner or

English-Afrikaner. Seven percent of the associations have members from all three white ethnic groups. Young people's groups are also ethnic bounded, with English in the Boy Scouts and Girl Guides, Germans in the *Pfadfinder* and *Deutscher Jugenbond*, and Afrikaners in the *Voortrekker* movement.

The *Otjiserandu* ("Red Band") is the only African ethnic association in Katutura.[13] It has existed for more than fifty years and functions as a mutual-aid and burial society. Only Herero are allowed to join, and there are troops of the movement in all towns and reserves with any sizable Herero population. Every year in August *Otjiserandu* troops from all over South West Africa congregate in Okahandja, north of Windhoek, and march past the graves of dead Herero chiefs and have a mass meeting in the Okahandja African township.

Sports clubs are the most numerous clubs in Katutura. Nine teams existed in 1968. Since most team members are recruited from among those who played soccer together in school, they are primarily ethnic group bounded. Teams are frequently called the Herero team, the Ovambo team, and so on, although they have other nonethnic names, such as the Blue Heaven. Many soccer teams have women's groups that play volleyball. Team members are mostly between twenty and thirty years of age. The only nonethnic bounded sports club is the Tennis Club, which has many African elites as members, such as teachers and nurses. There are no business or cultural organizations in Katutura. Many Katutura teachers belong to the multiethnic South West Africa Teacher's Association.

In the past there were more voluntary associations among the African population. Today many people are reluctant to join associations or form new ones for fear of being accused of political or antigovernment activities. The socioeconomic condition of the African population is also a factor. In addition, the rise of African political parties has created interest groups where people who seek political involvement can associate themselves with others who share similar concerns.

Political Parties

African political parties have only recently emerged in Namibia.[14] The South West African National Union (SWANU) was the first African political party founded in the country. Since its founding there have been many leadership disputes, splits, and the establishment of other political parties. The present situation is that all major ethnic groups in the country have their own political parties. The South West African United National Independence Organization (SWAUNIO) is a Nama party, the South West African People's Organization (SWAPO) is primarily Ovambo, the National Union Democratic Organization (NUDO) is Herero, and *Die Stem van die Volk* ("The People's Voice") is a Damara party. There are other political parties in addition to the major ones listed above. Attempts have been made to form a national organization of nonwhite political parties, and several meetings have taken place in recent years. Due to leadership disputes and rivalries, they have been ineffective so far. Windhoek is a center of African political party activity.

The dominant white political party is the Afrikaner-oriented National Party. A majority of Afrikaners belong to it, and so do most Germans. The United Party has both English and Afrikaner members. Coloured people have their own political parties.

School

Separate schools are maintained for all the major ethnic groups in the Windhoek urban area. Within Katutura there are primary schools for the Ovambo, Nama, Damara, and Herero. Schools for coloured people are in Khomasdal, and in Windhoek there are primary and secondary schools for Afrikaners, English, and Germans. Instruction is in ethnic group languages, with Afrikaans and English taught as second languages.

A consequence of this policy is that not only is the classroom situation ethnically bounded, but all the activities connected with school attendance such as recreation, making friends, and, in general, secondary socialization are with coethnics.

Religion

Group boundaries are also reinforced by church affiliation. Separate churches are located in Windhoek, Khomasdal, and Katutura. Both the Nama and Herero have independent churches, having broken away from the Lutheran Mission Church.[15] The Nama affiliated with the African Methodist Episcopal Church (A.M.E.), and the Herero have their own independent church, the *Orwano* (Unity) Church. Separate services are held at the Lutheran Church for the Damara and Ovambo, and many Ovambo also attend the Catholic Church.

Most Afrikaners belong to one of the three Dutch Reformed Churches, Germans attend the Lutheran or Catholic Churches, and English belong to the Anglican Church. Attendance at church is usually with members of one's own ethnic and racial group, and church-related activities such as men's, women's, and youth groups are consequently ethnic bounded.

Economics

Ethnic influences are also present in the economic sector. It is locally thought by white employers in Windhoek that Africans of certain ethnic groups perform particular jobs better than Africans of other ethnic groups. The common stereotypes are that Herero men prefer office jobs and jobs connected with automobiles, such as drivers. Damara men are thought to prefer work as artisans' assistants. Town-Ovambo men often work on semiskilled jobs such as stock clerks, shop assistants, and policemen; but contract-Ovambo men are employed as domestic servants and as unskilled laborers for the municipality, administration, and private businesses, especially construction companies. Women in all African ethnic groups do domestic servant work, but Herero women will only do washing and ironing. Stereotypes about whites include Germans and English as businessmen and professionals, while Afrikaners are civil servants in the government, administration, and municipality.

A study of occupational prestige yielded the interesting result that Africans ranked job categories much the same as people all over the world without any statistically significant ethnic differences being found.[16] Whether they occupy particular jobs or not, they are aware of the differential prestige and rewards of skilled and professional jobs over unskilled ones, to which they are largely confined. This finding would seem to indicate that if Africans were able to participate in the job market more equitably, they would soon occupy a full range of different occupations. If this occurred, one development might be that a socioeconomic stratification of the African population would result that might crosscut ethnic interests and help break down the strong boundaries around ethnic groups. At present the stratification of the African population is limited to elites, who are primarily businessmen and intellectuals (such as teachers and ministers), and the majority of the population, which is poor and unstratified.

CONCLUSION

Ethnicity in the Windhoek urban area is strengthened by patterns of interaction, aspects of social structure, and stereotyped patterns of ethnic group identification in cues and values. The ritualized patterns of interaction between members of different racial and ethnic groups create social distance, which maintains boundaries between people. These patterns were described for formal interaction in public places, conjugal partner choice, friendship networks, voluntary association membership, political party membership, and church affiliation. Eating, drinking, sexual intercourse, and leisure time activities, by choice and as a result of structural constraints, usually take place with members of the same racial and ethnic group.

The cues for ethnic and racial group membership are well known and stereotyped. They enable all people in the society to quickly and accurately identify to which group people belong. These cues and the value orientations associated with each group provide the primary data for interaction with strangers and maintain social distance between people who see each other in formal interaction situations often.

The structure of the social system with its many racial and ethnic boundaries creates many constraints on interaction. These boundaries do not usually prevent intergroup interaction, but they make it difficult. The separation of racial groups into different townships, the ethnic segregation in Katutura, and segregated education—all create external boundaries that must be crossed if people from different groups are to become friends. The coincidence of so many boundaries for each group increases group consciousness and the feeling of belonging to an exclusive group.

Economic factors also establish group boundaries. The society is economically stratified with Africans on the bottom, coloureds in the middle, and whites on the top. Economic constraints keep most Africans in unskilled, low-paying jobs. In addition to the poverty this creates, it also increases African ethnic group solidarity. It is no coincidence that elite voluntary associations such as the Tennis Club and the Teacher's Associations are multiethnic, while most other African associations are ethnic group bounded. The same pattern exists in the white strata of the society; professional and elite associations are multiethnic, while those with members from the lower socioeconomic classes are primarily ethnic group bounded.

The cumulative effects of the structure and patterns of interaction create a society that is racially stratified with aspects of both caste and class. Africans, coloureds, and whites belong to caste groups that are occupationally specialized into unskilled, semiskilled, and skilled occupations, endogamous, and separated by social distance mechanisms. The fear of racial group pollution manifests itself in the absence of physical touching, sexual intercourse, and eating and drinking together.

There is a socioeconomic stratification within each racial group. The white population appears to have four socioeconomic classes: at the top are political and administrative elites, who are all Afrikaners; next is a business and professional elite of mostly Germans and English; in the middle are members of all ethnic groups; and on the bottom are mainly unskilled Afrikaners. The African population has a few businessmen, teachers, ministers, nurses, and skilled workers from different ethnic groups; but most people are poor and unstratified. They are drawn together in ethnic groups out of necessity.

Ethnicity is a serious problem in Namibia, and when the country eventually achieves independence and self-government, it will assume even more critical proportions. The Ovambo will not trust the Herero, Africans will not trust coloureds, and whites will not trust nonwhites. This lack of trust is reflected in the fact that members of one group will not allow members of other groups to make political decisions for them; each group feels that other groups will only act out of self-interest. Unless steps are taken to decrease the importance of ethnicity and increase intergroup interaction and confidence, the consequences could be serious ethnic group conflicts in the post-independence period.

NOTES

1. For a more detailed description and analysis of the Windhoek social system, see Wade C. Pendleton, *Katutura: A Place Where We Do Not Stay* (San Diego: San Diego State University Press, 1974). Unless otherwise indicated, all factual information is from this book and refers to the time period 1968-1972.

2. J. Clyde Mitchell, *The Kalela Dance*, Rhodes-Livingstone paper 27 (Manchester: Manchester University Press, 1956).

3. Fredrik Barth, *Ethnic Groups and Social Boundaries* (Boston: Little, Brown, 1969).

4. J. Clyde Mitchell, "Theoretical Orientations in African Urban Studies," in *The Social Anthropology of Complex Societies*, ed. M. Banton (London: Tavistock, 1966).

5. Abner Cohen, *Urban Ethnicity* (London: Tavistock, 1974), pp. ix, x.

6. *Namibia* is the name for South West Africa that has been adopted by the United Nations. The country is more generally known as South West Africa. Both names have been used in this chapter.

7. *Katutura* is the name adopted by white authorities for the new African township. It is a Herero word and means "place where we do not stay." The name must have been suggested to reflect the widespread protest over the forced relocation of Africans in the new township; it is ironic that white officials accepted it without apparently knowing what it meant.

8. All these restrictive measures were first imposed during the German colonial period. Some were copied from existing regulations in practice in South Africa. See Helmut Bley, *South West Africa Under German Rule 1884-1914* (London: Heineman, 1971), pp. 171-173.

9. The southern section is that portion of the country (locally called the "police zone") where whites may own land and settle. It comprises about three-quarters of the territory and contains about one-half of its population. No white settlement is allowed in the northern section.

10. Peter Banghart, "The Effects of the Migrant Labourer on the Ovambo of South West Africa," unpublished manuscript, 1971. Some of the negative effects of migrant labor are a disruption of marriage and homestead life, sometimes leading to divorce but often causing conflicts, frequent heavy drinking, prostitution, and adultery. For more details see R. Voipio, *Kontrak soos die Owambo dit sien* (Johannesburg: Christian Institute of Southern Africa, 1972), and Pendleton, *Katutura*, pp. 17-18.

11. See Wade C. Pendleton, "Social Categorization and Language Usage in Windhoek, South West Africa," in *Urban Man in Southern Africa*, ed. C. Kileff and Wade C. Pendleton (Gwelo and Chattanooga: Mambo Press, 1975). A speech variety is an alternative form of a language with different grammatical and vocabulary usage. It differs from a dialect because it is not identified with a geographical area.

12. Density is a measure of the extent to which people in a network know each other. A density of 100 percent means that all people in the network know each other; a density of 50 percent means that only one-half of the total possible links

among people in the network existed.

13. For a more detailed discussion of the *Otjiserandu*, see Wade C. Pendleton, "Herero Reactions: The Pre-Colonial Period, the German Colonial Period and the Period of South African Colonialism," in *African Responses to European Colonialism in Southern Africa*, ed. David Chanaiwa (Northridge: California State University Foundation, 1977).

14. More information may be found on political parties in I. Goldblatt, *History of South West Africa* (Cape Town: Juta, 1971); F. J. Kozonguizi, "Historical Background and Current Problems in South West Africa, in *Southern Africa in Transition*, ed. J. Davis and J. Baker (New York: Praeger, 1966); and Pendleton, *Katutura,* pp. 117-120.

15. For more details about the *Orwano* and A.M.E., see E. Kandovazu, *Die Oruuano-Beweging* (Karibib: Rynse Sending Drukkery, 1968); and K. Schlosser, *Eingeborenenkirchen in Süd- und Südwestafrika* (Kiel: Mülau, 1958).

16. More details of the occupational prestige study may be found in Wade C. Pendleton, "Ethnicity As A Factor in Occupational Prestige," *ASSA Sociology Southern Africa 1973* (Durban: University of Natal, 1973); and J. Kelley and Wade C. Pendleton, "Structure, Culture and Occupational Prestige: Data from Sub-Saharan Africa," *American Journal of Sociology*, in press.

SELECTED READING

Esterhuyse, J. H. *South West Africa 1880-1894*. Cape Town: Struik, 1968. A good history of the early period of the German colonial era, although not based on any oral-historical research.

Pendleton, Wade C. *The Peoples of South West Africa*. Cape Town: David Philip, 1977. This book has chapters on the origins and migrations, prehistory, colonial history, ethnography, and the current situation of the people who live in this disputed territory.

Wellington, J. H. *South West Africa and Its Human Issues*. London: Oxford University Press, 1967. The first half of the book is on the geography of the country, and the second part deals critically with the administration of the territory by the South African government.

8

Ethnicity, Neighborliness, and Friendship among Urban Africans in South Africa

Brian M. du Toit

The concept of social distance is not new. More than four decades ago Emory Bogardus, in a series of papers, developed it into a research tool. Since then it has been used in sociology, social psychology, and anthropology. In this study I will discuss its adaptation to urban Africans, mostly Zulu-speakers, living in an urban center in Natal.

THE SOCIAL DISTANCE CONCEPT

The concept of social distance is most clearly expressed where there are boundaries that divide people. The more marked such barriers, the more significant it becomes to transcend them. Thus geographical boundaries have all but lost their significance due to improved transportation and communication, and we frequently remark that the world is "shrinking." But linguistic, cultural, and ethnic boundaries may increase in significance and constitute divisions that rarely are transcended.

In modern Africa all these barriers are present as the concepts "tribe," "nation," or "race" have gained in importance. Every individual has, first of all, the home neighborhood, whether this is a small village or a city block. He also has a kingroup constituted by parents and an extended family. In most cases the traditional setting presents an overlap between these two groups, but in modern and particularly urban living arrangements the kingroup or even the linguistic and cultural group may be far removed from the immediate vicinity. The modern African city might have members of four or five linguistic groups living on the same block, and the modern urban African may have daily contacts with even more groups.

One important question that has concerned researchers is the selection a person might make when confronted with a decision-making situation. When a person selects a good friend, when he needs some person to depend on, when there is an immediate need of a loan, to whom does a person turn? Is it the next-door neighbor or the kinsman in the next block? Does a recent migrant call on a coworker or a person who comes from the same home area? Harries-Jones has shown how "home-boy" ties are important in various voluntary organizations and the political aspirations of people on the Zambian Copperbelt.[1] Thus the Bemba *Abakumwesu* are not only persons from the same linguistic district, but "the 'home-boys' who are important for maintaining his ties with his rural home are usually near kin."[2] Much the same picture has been sketched by Mayer[3] as well as Wilson and Mafeje[4] for Xhosa urban migrants, among whom the *amakhaya*, or "home people," are of basic importance.

Such cliques or reference groups—which are based on a combination of ethnic and geographical factors—can be interpreted at different levels. They may refer to the immediate group of interaction, i. e., the people with whom a person carries on interaction or with whom he forms voluntary associations. At a higher level this reference group forms the basis of the "we-they" dichotomy. Here the linguistic group or ethnic group is factually compared with other groups in an attempt to discover similarities or contrasts. They then form the "insignia in terms of which alignments . . . may be formed."[5] But they also are at the basis of categorical relationships where persons of other ethnic, linguistic, or color groups, with whom relatively little contact is experienced, are typified. Such typification in many cases is completely unfounded or rests on hearsay. As Mitchell pointed out, "Categorical relations, because of their superficiality, imply stereotypes."[6]

We might expect that as an African urban elite develops, stereotypes might disappear and ethnic heterogeneity gain in importance in the reference group. People might prefer to interact with fellow educated or elite persons with the result that "elite of different tribal backgrounds are woven into this single network, in which tribal or ethnic differences are overridden by common status as elite."[7] It is equally possible that a person might continue to make a selection among other elite persons on the basis of the criteria discussed.

When we turn from the interaction set to more meaningful relations, social distance is really tested. For rather than looking at friendship and interaction, we might deal with coresidence or marriage. The question then changes from a generalized reference to a very personal one, from relations that usually occur in the group context to the intimate relations of two persons. This does not deny that all friendships are based on dyadic relations, only that we do face here degrees of intimacy and duration of contact.

Bogardus distinguished between "personal distance," the degree of sympathetic understanding between two persons; "personal-group distance," the degree of sympathetic understanding between a person and each of the social groups of which he is a member; and "group distance," the degree of sympathetic understanding between any two social groups. "Each person, therefore, continually plays a three-fold social distance role:

one, as person to person; two, as person to social group; and three, as a member of a group in its relations to other groups."[8] It is particularly this last aspect that Clyde Mitchell utilized when he wrote about "tribal social distance." In this study Mitchell employed descriptions of contrived situations, much as Bogardus had, to represent grades of intimacy or distance.[9] These ranged from marrying or admitting near kinship to allowing such a person to live in the same country or tribal area.

Within the South African context, the social distance concept was first employed by MacCrone in his study of *Race Attitudes in South Africa*.[10] He applied his questionnaire to a representative sample of whites only, testing their reaction to Belgians, Scotsmen, Indians (Asiatic), English-speaking South Africans, Cape Coloured, Portuguese, Native (Bantu), Hollanders, Afrikaans-speaking South Africans, and Englishmen.

During the early 1960s two studies were published—both based on research at the University of Natal. In 1960 Pettigrew published his study of the attitudes of 627 white students questioned during 1956.[11] Two years later Pierre van den Berghe reported on his 1960 study of white and nonwhite students at the University of Natal, a local nursing school, and two technical colleges.[12] Up to this point, then, the research had been almost exclusively on whites.

Five years later, Grove analyzed the results of his research on social distance between coloureds and Indians in Pretoria.[13] This was the first systematic use of the social distance concept relative to population groups other than white South Africans. I am here not referring to such studies as Brett, which dealt with the attitudes of Africans about whites but did not include the social distance concept quantification.[14]

The first systematic study on the attitudes of urban Africans with special reference to stereotyping and social distance was that of Edelstein.[15] His sample population consisted of 200 African matriculation (twelfth-grade) pupils whose families resided in Soweto, the large African satellite city near Johannesburg. These pupils' attitudes were tested toward a number of African linguistic groups as well as toward non-Africans, namely, coloureds, Indians, Jews, English-speaking white South Africans, and Afrikaans-speaking white South Africans. In general, his conclusions are the same as those represented in table 8.4 below.

For a critical discussion of the original Bogardus tests and the need for situational adaptation, the reader is referred to Banton[16] and Edelstein.[17]

RESEARCH METHODOLOGY AND SAMPLING

The research on which this chapter is based employed six of the original seven basic contrived situations mentioned above.[18] These situations implied degrees of contact or social distance. The question also referred to a variety of ethnic groups—based, we thought, on both geographical and social distance. Where the questions needed explanation or elaboration, we employed the same examples for all subjects, thus limiting the kinds of cues given to different subjects.

This study of social distance employed two basic questions. The questions for the first were arranged on a grid pattern. Along the horizontal axis were listed the names of six different Bantu-speaking African groups as well as coloured, Indian, and white. Along the vertical axis was a list of questions implying different degrees of intimacy or closeness. These were:

1. Would you marry?
2. Would you work alongside?
3. Would you have several families live near you?
4. Would you have as a regular friend?
5. Would you share a meal with?
6. Would you visit or have visit you?

The sequence represented what I thought to be a realistic ordering of distance vis-à-vis ethnic groups and regarding social distance criteria. For each question the researcher conducting the interview would in turn apply the names of the different ethnic groups. The African group included Swazi, Xhosa, and Ndebele, who are linguistically and ethnically close to the Zulu, as well as Sotho, Tswana, and Tsonga, who are more distant. Contrasting with this linguistic and ethnic relationship is the fact that the Sotho, who live basically in and around Lesotho, are geographically much closer than the Tswana. Familiarity, it was argued, is a factor of geographical distance, experience, and intelligibility.

The second question actually goes beyond the limited social distance test and asks the subject to name three persons with whom he or she visited or received visits from, with whom they went out socially, with whom a mother shared child care or a man relaxed outside the home. We also asked which one of these three persons the subject would turn to in time of personal crisis and to which he would entrust a money loan. Each person was identified in terms of specific place of residence and of kinship to ego.

The research sample was a 3 percent random sample of households in each of the eleven residential neighborhoods of Kwa Mashu, a large African satellite city northwest of Durban in South Africa. Owing to its geographical location, the residents are largely Zulu-speaking. A breakdown of the sample population in this study is presented in table 8.1. In an attempt to minimize variation in responses due to ethnic and cultural differences, this study will deal only with the 362 Zulu speakers.[19]

In terms of the social distance between Zulu and members of other ethnic-linguistic groups, it is interesting to note the ethnic distribution in the first table. Xhosa, Swazi, and Sotho are best represented in our random sample, while the other three African groups are present, though in smaller numbers. Males are present in larger numbers than females. On a

TABLE 8.1

Ethnic Distribution by Sex

Ethnic Group	Male	Female	Total
ZULU:	219	143	362
NON-ZULU:			
Xhosa	7	9	16
Swazi	5	8	13
Sotho	7	6	13
Hlubi	4	4	8
Baca	6	1	7
Tsonga	4	1	5
Pondo	3	0	3
Tswana	2	0	2
Pedi	1	0	1
Ndebele	1	0	1
Karanga	1	0	1
TOTAL	260	172	432

residential basis the Zulu do meet and interact with other non-Zulu.

The sketch map is included merely to give a general impression of the relative geographical distance among the different African ethnic-linguistic groups. The fact that some Zulu live adjacent to some Swazi in the north or that some Zulu in the south live adjacent to Xhosa is self-evident. The graph that accompanies the sketch map is an attempt to compare three kinds of distance. The first bar for each ethnic-linguistic group represents socio-linguistic or cultural distance. This is an arbitrary evaluation in which I considered language and cultural factors. There might be a way of developing a better-controlled measure, but for this discussion the evaluation is my own. The second bar for each ethnic-linguistic group represents geographical distance. I have counted each small square the equivalent of 10 miles. Thus the Swazi live 200 miles from the Zulu with whom we are dealing. Since coloured, white, and Indian differ in critical linguistic and cultural aspects and live adjacent to or work with the Zulu, these two aspects were not measured for them. The third bar in each case represents the actual social distance as we measured it in this study. Thus each five negative responses in table 8.4 (see page 151) were represented here by one shaded square. Thus the Swazi received 95 negative votes represented by 19 shaded squares, whereas the Indians received 532 negative responses, represented by 106 shaded squares.

For years, social scientists have been writing about the fact that a people's horizons expand as they receive more education, as they have interethnic experiences, and as they have the opportunity of being exposed to other cultures and values. Some writers have even gone so far as to suggest measures of "urbanization," "detribalization," or urban commitment.[20] In spite of the fact that we have now theoretically and methodologically gone beyond such measures, the criterion of education remains of critical importance. It is also one of the indexes to be used here

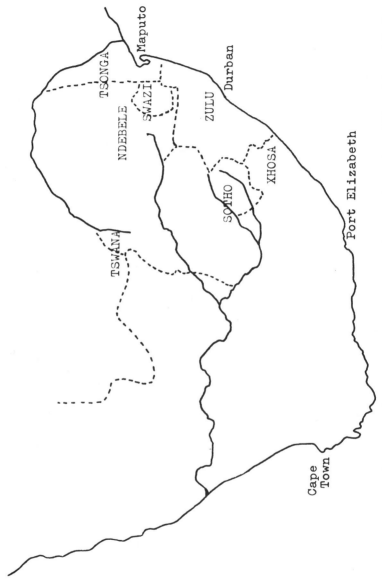

Figure 8.1 Geographical distribution of aboriginal ethnic-linguistic groups in southern Africa. Coloured, Indian and white South Africans live throughout the region and have not been localized on this map.

TABLE 8.2

Education by sex by age

Years of Schooling

MALES:	NR	None	below 2	2-7	8-10	11-12	over 12	Total
below 25	1	0	0	0	1	1	0	2
25-34	0	0	0	2	6	6	4	18
35-44	0	10	0	32	24	14	16	96
45-54	0	17	1	32	8	0	6	64
55 and over	0	17	3	13	3	1	1	38
TOTAL	1	44	4	79	42	22	27	219
FEMALES:								
below 25	0	0	0	1	2	0	0	3
25-34	0	0	0	13	11	6	4	34
35-44	0	4	1	20	11	5	4	45
45-54	0	5	1	16	8	2	3	35
55 and over	0	6	1	13	4	2	0	26
TOTAL	0	15	3	63	36	15	11	143

in marking persons on a social distance scale. The sample of Zulu-speakers to be used in this study includes persons who have had no formal schooling as well as others who have completed university degrees. This information is contained in table 8.2.

The cutoff points in years of schooling used here should be explained. Less than two years of schooling really has no significance and would hardly make a child literate. It has been used here because at least the person has been exposed to some elementary education. The category that includes two through seven years of schooling would include the largest number of children who have at least been in or through junior school, terminating in the South African system with standard five. The next two categories cover persons in junior high, i. e., standard six through eight, and the matric years are standards nine and ten, or the equivalent of eleventh and twelfth grades in the American system. The last category would cover all persons who received some tertiary education—just over 10 percent of our Zulu sample.

We should of course keep in mind that these figures reflect the education of the respondent, the subject who was interviewed and not necessarily the breadwinner of the family. Forced as we were to work basically during daytime, we may have missed many better-educated and professional people who were at work. Care should be taken in attributing the parameters of this sample to the larger population.

A further criterion frequently mentioned as a possible indicator of urbanism or of social distance is length of residence in the urban area. We have not included a table here to reflect this data, since there is not sufficient variation on this subject to allow significant comparison. For the males interviewed, 170 had been living in the city for twenty years or more or alternately for their entire lives, while 87 of the female respondents were in this same category—thus 71 percent of our total sample. On the other hand, only 8 percent had been living in the city or urban environment for less than twelve years.

Income, in most cases, is a product of education or experience or both. Table 8.3 shows the distribution of our sample as age and income relate. The largest number of persons, just over 56 percent of the males and just over 62 percent of the females, earned below R80 per month. This would include most of the unskilled workers and laborers. With reference to our previous table, it would probably include most of the persons in the educational level below eight years of schooling. While salaries in South Africa have recently been adjusted to the cost of living, this salary category in 1973 included most of the Africans working for the Durban Corporation. Elsewhere I have discussed the purchasing power of such an income during 1973.[21]

TABLE 8.3

Income by sex by age

Age	below R80 p.month			R81–R100 p.month			R101–R150 p.month			over R151 p.month		
	M	F	Total	M	F	Total	M	F	Total	M	F	Total
Under 34	11	17	28	1	2	3	3	11	14	5	7	12
35–44	38	26	64	13	7	20	22	5	27	23	7	30
45–54	41	23	64	10	4	14	10	4	14	3	4	7
over 55	32	23	55	4	0	4	1	2	3	1	1	2
N/R	1	0	1	0	0	0	0	0	0	0	0	0
Totals	123	89	212	28	13	41	36	22	58	32	19	51

Educational facilities and programs as well as employment opportunities for educated Africans have greatly improved in recent years. This is clearly discernible in the number of young well-educated men and women who earn fairly good salaries, while these numbers decrease with age. One also notices that in the older age brackets more men earn good salaries, while younger women are competing very successfully with men for good salaries.

One of the central questions in any social distance study is whether criteria of age, sex, education, income, and similar factors influence attitudes. In other words, do these criteria act as differential factors? If so, what are the likely explanations?

A STUDY OF RESPONSE PATTERNS

The interviewer in every case requested the subject to respond positively or negatively to the question and, wherever possible, to give a reason for a negative response. Table 8.4 contains the responses as given by females and by males. As stated before, respondents are all Zulu-speaking.

TAFLE 8.4

Negative Responses to Social Distance Questions*

	SWAZI	SOTHO	XHOSA	TSWANA	NDEBELE	TSONGA	COLOURED	WHITE	INDIAN	TOTAL
FEMALES:										
Work with	3	4	7	6	9	8	9	4	13	63
Share meal with	1	3	4	4	7	5	5	11	24	64
Visit and allow to visit	1	3	5	5	6	5	4	10	18	57
Have as regular friend	6	9	11	12	15	13	20	21	28	135
Live adjacent to	9	13	15	16	14	13	28	29	35	172
Marry	38	49	49	58	65	64	122	118	127	690
TOTAL:	58	81	91	101	116	108	188	193	245	1181
MALE:										
Work with	0	3	1	0	0	2	5	3	15	29
Share meal with	3	2	4	5	3	6	8	9	26	66
Visit and allow to visit	3	4	6	6	5	6	16	16	23	85
Have as regular friend	3	4	4	7	7	9	14	19	26	93
Live adjacent to	5	7	5	9	8	9	22	28	41	134
Marry	23	26	37	34	40	46	124	152	156	638
TOTAL:	37	46	57	61	63	78	189	227	287	1045
TOTALS:										
Work with	3	4	8	9	9	10	14	7	28	92
Share meal with	4	5	8	9	10	11	13	20	50	130
Visit and allow to visit	4	7	11	11	11	11	20	26	41	142
Have as regular friend	9	13	15	19	22	22	34	40	54	228
Live adjacent to	14	20	20	25	22	22	50	57	76	306
Marry	61	75	86	92	105	110	246	270	283	1328
TOTAL:	95	124	148	165	179	186	377	420	532	2226

*Note that in this table the social distance criteria have been ordered in terms of their relative importance. The ethnic groups are also listed here as from those who are least to those who are most distant.

Several response patterns deserve special mention. Merely working with a person of another ethnic group contains no problems, since people are used to this arrangement—though males react less strongly than females. Much the same response is found in the question about sharing a meal. Subjects repeatedly pointed out that they would eat with a person of another ethnic group provided this did not entail sharing the same pot. Africans in traditional life-style prepare a large pot of the staple food, and this is shared by all who share the meal, using in many cases, the thumb and first two fingers on the right hand.

Visiting and allowing to visit, an action we originally considered to be of minor significance, evoked surprising reaction. This was not directed against any particular ethnic group, although the Indians were singled out much more frequently than any other group. As some subjects explained it, the very idea of a non-Zulu in their home was a strange idea. When on a number of occasions I pointed out that I was white, they would laugh and retort, "Yes, but you're working!"

When we move to the last three questions, we find that the idea of having a non-Zulu as a close personal friend is reacted to relatively strongly, especially by women who are more housebound and do not move in interethnic situations as much as their menfolk. Regular friendship of course implies all three of the previous questions, and one might expect that persons who responded negatively to the first three might very well have replied in the negative here as well. Once again the reaction here is not against a particular ethnic group.

The last two questions deal with the most intimate contact. The first implies association on a neighborly and daily basis with persons who speak a different language, differ culturally, eat other foods, and finally are not Zulu. We find here almost the same number of males and females reacting against the idea of living with non-Africans, but again it is the women who respond most strongly against living with other non-Zulu-speaking Africans.

When we get to the question of marriage, we are dealing with a very delicate subject. Some subjects reacted with disgust at our questions, while others laughed at the thought of marrying a non-Zulu. As regards marrying members of other color castes, we clarified that this presumed the abolition or absence of any legal restrictions—thus complete freedom for a person to marry any other person willing to do so. The recorded responses show an interesting patern, with a higher frequency of negative responses for Zulu women as regards other African linguistic groups. When we turn out attention to the non-Africans, we find that males respond much more negatively. This might be due to various factors, two of which are: a greater national consciousness and pride among males, and the fact that African women are very frequently domestics. In this role they are in homes, hotels, and boarding houses, and generally in closer contact with whites, especially white males.

But this table of negative responses also shows—and this is central to the topic of this paper—that Zulu men and women consider members of certain ethnic groups more acceptable than others.

Zulu and Swazi speak languages that belong to the same branch of Nguni and that differ from the southern Nguni languages, which include Xhosa. The Swazi also live much closer to the traditional Zulu rural areas than any other group. It is thus predictable that there will be a minimum of negative responses to the Swazi. The Sotho, who live mostly in and around Lesotho, are a different matter. Their language belongs to a different family of Bantu, and they differ culturally in some major respects. Yet, during the fieldwork period informants frequently explained that during Shaka's rise to power numerous "Zulu" clans fled Natal and that a number joined Moshesh in the Maluti mountains and beyond, "so, you see, they [the Sotho] are really Zulu!" To what extent this argument affected the responses is unknown, but relatively few people find fault with marrying a Sotho-speaker. We should also keep in mind that Lesotho is an exceptionally poor country and that large numbers of Sotho are forced to

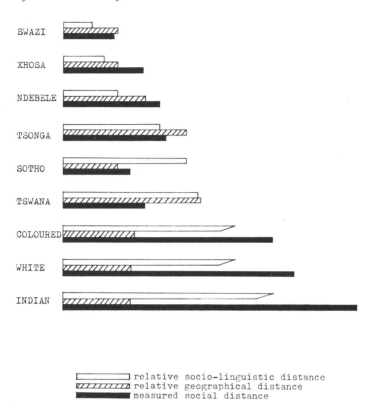

Figure 8.2 Relative socio-linguistic, geographical, and social distance between urban Africans and other South Africans.

seek employment outside their country. Zulu therefore have a fair acquaintance with Sotho-speakers.

Xhosa, of course, ought to be seen as cousins, much as Swazi are. But Xhosa women are different from Zulu women. One fact that was mentioned is the practice of pipe smoking, another is the fact that Xhosa are so "clever." They would make a Zulu feel "lost" and "inferior," or else they would cheat him out of his belongings.

In geographical distance the Tswana of Western Transvaal and Botswana are quite far removed from the Zulu. Once again, as can be expected, Zulu women respond much more negatively than their menfolk. Men are of course employed in various jobs that permit them to travel outside the Durban area and Natal, and many young Zulu have a stint on the gold mines in the Orange Free State or the Transvaal before settling down. They also spend some free time in beer halls, eating houses, and bars, where an ethnically heterogeneous population drowns the boredom of poverty and frustration.

The Ndebele and Tsonga are both related to Nguni speakers, but both are fairly distantly resident. As a result, subjects frequently responded negatively and then explained that they really did not know the Ndebele or the Tsonga. Because they did not know them, they also presumed sociocultural differences that might separate them. In both cases, as had been evident with regard to other ethnic groups, we find a much higher percentage of negative responses with women than with men.[22]

When we turn our attention to the three non-African ethnic groups, the total number of negative responses doubles for women and triples for men. The idea of marrying a person who is not black shocked many subjects. There is no excuse here of not knowing the coloureds, whites, or Indians. There also is little reason to say that linguistic barriers would be insurmountable, for anybody who has lived in Natal for a couple of years at least tries elementary Zulu, and our subjects are all at least bilingual and some trilingual. English is the lingua franca both in education and in commerce.

One unexpected fact is the relative position of white versus Indian. It is known that very deep resentments separate Africans and Indians in South Africa and that these were part cause of the 1949 riots in Durban. We expected, however, that Indians would be more acceptable than whites and thus closer on the social distance scale. The figures presented here illuminate these attitudes. We had expected Xhosa to be closer to Zulu than Sotho, and Ndebele closer than Tswana. The attitudes of Zulu subjects must be seen in a wider context.

During the research it was decided that subjects should explain their motivation for any negative response. This should not suggest that every response is a conscious evaluation or that a subject weighed his motivation before answering. It should also not suggest that we insisted on a reason for every negative response.

REASONS UNDERLYING RESPONSES

Every time a person voiced a reason for a negative response, we recorded it. There was no attempt to insist on logical reasoning or to influence in any way the decision or the answer, with the exception that when a person was asked whether he or she would marry a white, we explained that this would imply an absence of the apartheid laws. We would explain that *if* there were no legal barriers and *if* it were possible and permissible, "Would you marry a white man/woman?"

The negative responses recorded for the first four questions were small in number, and for that reason their analysis here would only occupy space and time. They followed in essence the pattern of the last two questions.

Table 8.5 contains the responses for question five: "Would you have several families live near you?" Once again the subject had to imagine a situation in which they as an African Zulu-speaking family would be surrounded by families representing other cultural, linguistic, and ethnic groups. These responses and reasons were noted down on the interview schedule by the interviewer. After my return from the field, all these diverse responses were transferred to cards. I then developed a code system by which each verbal response could be classified into one of the fifteen categories.

TABLE 8.5

Reasons for not living adjacent to other ethnic groups

		Swazi	Sotho	Xhosa	Tswana	Ndebele	Tsonga	Coloured	White	Indian	TOTAL
A.	Ethnic-national pride	1	1	1	1	1	1	3	2	3	14
B.	Ethnic-racial pride	-	-	-	-	-	-	2	1	1	4
C.	Socio-Cultural	1	1	1	1	1	2	4	7	9	27
D.	"I don't like them"	-	-	-	-	1	-	-	2	8	11
E.	Superstitions	1	1	-	2	-	-	3	2	2	11
F.	"I don't know them"	-	-	-	1	1	2	-	-	-	4
G.	Children would be mixed	-	-	-	-	-	-	-	-	1	1
H.	Linguistic reasons	1	3	1	2	2	2	2	2	4	19
I.	Fear	-	-	-	1	-	1	-	3	1	6
J.	Social isolation	-	-	-	-	-	-	2	2	2	6
K.	"They are too clever"	-	-	1	-	-	-	-	-	-	1
L.	Apartheid	-	-	-	-	-	-	-	-	-	0
M.	"They are backward"	-	-	-	-	-	-	-	-	-	0
N.	Ancestors - religion	-	-	-	-	-	-	-	-	-	0
O.	"I would feel inferior"	-	1	-	1	-	-	-	-	-	2
	TOTAL:	4	7	4	9	6	8	16	21	31	

Responses referring to "my race" or "not black" were scored as *ethnic-racial*, but responses dealing with "my people," "not Zulu," or a "different nation" were scored as *ethnic-national*. A large category of responses are those in which a respondent simply states that "I don't like them." This is a gut reaction, not logically thought out and hard to explain. A very wide range of responses that are clearly illogical, inconsistent, or unfounded were classified as *superstitions*. Such evaluations are frequently founded on unfamiliarity with the group in question, resting on hearsay or superficiality, and leading to stereotypes. They may also be contrived excuses for other kinds of real social distance. Some of these will be discussed in the following section of this chapter. In many cases informants explained that they "really did not know" the people in question.[23] This was seen as either an easy answer or as being based on actual unfamiliarity due basically to geographical distance resulting in social distance. In a number of cases, and this applied almost exclusively to the question of marriage and the resultant progeny, subjects feared for the fact that their *children* would be either not Zulu or not black. A reason that affected persons who were not as fluent in languages other than Zulu as they might be, was one that raised various *linguistic* considerations. Either the reason given was that the other ethnic group spoke a "foreign language," or it was explained that "we Zulu are very stubborn about our language." As unfounded and inexplicable as the category of superstitions is one in which subjects explained that they would be afraid—the *fear*, it seemed, was partly due to the unknown rather than any empirical fact or cultural practice, and this category might well be joined with that which claims that the group is unknown to the subject.

Living among or being married to a member of another African ethnic group would not really affect the person's contact with fellow Zulu—after all, many interethnic marriages do occur. But the factor of *social isolation* enters as a likely possibility should a family live among non-Africans, and particularly so should he or she be married to a non-African. In a number of cases, in spite of the hypothetical situation we sketched, informants would point out that if they married a white or an Indian they "would have to go overseas—away from my family." Most people know of or have read about persons who have married across these ethnic boundaries and have been forced by the Immorality Act to leave South Africa. This response is very closely related to one in which the apartheid laws are given as reason for a negative response. It seems that in many cases people are psychologically so conditioned that they cannot even imagine a situation in which this would be different. Mostly these persons were women who had not been in the urban area for any length of time.

The reason that they are *too clever* was given with reference to the Xhosa only—save one person who ascribed it to the Ndebele. The basis for this evaluation is unknown, but we found it in various contexts. Being clever might be a positive trait, but being too clever had the connotation of looking down on or despising others. This response also is closely related to superstitions about Xhosa—a topic to be discussed in the next section of this chapter.

The last three negative responses were not recorded in regard to residential arrangements but only as regards marriage. Table 8.6 shows a small number of persons who would not marry Ndebele or Tsonga because the latter two ethnic groups are thought to be *backward*. The same number of persons felt that by marrying non-Africans, the *ancestors* would become incensed and possibly affect them detrimentally. We should of course keep in mind that the ancestor cult is very strong among the Zulu and that a very large percentage of urban residents retain an active recognition of the ancestors. In some cases people may classify themselves as traditionalists, but in many more the ancestor cult exists side by side, or in fact may be integrated, with various Christian or neo-Christian church practices.[24]

The last, and very small, topic of response is the feeling that by marrying such a person, and this was restricted to whites and Indians, the subject would "feel inferior."

With reference to the previous two tables it would be valuable to look at the contents of some of these negative response categories. It will be possible only to refer to some of the most interesting of these, but they do show some of the stereotypes that exist and that are perpetuated by hearsay. As persons become better informed or gain in mobility, these

TABLE 8.6

Reasons for not marrying other ethnic groups

		Swazi	Sotho	Xhosa	Tswana	Ndebele	Tsonga	Coloured	White	Indian	TOTAL
A.	Ethnic - national pride	16	3	20	22	18	6	68	61	77	291
B.	Ethnic - racial pride	2	9	2	8	9	10	60	77	68	240
C.	Socio-Cultural	4	19	7	7	13	29	29	30	36	174
D.	"I don't like them"	2	4	5	4	9	8	29	30	35	126
E.	Superstitions	7	5	19	10	7	10	10	14	20	102
F.	"I don't know them"	5	3	6	19	30	27	3	4	3	100
G.	Children would be mixed	-	-	-	-	-	-	17	24	25	66
H.	Linguistic reasons	2	9	2	8	3	5	4	3	5	41
I.	Fear	-	-	-	1	1	2	5	8	2	19
J.	Social Isolation	-	-	-	-	-	-	5	4	6	15
K.	"They are too clever"	-	-	12	-	1	-	-	-	-	13
L.	Apartheid	-	-	-	-	-	-	-	6	1	7
M.	"They are backward"	-	-	-	-	1	3	-	-	-	4
N.	Ancestor - religion	-	-	-	-	-	-	1	1	2	4
O.	"I would feel inferior"	-	-	-	-	-	-	-	3	1	4
	TOTAL:	38	52	73	74	92	100	231	265	281	

superstitions and stereotypes are challenged and finally change.

SUPERSTITIONS AND STEREOTYPES

As suggested before, the greater the familiarity due to interaction or knowledge, the less likely a person will be to have superstitions and stereotypes regarding the members of another ethnic group. In fact, at one point in the study I formulated a working hypothesis that suggested that: "As F (in table 8.6) decreases, so will E, I, K, and M." But this only partly proved to be the case, as we can see by looking at Xhosa: six negative responses in F versus thirty-one negative responses in the other four categories. The Tswana show nineteen negative responses in F versus eleven, and the Ndebele thirty versus thirteen. This was puzzling until I recalled the findings of Gluckman, Marwick, Nadel, and Wilson in their discussions of the sociology of conflict, particularly pertaining to witchcraft and sorcery accusations.[25] If such accusations are a reflection of strains in social relations or threats to status, then obviously strangers would not be accused because they are outside the social system of the accuser. Basically the same argument operates here. The less known another social group is, the less threat or competition, and, conversely, the less is known about them to use as an excuse for social distance.

Zulu would not have superstitious excuses for not liking the Mongo, Diola, or Edo because they most likely never even heard about them. In other words rather than social distance we would be dealing with a neutral no-response category. The better people are known, the easier it is to find some reason for not liking them or for liking them. The same principle is at work here. The result is almost the opposite of what I had originally expected to find. Tswana, Ndebele, and Tsonga, who live a long way off and with whom the average Zulu-speaker would have absolutely minimal contact, stand respectively in a 19/11, 20/15, and 27/15 relation as regards people who say they do not know them versus excuses in categories E, I, K and M. However, coloured, white, and Indian (who are well known and with whom a great deal of interaction takes place) stand respectively in a 3/15, 4/22, and 3/22 relation in these same responses.

Superstitions operate particularly with regard to marriage. As was done in the previous section, I will discuss each ethnic group as they were evaluated in terms of the social distance scale.

The Swazi are said to "pick on you if you are not of their nation," while other subjects stated that they are "anti-Zulu," that they are "not clean," and that they are "poor." But on the whole very few responses contrast the two ethnic groups, two nations that have a solid knowledge of and high esteem for each other.

Sotho, who are almost as well known but who differ culturally, are said to "live a very fast life and they say Zulu eat horses," Subjects also complained that they are "to stingy," that they are "poor," and that they have "very low morals."

But Zulu really get their ire up when they refer to the Xhosa. First of all, the latter are "too clever," but others saw them as "crooks," as "scheming," or as "very cunning" and "crafty." One old lady stated emphatically that if she had anything to say about the matter, she would not have even one as a daughter-in-law. Better-educated women complained that Xhosa men are "stiff-headed. If you are their woman, you must act like a woman and be suppressed." Or, responded a third, "they treat wives like they treat goats."

Separated by linguistic, cultural, and geographical distance are the Tswana, for whom there are actually relatively few superstitions. One person started his reasoning by explaining, "I have heard that. . . ." In other words, the negative response is not based on firsthand experience.

Ndebele are seen as *isilwane*, or "animals," as having "funny customs," and as "backward because they have never been in cities or schools."

Prototypes are frequently based on a single experience or an event that is seen as typical and representative. When the same response is found among diverse subjects, it is likely that such prototypes have been communicated to others. Thus the Tsonga have a reputation for "specializing in witchcraft" and for being "witches and wizards." They are also said not to "dress decently," and since they "don't conduct themselves properly, they usually are harlots."

All the previously discussed groups are African. When we consider the last three ethnic groups, the Zulu subjects were discussing peoples with whom they had a great amount of social contact but who differ from them cuturally, linguistically, and phenotypically.

Regarding the coloureds, we were told that Zulu would not marry them because they are "loafers" and "fond of drink." One subject stated that they are "stubborn," but it was the Zulu males who felt most strongly on the subject of the complexion. One man who obviously has a rural prototype stated that a coloured woman "will refuse to work in the fields because she would not want to lose her light complexion." Another feared that a coloured woman "easily registers wrinkles."

Some of the same reasons were suggested for not marrying a white woman, for instance her unwillingness to work in the fields. But it was also felt that even if she were willing, she would be too weak to stand "the strain of the work." One man, who might have experienced a woman whose tastes were beyond his income, felt that white women are "too expensive to maintain," while another pointed out that they "lead an expensive life." Recalling a recent statement by Mrs. Verwoerd,[26] widow of the former prime minister, regarding the "smell" of black Africans, it is interesting that two subjects remarked that whites "have a bad smell," one singled out particularly "in the morning."

The ethnic group with whom greatest social distance was recorded (the Indians) was also said to "have a nasty smell" or a "peculiar smell." Other subjects felt that Indians were "untidy," that their "food is strange," or that they are weak in health. Due to their involvement in the retail trade in Natal, the Indians are necessarily tied in with financial matters. One subject explained that Indians "cheat too much," and a second called them all "swindlers." Once again the Zulu males who had the colorful silk sari in mind remarked on the expense of maintaining an Indian woman, while one noted that they "register wrinkles" at a much younger age than Zulu women do.

These, then, are some of the reasons for not being willing to marry non-Zulu. Obviously many of the same reasons underlie other forms of social distance.

The previous three sections of this chapter have looked at the patterns of response and the content of these responses. The logical question is how any of these criteria find significance in the social distance survey. To express this quantitatively and in easily comparable form, we have drawn up an index that allows one to deal with percentages rather than numbers of positive or negative responses.

SOCIAL DISTANCE MATRIX INDEX

These "index" figures are percentages, which can be compared directly with other such percentages. Higher numbers indicate greater perceived social distance as reflected by the matrix. To obtain the figures for each ethnic group, the total of negative responses registered was divided by the total possible responses for that particular group of persons.

In the following calculations, the sample population was divided on the basis of a number of factors, some of which we have already discussed, and the Social Distance Matrix Index (hereafter referred to as SDMI) was calculated on each subgroup. The sample population considered here is the subpopulation of Zulu respondents (N=362). The sample population was

TABLE 8.7

S.D.M.I. by Sex

	MALE	FEMALE	TOTAL
Response to all ethnic groups	.0884	.1527	.1136
Response to Swazi	.0282	.0686	.0440
" " Sotho	.0327	.0957	.0574
" " Xhosa	.0434	.1076	.0685
" " Tswana	.0487	.1194	.0764
" " Ndebele	.0479	.1371	.0829
" " Tsonga	.0594	.1266	.0861
" " Coloured	.1438	.2009	.1662
" " White	.1728	.2281	.1944
" " Indian	.2184	.2896	.2463
Response to work with	.0147	.0496	.0284
" " share meal	.0355	.0504	.0401
" " visit	.0431	.0449	.0438
" " have as friend	.0472	.1064	.0704
" " live adjacent to	.0680	.1355	.0944
" " marriage	.3237	.5296	.4043

M=219   F=141*   N=360

*Two female respondents did not reply to the social distance question.

divided by sex, education, income, age, and length of urban residence for cross-tabulation with their response to the questions dealing with the intensity of social contact or social distance. Table 8.7 expresses the SDMI as measured by the criterion of sex. As can be expected, women recorded a higher incidence of negative responses. This is due to a major extent to the degree of exposure and contact males have with persons of other ethnic groups. Since they know them better or have worked with such culturally heterogeneous persons, they are much more willing to accept them as friends or even live with them. The major negative response, more than four times any other response, is to the idea of marrying such persons. The criterion of sex must thus be seen as the qualifier for the concomitant factors that in fact produce the response pattern. These factors are partly due to the activity realm and job potential of males and females, but they are largely the result of the sociocultural background of the Zulu. While the variation is small, thus possibly resulting from random variation, it is nevertheless interesting to note that women are less conservative when it comes to visiting than on any other criterion. After all, they visit around the town, and every day many find themselves in homes where they work as domestics.

But what happens when these same people are looked at in terms of educational achievement—or years of schooling, to be more precise? In

TABLE 8.8

S.D.M.I. by Education for Males

	0-7 years	8-10 years	11-12 years	more than 12
Response to all groups	.0864	.0967	.0648	.1029
Response to Swazi	.0227	.0147	.0152	.0432
" " Sotho	.0253	.0556	.0152	.0432
" " Xhosa	.0307	.0293	.0152	.0617
" " Tswana	.0400	.0778	.0227	.0617
" " Ndebele	.0440	.0630	.0303	.0556
" " Tsonga	.0547	.0704	.0227	.0926
" " Coloured	.1587	.1185	.1212	.1358
" " White	.1840	.1481	.1288	.1975
" " Indian	.2173	.2148	.2121	.2346
Response to work with	.0160	.0148	.0074	.0082
" " share meal	.0320	.0198	.0353	.0617
" " visit	.0453	.0519	.0202	.0370
" " have as friend	.0436	.0469	.0253	.0741
" " live adjacent to	.0569	.0864	.0296	.0947
" " marriage	.3209	.3605	.2323	.3498

N=219   N/R=0

tables 8.8 and 8.9 we calculate the SDMI according to education for males and females, respectively. We have compressed the years of schooling into four categories. The first, zero through seven years, would include all persons up to the end of standard five (provided they passed every year) or conclusion of the junior school level. The next category would include the middle school years, then the two matriculation, or final, years at high school, and finally tertiary education.

Regarding the men, the same social distance pattern already alluded to emerges here, but it is most important that the greatest social distance is found among the best educated, both as regards other ethnic groups and as regards degrees of contact with all non-Zulu. The lowest, on the other hand, is among persons with a senior high school education. It will be recalled from the second section of this chapter that most of the men in both educational categories were between 35-44 years of age. I would suggest that as persons were exposed to tertiary education and competed for somewhat limited highly salaried jobs, they became more conscious of competition in this "Zulu region" from others who are really foreigners. But it seems likely that with modern "homeland" development their best chances are in excluding those who do not really belong. In addition it must be stated, as became clear in our discussion of the reasons for negative responses, that the Zulu have a highly developed ethnic consciousness and ethnic pride.

Among the females we find a different response pattern, and we are also dealing with almost an equal number of persons in the two educational

TABLE 8.9

S.D.M.I. by Education for females

	0-7 years	8-10 years	11-12 years	more than 12
Response to all groups	.1556	.1545	.1802	.0859
Response to Swazi	.0759	.0857	.0222	.0303
" " Sotho	.0992	.1095	.1111	.0152
" " Xhosa	.0928	.1095	.2444	.0303
" " Tswana	.1245	.1333	.1333	.0303
" " Ndebele	.1269	.1524	.2222	.0455
" " Tsonga	.1287	.1286	.1889	.0455
" " Coloured	.2152	.1857	.2000	.1212
" " White	.2447	.2143	.1556	.2273
" " Indian	.2932	.2714	.3444	.2273
Response to work with	.0534	.0349	.1037	.0
" " share meal	.0619	.0413	.0370	.0202
" " visit	.0492	.0540	.0370	.0
" " have as friend	.0985	.1429	.1037	.0303
" " live adjacent to	.0999	.1302	.3185	.1414
" " marriage	.5710	.5238	.4815	.3232

N=140   N/R=3

categories and the three age categories between twenty-five and fifty-four years of age. Here those persons educated through senior high school have the highest negative response pattern while the best educated have the lowest index of negative responses. My suggestion is that given the job situation in South Africa and the Zulu social structure (which relegates women to a secondary position), we are dealing here with a case of delayed ethnic consciousness. The same consciousness, pride, and potential threat felt by males with a tertiary education are experienced by women with a secondary education. Their educational status is a relatively recent achievement, and a diversity of high-paying jobs are now opening up for them.

The best-educated women are interested in their professions and the best match they can make in marriage. They select their friends on bases other than ethnicity and are thus really the "women of the world." They had no negative feelings about visiting or working with non-Zulu, because they probably did it daily, and even the other aspects of this index are low for them. In contrast to their menfolk, who had to uphold Zulu pride and the Zulu nation, they were out to improve their positions. Among women, then, ethnicity and ethnic social distance decreases as education increased.[27]

On the whole there is a positive correlation between years of education and income. the SDMI according to income is given in tables 8.10 and 8.11 for males and females, respectively. Males in the lower income brackets, e.g., below R80 (U.S. $120) per month would also be persons with a poor

TABLE 8.10

S.D.M.I. by Income for males

		below R80 p.m.	R81-R100 p.m.	R101-R150 p.m.	Over R151 p.m.
Response to all groups		.0997	.0668	.0808	.0706
Response to	Swazi	.0339	.0060	.0185	.0260
"	" Sotho	.0461	.0060	.0185	.0260
"	" Xhosa	.0515	.0060	.0741	.0417
"	" Tswana	.0650	.0119	.0278	.0260
"	" Ndebele	.0637	.0179	.0278	.0365
"	" Tsonga	.0772	.0289	.0417	.0417
"	" Coloured	.1585	.1369	.1389	.0990
"	" White	.1843	.1786	.1713	.1250
"	" Indian	.2168	.2083	.2083	.2135
Response to work with		.0136	.0238	.0062	.0139
"	" share meal	.0470	.0198	.0154	.0174
"	" visit	.0596	.0357	.0154	.0069
"	" have as friend	.0678	.0238	.0185	.0660
"	" live adjacent to	.0777	.0476	.1142	.0104
"	" marriage	.3442	.2381	.2870	.3090

N=219   N/R=0

education and frequently recent urban migrants. Those with the highest incomes are persons who are serving in select jobs, positions that qualify them to compete with others on a national level. This and the role of males in this patrilineal society produce conflicting themes—an increasing interethnic network of associates with decreasing social distance against the role of men in a Zulu nation. As in previous cases, the differences between social or even residential contact and marriage are critical.

When we look at the figures for females, we again find the general tendency noticed in the educational breakdown above. The higher the woman's income (and this no doubt presupposes a positive correlation with

TABLE 8.11

S.D.M.I. by income for females

		below R80 p.m.	R81-R100 p.m.	R101-R150 p.m.	Over R151 p.m.
Response to all groups		.1583	.1453	.1305	.1404
Response to	Swazi	.0506	.0897	.0455	.0965
"	" Sotho	.0955	.0769	.1061	.0877
"	" Xhosa	.1030	.0769	.0682	.1228
"	" Tswana	.1217	.0897	.1136	.0965
"	" Ndebele	.1479	.0769	.1288	.0965
"	" Tsonga	.1348	.0769	.1364	.0965
"	" Coloured	.2116	.2436	.1667	.1754
"	" White	.2360	.2821	.1515	.2193
"	" Indian	.3240	.2949	.2576	.2719
Response to work with		.0612	.0256	.0303	.0058
"	" share meal	.0642	.0427	.0253	.0351
"	" visit	.0612	.0256	.0152	.0409
"	" have as friend	.1086	.1368	.1010	.0643
"	" live adjacent to	.1086	.0684	.1364	.2749
"	" marriage	.5293	.5641	.5101	.4152

N=141   N/R=2

relative youth and a better-than-average education, the slighter the perceived social distance. This does not mean that they are less conscious of these factors than their menfolk; in fact, in most categories their lowest response percentage is still higher than the highest percentage for males. But compared to other women in the sample, they are less negative,. In 1973, when this study was conducted, R100 (about U.S. $150) per month was a fairly good salary for a single or a married woman. It would of course not have been so for a woman with children who was a household head. If we compare those who earned less than this figure, namely, the first two columns, with those who earned more than R100, the contrast is quite clear. It should also be mentioned, though I do not have figures here to support it, that in our sample a larger percentage of the females were single or living alone than the males. These would be staff nurses or matrons, legal secretaries, saleswomen, or teachers. This of course is something new among the Zulu. Traditionally a woman would be expected to get married, but with urban conditions, a rise in women's status, and economic independence, this expectation is changing. In a study conducted at the same time in the same urban area, Shelley found that school girls had their sights on such professions as doctors, dentists, and X-ray specialists, 72 percent planned not to marry before they were at least twenty-five years old, and 4 percent planned not to marry at all.[28] These are the women who have found a freedom their mothers never knew, a freedom that includes associating with other educated and professional persons, both Zulu-speaking and members of other ethnic groups. This freedom also applies to the possibility of marrying someone who is not a Zulu.

Many of the males in our sample were either polygynists or openly kept girlfriends. This again is in keeping with the traditional expectations for a man, particularly a man of achievement. The Zulu Don Juan was not left behind in the rural area; he now finds expression through new avenues. Rather than the great warrior, the agile dancer, or the accurate hunter, he now frequently is in the high income bracket, dresses well, and knows his way about the city, geographically and administratively.

Tables 8.12 and 8.13 present the Social Distance Matrix Index according to age for males and females, respectively. The highest incidence of negative responses among both male and female respondents is among persons fifty-five years of age and older. This category also includes the largest ratio of persons in the lowest income group.

The variations evident in the Social Distance Matrix Indexes occurred independently of the factor of length of urban residence, since there was very little significant variation in length of residence among respondents. On the whole, social distance decreased as people rose in the educational and income hierarchy. We also found that the middle age group responded most favorably to interethnic contacts. Women in the better educated and higher income category responded less negatively than their male counterparts to contact with other ethnic groups.

TABLE 8.12

S.D.M.I. by age for males

	under 34 years	35-44 years	45-54 years	over 54 years
Response to all groups	.0852	.0801	.0820	.1130
Response to Swazi	.0333	.0260	.0159	.0385
" " Sotho	.0417	.0313	.0159	.0470
" " Xhosa	.0333	.0382	.0370	.0641
" " Tswana	.0417	.0313	.0423	.0726
" " Ndebele	.0417	.0434	.0370	.0726
" " Tsonga	.1167	.0486	.0529	.0610
" " Coloured	.0917	.1285	.1481	.1624
" " White	.1333	.1615	.1640	.2051
" " Indian	.2333	.2118	.2249	.2906
Response to work with	.0222	.0046	.0123	.0285
" " share meal	.0333	.0208	.0370	.0598
" " visit	.0278	.0301	.0423	.0712
" " have as friend	.0278	.0347	.0600	.1054
" " live adjacent to	.0944	.0637	.0564	.0940
" " marriage	.3167	.3264	.2840	.3333

N=218   N/R=1

TABLE 8.13

S.D.M.I. by age for females

	under 34 years	35-44 years	45-54 years	over 54 years
Response to all groups	.1524	.1469	.1471	.1711
Response to Swazi	.0476	.0667	.0972	.0600
" " Sotho	.1143	.0889	.1204	.1200
" " Xhosa	.1095	.0852	.0926	.1267
" " Tswana	.1143	.1037	.1157	.1600
" " Ndebele	.1429	.1148	.1296	.1733
" " Tsonga	.1143	.0889	.1065	.1800
" " Coloured	.2095	.2185	.1898	.1933
" " White	.1857	.2667	.2269	.2333
" " Indian	.3286	.2778	.2454	.3000
Response to work with	.0381	.0420	.0216	.0978
" " share meal	.0667	.0642	.0185	.0533
" " visit	.0603	.0469	.0340	.0533
" " have as friend	.1048	.0938	.1265	.0889
" " live adjacent to	.2000	.0938	.1389	.1200
" " marriage	.4444	.5407	.5432	.6133

N=141   N/R=2

FRIENDS AND NEIGHBORS

The second question we dealt with (see above) was that of friendship. From a number of items employed we used three as indicators of levels of friendship. Each subject was asked to mention three people with whom he or she visited reciprocally. These persons were identified as to the specific location of their residence and whether they were kin, as well as to their origin if they were migrants to the city. Our intention here was to discern long-standing links that were continued in the city or potential links (e.g., home district or clan membership) that were activated in the city. Along with the question dealing with friendship, we posed two further questions in the following way: (1) If you found your house burned down or if some such disaster struck your family, to which *one* of these three persons would you turn? Why? (2) If you arrived home on Friday with your paycheck and these three friends with equally good reasons requested a substantial loan, to which *one* would you lend the money? Why?

Several research subjects explained that it would be impossible to select one, while others had no problem deciding or justifying their choice. But what we wanted to learn was whether these choices would be randomly distributed or whether they would favor kinsmen, "home boys," close neighbors, or any other category. If a person had relatives in the same area and selected a neighbor or a friend, this would be significant. If the person selected a kinsman living further away but ignored a next-door neighbor, this would show the retention of kinship ties.

Table 8.14 deals with our first question, namely, the person to whom one can turn in a crisis. Under the first set of alternatives, those of proximity of residence, we find the expected pattern of turning firstly to persons who live close by, preferably within the same section of the satellite city. Women are slightly more likely to turn to people living close by, partly because they spend more of their time in or near the home. Men, because they work and visit away from the home, select close friends on criteria other than residential proximity.

The second part of the table under discussion analyzes the relationship to the respondent of the person selected. Once again we find that women on

TABLE 8.14

Person chosen for "Call on in a crisis"

Variable	Male Number	%	Female Number	%	Total Number	%
Proximity of Residence:						
Same Section of K M	97	47.8	66	50.4	163	48.8
Different Section of K M	77	37.9	42	32.1	119	35.6
Another part of Durban	19	9.4	15	11.4	34	10.2
Outside the Durban area	10	4.9	8	6.1	18	5.4
Relationship to Respondent:						
Relative	71	35.0	57	43.5	128	38.3
Neighbor	64	31.5	36	27.5	100	29.9
Friend - no further information	46	22.7	16	12.2	62	18.6
Church Friend	5	2.5	12	9.2	17	5.1
Home Boy	8	3.9	6	4.6	14	4.2
Co-worker	9	4.4	4	3.0	13	3.9
Kinship:						
Not related	132	65.0	74	57.3	206	61.7
Related:	71	35.0	57	43.5	128	38.3
Extended	37	52.1	27	47.4	64	50.0
Affinal	21	29.6	14	24.6	35	27.3
Nuclear	13	18.3	16	28.0	29	22.7

N: M = 203, F = 131

the whole turn more to relatives than do men. The latter would be much more likely to turn to a friend. Although women do in some cases turn to friends, we find that "church friends" are singled out as a significant category. These women frequently meet on Thursdays, which is the day set aside for ladies' meetings, prayer groups, and church meetings. For women this day is almost as important as Sundays in terms of their religious activities and interaction with fellow believers.

All those persons designated as relatives in the second part of the table are further subdivided in the third section under kinship. Of the fifty-seven women who selected a kinsman as the person to whom she would turn in a crisis, 28 percent indicated a person who was a member of the nuclear family, i.e., parent or sibling.

Much the same picture emerges from table 8.15, in which we analyze the persons who are selected for a cash loan. We should of course realize that lending money is a very serious matter, and the less one has the harder it will be even to agree to make a loan. Several persons refused to make the selection, simply stating that they earned too little ever to make a loan—even to to a parent. From previous studies along these lines, we know that recent migrants to the city were unwilling to allow their money out of their sight and if they did, it was by giving it to persons over whom they had some control, e. g., kinsmen or persons from the same point of origin.[29] This then would mean a greater selectivity of the person singled out to be the recipient of the limited hard-earned cash.

The point was not stressed quite as heavily as we expected, but it is clear

TABLE 8.15

Person chosen for "Entrust money to"

Variable	Male Number	%	Female Number	%	Total Number	%
Proximity of Residence:						
Same section of K M	80	40.6	59	45.0	139	42.4
Different section of K M	82	41.6	49	37.4	131	39.9
Another part of Durban	25	12.7	14	10.7	39	11.9
Outside the Durban area	10	5.1	9	6.9	19	5.8
Relationship to Respondent:						
Relative	68	34.5	55	42.0	123	37.5
Neighbor	53	26.9	33	25.2	86	26.1
Friend - no further information	54	27.4	22	16.8	76	23.2
Church friend	5	2.5	12	9.2	17	5.2
Home Boy	8	4.1	5	3.8	13	4.0
Co-worker	9	4.6	4	3.1	13	4.0
Kinship:						
Not Related	129	65.5	76	58.0	205	62.5
Related	68	34.5	55	42.0	123	37.5
Extended	37	54.4	26	47.3	63	51.2
Affinal	18	26.5	14	25.4	32	26.0
Nuclear	13	19.1	15	27.3	28	22.8

N: M = 197, F = 131

that there are already changes in the section dealing with the proximity of residence. Both women and men would be less likely to deal with somebody simply because he or she lived in the same neighborhood. In fact our urban residents would be more likely not only to go outside the particular satellite city in which they live but even outside of the Durban area in selecting a person they could trust. This need not necessarily be a relative, because more of them now would trust friends—in fact there is an overall increase of 5 percent for this category.

When we look at the breakdown of the kinship ties between the respondent and the person selected, the picture is almost identical to the former table. There is then an emphasis on persons who can be trusted and be depended on to return the loan. Such persons are more likely to be friends than relatives.

Turning to this question from a different angle, we were interested in seeing whether there were any differences in the kind of person who was selected as a first friend, or whose name came to mind first when the topic of friendship was raised. In all cases we are taking up kinship, residential proximity, and similar factors relative to the respondent. Since there was very little difference between "second" and "third" friends, we have deleted the table dealing with the latter.

Tables 8.16 and 8.17, respectively, deal with those persons selected as "first" and "second" friends. Among the male respondents, we have almost half in both categories being selected in the same section of the residential area, and a little over a third in both cases would still be in the same satellite

TABLE 8.16

Person chosen as "First Friend"

	Male		Female		Total	
	Number	%	Number	%	Number	%
Proximity of Residence:						
Same Section of K M	98	47.3	67	51.1	165	48.8
Different Section of K M	76	36.7	40	30.5	116	34.3
Another part of Durban	23	11.1	17	13.0	40	11.8
Outside the Durban area	10	4.8	7	5.3	17	5.1
Relationship to Respondent:						
Relative	51	24.6	45	34.4	96	28.4
Neighbor	72	34.8	41	31.3	113	33.4
Friend - no further information	57	27.5	26	19.8	83	24.6
Church friend	4	1.9	14	10.7	18	5.3
Home Boy	12	5.8	3	2.3	15	4.4
Co-worker	11	5.4	2	1.5	13	3.8
Kinship:						
Not related	156	75.4	86	65.6	242	71.6
Related	51	24.6	45	34.4	96	28.4
Extended	26	51.0	17	37.8	43	44.8
Affinal	16	31.4	14	31.1	30	31.2
Nuclear	9	17.6	14	31.1	23	24.0

N: M = 207, F = 131

TABLE 8.17

Person chosen as "Second Friend"

	Male Number	%	Female Number	%	Total Number	%
Proximity of Residence:						
Same section of K M	102	49.3	51	38.9	153	45.3
Different section of K M	72	34.8	57	43.5	129	38.2
Another part of Durban	25	12.1	15	11.5	40	11.8
Outside the Durban area	8	3.9	8	6.1	16	4.7
Relationship to Respondent:						
Relative	55	26.6	43	32.8	98	29.0
Neighbor	75	36.2	29	22.1	104	30.8
Friend - no further information	59	28.5	34	26.0	93	27.5
Church friend	5	2.4	11	8.4	16	4.7
Home Boy	4	1.9	9	6.9	13	3.8
Co-worker	9	4.3	5	3.8	14	4.1
Kinship:						
Not Related	152	73.4	88	67.2	240	71.0
Related	55	26.6	43	32.8	98	29.0
Extended	35	63.6	27	62.8	62	63.3
Affinal	11	20.0	7	16.3	18	18.4
Nuclear	9	16.4	9	20.9	18	18.4

N: M = 207, F = 131

city. Among the women, a "first" friend would in more than half the cases live in the same residential section, but only 38.9 percent of the "second" friends would. There is, then, a great emphasis for women to have very close relations with next-door neighbors or persons on the same street. A great deal of their time is spent at home, running to the corner store, doing the family laundry in the backyard, or caring for a neighbor's infant. Men spend the greater part of the day at work, and when they return it is common to see them walking to visit with a friend, going to the canteen or more frequently a *shebeen* for a social drink, or taking off for a professional soccer game at the stadium. These men on the whole do not spend much time at home.

Turning to the second part of the two tables under scrutiny, we notice the importance of "church friend" for women, but the first friend of a man is statistically more likely to be a "home boy." We also find in both cases that the category "neighbor" is more important than the mere fact of being a relative. In fact, as second friend these urban residents are almost equally likely to select a neighbor, a relative, or a friend.

If we turn to the factor of kinship, which is almost identical in its overall importance for the two categories of friends, we notice that although it becomes more important for males when selecting the second friend, it decreases in importance for females.[30] In spite of this, it is still of greater importance to women than to men. However, all forms of kinship are not equally dependable or close. When selecting a first friend, only 24.6 percent of the males selected a kinsman. Of those that did, a full 51 percent indicated a clansman, while only 17.6 percent mentioned a member of their nuclear family only slightly more frequently than an affine or a member of their own nuclear family. In considering a second friend, we find that when subjects selected a kinsman to fill this category, men and women pointed to their extended kin in 63.6 percent and 62.8 percent of the cases, respectively. There is also a corresponding decline in the importance of the close and affinal relatives.

Overall, the expected significance of the "home boy" category did not materialize, and a new category, namely, "church friend," emerged—particularly for women. Neighborliness is important for those who spend a great deal of time in the neighborhood, mainly women, but the coworker, the "home boy," or the friend figures much more prominently in the selection pattern of males.

In table 8.18 these three aspects, namely, the residential proximity of friends or relatives who were selected have been cross-correlated. The upper part of the table pertains to "first" friend and the lower part to "second" friend. Since we have included a great deal of material in the table, it requires some explanation.

TABLE 8.18

"Friends"/"Relatives" classified according to proximity of residence

		Same Section of K M	Different Section of K M	Another part of Durban	Outside the Durban Area
	FIRST FRIEND				
MALE	FRIEND	81 – 39.1 81.8%	53 – 25.6 70.7%	16 – 7.7 69.6%	6 – 2.9 60.0 %
	RELATIVE	18 – 8.7 18.2 %	22 – 10.6 29.3%	7 – 3.4 30.4%	4 – 1.9 40.0%
FEMALE	FRIEND	49 – 37.4 73.1%	25 – 19.1 73.5%	8 – 6.1 34.8%	4 – 3.1 57.1%
	RELATIVE	18 – 13.7 26.9%	9 – 6.9 26.5%	15 – 11.5 65.2%	3 – 2.3 42.9%
TOTAL	FRIEND	130 – 38.5 78.3%	78 – 23.1 71.6%	24 – 7.1 52.2%	10 – 3.0 58.8%
	RELATIVE	36 – 10.7 21.7%	31 – 9.2 28.4%	22 – 6.5 47.8%	7 – 2.1 41.2%
	SECOND FRIEND				
MALE	FRIEND	84 – 40.6 82.4%	50 – 24.2 69.4%	13 – 6.3 52.0%	3 – 1.4 37.5%
	RELATIVE	18 – 8.7 17.6%	22 – 10.6 30.6%	12 – 5.8 48.0%	5 – 2.4 62.5%
FEMALE	FRIEND	41 – 31.3 82.0%	33 – 25.2 56.9%	9 – 6.9 60.0%	4 – 3.1 50.0%
	RELATIVE	9 – 6.9 18.0%	25 – 19.1 43.1%	6 – 4.6 40.0%	4 – 3.1 50.0%
TOTAL	FRIEND	125 – 37.0 82.2%	83 – 24.6 63.8%	22 – 6.5 55.0%	7 – 2.1 43.8%
	RELATIVE	27 – 8.0 17.8%	47 – 13.9 36.2%	18 – 5.3 45.0%	9 – 2.7 56.3%

Males as a category of respondents made a certain number of selections of persons who were friends only, while others were designated as relatives. If we look at the category "friends," we find that eighty-one, or 39.1 percent of all such persons, live in the same section of the satellite city as the respondent and could thus classify as a neighbor. Fifty-three, 25.6 percent of all such persons, live in a different section but still in the same satellite city; sixteen, or 7.7 percent, live in another part of Durban; and six, or 2.9 percent, live outside the urban area; the final category for males who are selected as "first friend" refers to four persons who are relatives, who live outside Durban and who form only 1.9 percent of the total category of friends and relatives. But the eighty-one persons who are only friends also form 81.8 percent of the total number of persons who live in the same section and who were selected, be they friends or relatives. The other eighteen persons who are relatives make up the rest, or 18.2 percent. (Since the percentages in all of these tables are rounded, it is possible that they may not always add up to exactly 100 percent in their rounded form.)

In the case of first friend, we can see that in general, then, persons living in the same section who are selected are likely to be friends rather than relatives. This is particularly so for males. When persons who are selected live in a different section, they are more likely to be a relative in the case of male respondents, even though they still select friends in 70.7 percent of the cases. As persons who live further away are selected, they are increasingly more likely to be relatives than friends. In fact, in the case of women 65.2 percent of persons living in another part of Durban were relatives.

Looking at the second part of the table, we notice that for males only 17.6 percent of those who are selected and who live in the same section are relatives. Persons living outside Durban who are selected, however, include 62.5 percent relatives. Also, in the case of women, there is a gradual decline in the importance of friends (and neighbors) as one deals with persons living further away or even outside the urban complex. While friends are more important in the total when first friends were selected, they are less significant in the selection of a second or third friend. This is well illustrated by the total number of persons who indicated their choices for a second friend. Here relatives in the same neighborhood accounted for only 17.8 percent of the choices, but relatives outside the urban area rose to 56.3 percent.

CONCLUSION

This chapter has treated two related questions, one dealing with social distance between Zulu and non-Zulu and, second, one dealing with the closeness of persons within the Zulu-speaking ethnic group.

One of the major distinguishing criteria is that of sex. The reason for differential responses between men and women, not only in regard to ethnic social distance but also regarding selection of friends, must be explained by reasons related to traditional culture and social structure, to modern urban conditions, and to the employment avenues available to men and women. Better education and employment opportunities have created a category of women who are more interethnic in their social networks than their male counterparts.

In spite of this, or perhaps because these women are relatively few in number, the selection of friends by women tends to focus on persons geographically or familialy close to the respondent. Thus, neighbors and kinsmen are of greater importance to women than they are to males. Here is the apparent contradiction: women simultaneously lean toward interethnic social network and possibly marriage while they also depend heavily on neighbors and kinsmen. Along with this pattern is the growing importance of the church friends who give support in various crisis situations.

This material has illuminated a life-style and action group quite different from the traditional Zulu described to date. While the conservative persons are present and while there are particular categories more likely to act in a conservative or traditional fashion, education, income, and job opportunities are producing urban Africans—who speak Zulu.

NOTES

1. P. Harries-Jones, " 'Home Boy' Ties and Political Organization in a Copperbelt Township," *Social Networks in Urban Situations*, ed. J. Clyde Mitchell (Manchester: Manchester University Press, 1969), pp. 297-347.

2. Brian M. du Toit, "Cooperative Institutions and Culture Change in South Africa," *Journal of Asian and African Studies* 4, no. 4 (October 1969): 303.

3. Philip Mayer, *Townsmen or Tribesmen* (Cape Town: Oxford University Press, 1961), pp. 99-100.

4. Monica Wilson and A. Mafeje, *Langa: A Study of Social Groups in an African Township* (Cape Town: Oxford University Press, 1963), p. 47.

5. J. Clyde Mitchell, "Tribe and Social Change in South Central Africa: A Situational Approach," in *The Passing of Tribal Man in Africa*, ed. Peter C. W. Gutkind (Leiden: E. J. Brill, 1970), pp. 93.

6. J. Clyde Mitchell, "Some Aspects of Tribal Social Distance," in *The Multitribal Society*, ed. A. A. Dubb (Cape Town: Rhodes-Livingstone Institute, 1962), p. 8.

7. David Jacobson, "Friendship and Mobility in the Development of an Urban Elite African Social System," in *Urban Growth in Subsaharan Africa*, ed. Josef Gugler (Kampala: Makerere Institute of Social Research, 1970), p. 57.

8. Emory S. Bogardus, "Social Distance and its Practical Implications," *Sociology and Social Research* 22 (May-June 1938): 462.

9. Emory S. Bogardus, "A Social Distance Scale," *Sociology and Social Research* 17 (January-February 1933): 265-271.

10. I. D. MacCrone, *Race Attitudes in South Africa* (Johannesburg: Witwatersrand University Press, 1937).

11. T. F. Pettigrew, "Social Distance Attitudes of South African Students," *Social Forces* 38 (1960): 246-253.

12. P. L. van den Berghe, "Race Attitudes in Durban, South Africa," *Journal of Social Psychology* 57 (1962): 55-72.

13. D. Grove, "Sosiale Afstand tussen Kleurlinge en Indiërs in Pretoria," (M. A. thesis, University of Pretoria, 1967).

14. E. A. Brett, *African Attitudes*, Fact Paper no. 14, (Johannesburg: South African Institute of Race Relations, 1972).

15. Melville L. Edelstein, *What Do Young Africans Think?* (Johannesburg: South African Institute of Race Relations, 1972).

16. Michael Banton, *Race Relations* (New York: Basic Books, 1967), pp. 313-333.

17. Edelstein, *What Do Young Africans Think?*, pp. 26-28.

18. This study is based on material collected during 1972-1973, while I was in the preliminary phases of field research supported in whole by P. H. S. Research Grant D. A. 00387 from the National Institute on Drug Abuse. My sincere gratitude is extended for this financial support. The findings are not necessarily shared by the granting agency or any person associated with it.

19. This results in an 83.8 percent Zulu sample. Sources suggesting Zulu to constitute 86 percent of the African population in Durban are quoted by Wade C. Pendleton, "Introduction," in *Urban Man in South Africa*, ed. C. Kileff and Wade C. Pendleton (Gwelo: Mambo Press, 1975), p. 11.

20. A number of these indexes have been discussed in Brian M. du Toit, "Cultural Continuity and African Urbanization," in *Urban Anthropology*, ed. Elizabeth M. Eddy, Southern Anthopological Society Proceedings, no. 2, (Athens: University of Georgia Press, 1968).

21. Brian M. du Toit, "Strike or You're in Trouble," in Kileff and Pendleton, *Urban Man in Southern Africa*, p. 11.

22. The difference between intertribal marriage, i. e., between members of different sociolinguistic groups, and "same origin" marriages is discussed by Mia Brandel-Syrier, in *Reeftown Elite* (New York: Africana Publishing Corporation, 1971), p. 85. In this she is analyzing the origin of husband and wife with regard to the urban-rural component. She concludes that in the light of the "large number of same origin marriages [among the elite], the number of same-'tribe' marriages was small."

23. We should of course allow for the possibility that both these negative reactions, namely, "I don't like them" and "I don't know them," are forms of withdrawal. They may in fact reflect an attitude of not wanting to disclose factual experiences or of not being willing to think about the questions.

24. African separatist churches with varying degrees of nativism have long been studied and recorded. For a cult group in this general area, see Brian M. du Toit, "Religious Revivalism Among Urban Zulu," in *Man: Anthropological Essays*, ed. E. D. de Jager (Cape Town: C. Struik (Pty) Ltd., 1971).

25. Max Gluckman, *Custom and Conflict in Africa* (Oxford: Basil Blackwell, 1965); M. G. Marwick, *Sorcery in its Social Setting: A Study of the Northern Rhodesian Cewa* (Manchester: Manchester University Press, 1965 (a)); idem., "Some Problems in the Sociology of Sorcery and Witchcraft," in *African Systems of Thought*, ed. M. Fortes and G. Dieterlen (London: Oxford University Press, 1965 (b)), pp. 171-191; S. F. Nadel, "Witchcraft in Four African Societies," *American Anthropologist* 54, no. 1 (1952): 18-29. Monica Wilson, "Witch Beliefs

and Social Structure," *American Journal of Sociology* 56, no. 4 (January 1951): 307-313.

26. E. Verwoerd, "Die Vrou Se invloed op goeie rassebetrekkinge," *Die Taalgenoot*, November 1973, p. 11.

27. See Kenneth Little's chapter in this volume.

28. J. Karen Shelley, "Township Teens: A Study of Socialization Processes Among Urban Black School Girls in South Africa," (M. A. thesis, University of Florida, 1975), p. 138. While Zulu does have a term for spinster—a term by the way which can also apply to an old bull—Bryant points out that spinsterhood "was always due to some special ugliness or deformity of person." *The Zulu People As They Were Before the White Man Came* (Pietermaritzburg: Shuter and Shooter, 1949), p. 567.

29. Du Toit, "Cooperative Institutions and Culture Changes in South Africa."

30. E. Preston-Whyte, "The Adaptation of Rural-Born Domestic Servants to Town Life," in *Focus on Cities*, ed. H. Watts (Durban: University of Natal, 1970), p. 278, found while dealing with a more limited sample of females that domestic servants are likely to turn to clan members for assistance and association.

9

Countervailing Influences in African Ethnicity: A Less Apparent Factor

Kenneth Little

Writing of the categorization of ethnic groups, Mitchell points out that there is often a discrepancy between the analyst's and the actors' view of the matter. This is because the category that the ethnographer may eventually establish for his analytical purposes must be of a different order of abstraction from the ethnic categories of the people themselves. He is likely to come up with a set of diacritica of an ethnographic category that is meaningless to the people concerned.[1] The likelihood of this happening for social and political reasons has been well documented, but for immediate purposes here, I am only concerned with "ethnicity." This—again mainly following Mitchell—I regard primarily as a construct of perceptual cognitive phenomena.[2] Therefore, although similarities of custom, religion, and language are generally a very cohesive factor, it does not matter for empirical purposes whether a particular individual or number of individuals actually share the cultural or other traits of most of the group with which they are associated. The relevant consideration is actual behavior in a given social situation, meaning that there should be mutual identification on the basis of supposed common origin in terms principally of ethnic group or tribe. In addition, the individuals concerned feel more loyalty toward each other than toward any other social group.

This point is made because the phenomena referred to are a particularly marked feature of what has elsewhere been termed the "modern" African city.[3] There, as Cohen has also pointed out, the division of labor is usually highly advanced, and the struggle for resources such as employment, housing, and political following is intense.[4] The consequence, if these urban populations are highly heterogeneous, is that the urban social structure tends to be constituted of many more or less separate ethnic groups. These, although interdependent in terms of the overall system, are by virtue of their own semiofficial headmen and tribunals almost autonomous units.[5]

It is generally considered that a contradictory social process opposes this pluralistic tendency. This process is becoming increasingly significant, and it derives from a quite different practice—that of alignment according to similarity in occupation, income, political rank, education, and life-style in general. The consequent stratification is important because the lines of social differentiation created cut across the total society, including its individual ethnic components. To the extent that this has the effect of making loyalty to "class" stronger than loyalty to ethnic group, it naturally diminishes the sociological significance of ethnicity as well.

It is not our objective here to decide whether any such "class" feelings are more pervasive than ethnic ones, but to consider the working of another social force. This, although different in origin, operates in a somewhat similar way to "class," and its momentum comes from the aims and attitudes of African women living in cities.

These aims and attitudes are directed principally toward the improvement of women's status, and it is not accidental that they develop most readily in urban areas. For one thing, it appears that especially for village women, migration to the larger cities is seen as offering more "freedom" than is available at home. There are also increasing numbers of women and girls who, mainly through the general cultural and educational process, have acquired a modern, Westernized conception of women's role. Such individuals consider that women are entitled to play a more prominent part in the affairs of the nation as well as the local community. Furthermore, although the wish for a more egalitarian relationship with the opposite sex is usually confined to the more educated category of women, there is widespread desire among women in general for more "independence." The difficulty is that in the urban economy women are educationally at a disadvantage in competition with men and that there is quite often male opposition to women engaging in wage employment at all. It is also the case that although the urban household may have reached the stage of managing its own budget, the financial support of kinsmen and relatives is still often needed. The effect of this combination of circumstances is to reinforce traditional ideas of male domination. In many cases it also tends to make the wife more economically dependent upon the husband than she would be as a breadwinner in the rural situation.[6]

Since men thus continue firmly to control most of the city's social institutions, only a small proportion of women are able on their own to challenge the male position. These are principally women from elite families owning wealth, who possess high professional and other advanced qualifications. However, the fact that, nevertheless, status in urban society can be gained by achievement does provide women with fresh opportunities in nearly every field. Disadvantaged she may be for educational and traditional reasons, but the urban woman has discovered strategies she can use, and she sometimes takes tactical action on her own.[7] Mostly, however, this action involves participation in voluntary associations, usually with other women, but sometimes together with men.

THE ARGUMENT

It is our thesis that women's employment of this kind of organization coupled with their manipulation of various religious, political, economic, and social situations, constitutes a countervailing factor to ethnicity. I term this factor *unapparent*, because its effect, being mainly indirect, is sociologically more difficult to uncover than the effect of social class alignment.

Let us begin, then, with the most predominant kind of group that the women themselves organize. This is a convenient way of starting because such associations are concerned specifically with matters of immediate interest to women as a whole, including their matrimonial position, the welfare of children, hygiene, and other domestic skills. In other words, since the primary emphasis in this regard is on women's own affairs, it is likely that other considerations, including ethnic ties, will have secondary importance. There is an almost countless number of such clubs, societies, leagues, unions, and fellowships, including many in the rural areas. The latter, being mostly based on local communities, naturally draw their members from the same tribe or ethnic group. Many such urban associations, too, are similarly composed, but the much greater participation in them of educated and literate women makes a significant difference. Not only are these women more articulate, but they are also in a better position to provide a more up-to-date kind of leadership. Instead, therefore, of trying merely to ameliorate their followers' domestic lot, these "professional leaders" seek a radical solution. A typical example, in Accra, is a Committee on the Status of Women, typical because its objectives include the study and the eradication of laws and customs that discriminate against women. This committee seeks to mobilize public opinion on the role of women and their participation in the planning of national development programs. It also aims to cooperate with governmental and nongovernmental bodies in these respects, as well as to make representations to the government on women's behalf. In addition, this committee organizes talks, seminars, and symposia on such matters.

These "progressive" women are therefore not only concerned with the interests of women also as a whole; they also perceive the advantages of creating a united front. This is shown in the deliberate grouping together of the multifarious women's societies and clubs under a common umbrella. This takes the form of national federations of women's associations, and one example among many others is the former West Cameroon. This is a small country; nevertheless, there were, by 1964, more than 125 registered women's organizations with a total membership of over 3,500. The framework is usually quite simple and consists of each affiliated society being represented in the federation itself by a number of members who elect an executive committee. Thus, in Ghana, the Ghana Assembly of Women includes representatives of ethnic and national associations and social clubs. Significantly, too, the first stated objective of the latter federation is to "co-ordinate women's work regardless of race, colour and creed for the promotion of their common interest."[8] In addition, not only does this assembly seek to promote friendly relations with international organizations whose aims and objectives are similar, but it sponsors conferences in conjunction with such bodies. In 1974, for example, a joint conference of the International Alliance of Women and of the Assembly was held in Accra. This brought together representatives of women's associations in most of the West African Anglophone and Francophone countries as well as delegates from some Western countries.

In Kenya, according to Audrey Wipper's interesting report, the National Council of Women performed a somewhat similar coordinating function. The movement included women's clubs, church groups, welfare associations, auxiliaries to tribal associations, the Girl Guides, the YWCA, and the Red Cross. *Maendeleo ya Wanawake* (Swahili for "Women's Progress") was the largest and most important of these organizations. It had a membership of more than 50,000, and, although concerned ostensibly with such matters as child care and improvement of domestic standards, *Maendeleo* was the backbone of the whole women's movement. It provided it with its more militant and outspoken leaders.[9] These women have not hesitated publicly to criticize what they regarded as male intransigence.[10] Since tribal feeling nationally inhibits the growth of feminist unity, they also urge Kenyan women to forget ethnic differences and to work together as a single body.

This does not mean that African women in general are unwilling to cooperate with the men. On the contrary, the women have seen the advantages of such an alliance, especially when they and the male section of the community have common interests. Perhaps the two most striking examples of this male-female collaboration are in South Africa and Guinea. The South African situation is particularly relevant bcause whereas in most of the other African countries women's main struggle is against traditionalism, in South Africa it is apartheid that the women regard as mainly responsible for their subordination. They feel, in other words, that simply to seek some improved status or right within the framework of the apartheid state is all but useless. The changes needed being so overriding, black women need the right to vote just as black men do. Since, therefore, without political power there is little hope of solving their own immediate problems of jobs, houses, and education, women choose to fight along with the men rather than against them.[11] The upshot, in consequence, is that the tendency of the individual non-European sections to unite in opposition to racialism includes women of different ethnic groups and tribes as well. This is shown most dramatically in the numerous public demonstrations that the women have staged on their own. One of those was in Pretoria in 1955, when some 2,000 women protested against the Pass Laws. Most were Africans, but the assembly included women of all the other races. A year later, some 20,000 women took part in a similar protest.[12] There were Indian women dressed in brilliant saris; and although Xhosa women wore their own distinctive ocher robes with elaborate headscarves, thousands of women bore no tribal insignia at all.[13] They simply sported the green and black blouse of the African National Congress. A further relevant development in this regard was the foundation of the multiracial Federation of South African Women. Since the strongest and largest group was the Women's League of the African National Congress, this had the effect of joining the demands of the African women to those of other groups of women. This is significant because although the strength of the Federation rested largely in such women's leagues in the black townships, not only Indian and coloured women but white women were also active in the Federation. Indeed, while its president was a black woman, its secretary was a white woman; and it is also pertinent to remember that women of all colors, including white, have suffered jail sentences and torture.[14] Thus, it is likely that in South Africa women's struggles for sexual and racial equality complement each other.

In Guinea, to which we now turn, the main reason for women's cooperation has been the creation of a social order more favorable to female interests. The opportunity arose in the 1950s out of the efforts of a would-be nationalist leader to unify the people of Guinea against the French colonial regime. He was Sekou Touré, and at the outset the party he formed—Parti Démocratique Guinée—had to compete with groups organized on ethnic and regional lines around certain members of the African elite. Most of these were little more than creatures of the colonial administration, but Sekou Touré perceived that for the women and the younger men there were mutual advantages in joining forces. This also was duly recognized on both sides, and the common interest was expressed in the slogan "Independence." This meant, in political terms, equality with the French. It also meant liberation for both youth and women from an ossified social system in which parental and male authority was complete. The idea, in other words, was that the women would support the party, and the party would emancipate the women; because—the argument ran—to build the nation required women having the same status as men. And so the women in particular rallied to cries of freedom and were in turn singled out continuously for special mention in the interests of overcoming regionalism. This was because, although party songs, poems, dances, and slogans could be used, the women cut across clan, ethnic, and other divisions. That was the really important thing. They signified unity and symbolized this by wearing dresses made out of identical cloth.[15]

However, although the Guinean women seem to have been attracted by

Sekou Touré's personality, their own outlook was pragmatic rather than ideological. They associated themselves enthusiastically with the PDG and were treated by the PDG as partners.[16] But, as explained, the women regarded this affiliation mainly as an instrument of social liberation with whose cooperation a number of important successes had been achieved. This does not mean that the struggle for female emancipation was not integrated with the total struggle or that women were not seen as a potentially revolutionary force elsewhere. On the contrary, they were offered this role in Ghana and in Guinea-Bissau, too, and they played it because it suited them as much as it did the radical politicians concerned to break down traditional barriers.[17] The fact that in the process ethnic divisions came under attack as well is integral to our whole thesis.

It must also be borne in mind that urban women in particular find the mass political party's egalitarian philosophy attractive for specifically economic reasons. In Sierra Leone, for example, the solemnly declared aims of the African People's Congress are "to create a welfare state... in which all citizens regardless of clan, tribe, colour or creed shall have equal opportunity, where there shall be no exploitation of men by men, by tribe, or class by class."[18] It is probably not accidental at all, therefore, that the APC's eventual rise to power was due quite largely to the activities of its women's wing as a propaganda agent and pressure group.[19] This organization, the National Congress of Sierra Leone Women (the NCSLW) consists mostly of women in commerce, working women, and self-employed women.

The consequence, therefore, is not only that the APC has many female supporters, but also that its ideal of ethnic equality and interethnic cooperation is correspondingly strengthened. Does, however, the APC possess the means as well as the will to implement such a policy? What appears to have happened is that—at least for political purposes—the open expression of ethnicity has been rendered impracticable. Thus, a group of women from a particular ethnic group went on their own to pledge their allegiance to the president, and this was promptly dealt with by the NCSLW's leader. She construed it as tantamount to forming a "party within a party," made a public issue of the matter, and deplored in the name of Congress "the attitude of some women who intend to organize on a *tribal* basis."[20] True, possibly this lady and other aspiring political leaders take an interethnic line mainly for reasons of expedience or convenience. Nevertheless, the fact that they do presumably means that ethnic platforms are no longer so useful for electioneering.

These, then, are some of the ways in which political forces are manipulated to serve female interests. It goes without saying, of course, that urban women also endeavor on their own to improve things for themselves and their children. For the very small well-educated minority, there is professional work, and sometimes senior civil service jobs are obtainable. Posts in teaching and nursing are also available for educated girls; but the most general outlet is commerce, and this, especially in West Africa, is usually confined to petty trading. In Accra and Lagos, for example, nearly nine out of ten women gainfully employed are traders, and most of the stallholders in markets are women. In Dakar, too, the percentage of women traders is almost as large, and so is the proportion in some of the small Senegalese towns. Women also dominate the markets in Zaire; and in Rhodesia, Zambia, and Malawi there are more women than men among the sellers. True, in East African towns, it is more customary for men than women to engage in commerce. Nevertheless, the situation varies, and in a study of Kampala markets, for instance, women formed some 41 percent of the 1,664 vendors counted.[21]

We have singled out commerce because the women engaging in it represent the less-educated section of the community. For the latter even petty trading is a full-time job, and the prime necessity is the purely practical one of making ends meet. Consequently, unlike the women who head national federations and the elite associations, few market women and women traders as a whole have any ideological interest at all in feminism per se. Yet paradoxically, it is these same lower-class and lower-middle-class women who in some respects provide directly or indirectly the greatest challenge to ethnicity.

There are two important reasons for this. First, although ethnic exclusiveness is usually strong among this section of the population, the woman trader may be obliged to develop a different orientation for the sake of her business. This is because, when success in selling depends on a customer's goodwill, his or her attitude has naturally to be taken into account. Thus, for example, not only do women beer-sellers in a Nairobi squatter settlement claim that it pays to welcome with "jolly talk" any customer who arrives at the door, but most of them also stress the necessity of not practicing "tribalism." Some even go so far as to insist that unless everyone in the room is of their own ethnic group, they speak mostly Swahili.[22] It appears, also, that a seller needing to check a customer's reliability seeks advice from those of her fellows in the best position to provide it, irrespective of their tribe.[23]

Furthermore, a market woman needs her fellow traders' cooperation for more important purposes. This, as implied above, is gained mainly through participation in voluntary associations. In addition to mutual benefit schemes, these groups regulate trading practices in the common interest. For example, they ostracize any trader undercutting prices, and they also assist their members through a system of "rotating credit" to raise capital and loans.[24] True, many of these associations consist of women from the same locality or tribe. The fact that the success of such schemes rests on personal guarantees, however, makes it likely that the question of individual members' trustworthiness matters more than their ethnic identity.

This point is the more significant because, according to the head of what is probably Nigeria's largest association of market women, its some 300,000 members include non-Nigerians, such as Sierra Leonean and Dahomean traders, as well as Fulani, Hausa, Ibo, Beni, and Urhobo. She stresses, in fact, that the aims of this massive association are as nonpolitical and nonsectarian as they are nontribalistic. They are simply to improve welfare and marketing amenities for their members and to cooperate to this end with government authorities.[25]

Another urban section of the female population with a commodity to sell are prostitutes. Admittedly, prostitution is not as characteristic of African cities as it is of Western urbanism. Nevertheless, African prostitutes are as much concerned as any others to safeguard their economic position. Consequently, they, too, regulate trading practices and apparently attach more importance to business interests than to ethnic differences.

In Brazzaville, for instance, there are associations of courtesans whose principal criteria of membership are elegance, deportment, and good looks. These carefully selected girls are expected to observe a proper code of professional conduct, and, like the market women, they are also discouraged from competing with each other.[26] This is also the case in the "red light" quarter of Abidjan, where Rouch and Bernus found that a local community of prostitutes had gone on strike. Among this group, which contained migrant prostitutes from a variety of tribes, were a number of Krobo. It appeared that a bitter quarrel had broken out between the latter and the other prostitutes, because the Krobo prostitutes had charged only 175 francs for a "service" instead of the 200 francs agreed by general consent. This breach of professional etiquette, not the Krobo women's different tribal affiliation, was the cause of the palaver.[27]

The other reason why lower-class and lower-middle-class women are an indirect threat to ethnicity is political. Since this has already been referred to within the context of the APC as a mass political party, it is necessary only to stress here the underlying motivation. In other words, many women who join the NCSLW as the APC's women's wing do so primarily for economic betterment. They believe that by becoming members of the Congress they can improve their situations by receiving some concessions in trade.[28] Other women join in the hope of securing employment through the efforts of the Congress. The NCSLW's own attitude is that any woman is welcome so long as she accepts Party policy, including naturally its interethnic aims.

Finally, there are large numbers of urban women who apparently consider that the "new" churches have more to offer than the indigenous religion. Consequently, in many churches women provide the bulk of membership, especially in churches of the Pentecostal kind and in syncretist cults that combine Christian with Islamic and pagan elements. Since Christianity's central dogma is the brotherhood of man, this sheer preponderance of women devotees might imply in itself additional opposition to ethnicity. The fact is, however, that the nucleus of many congregations is often provided by specific local ethnic communities within the town. Moreover, not only do the congregations concerned consist of members of the same ethnic group—Yoruba, Chewa, or Fanti—but they also take on the characteristics of ethnic associations. Few women, furthermore, have much say in the administration of the orthodox churches, where women's role, as in Western countries, is largely ancillary.

However, although in this case women may not be in a position greatly to influence personal conduct and group activities, the position in the "independent" churches and in the new syncretist cults is different. The latter, in particular, are popular among the masses of women because they are thought to have a cure for the barrenness and sterility supposedly caused by witchcraft. There is in consequence much more scope for women as leaders; and, indeed, some of the messianic movements favor women evangelists over men. One of the best known of these "prophetesses" was Alice Lenshina, whose Lumpa church began to flourish in what is now Zambia in the 1950s. This was an antisorcery cult as well as a Christian church because Alice claimed the ability to absolve people from the power of sorcery through baptism. By the end of 1955, Kasomo, her native village, had been visited by something like 60,000 pilgrims who carried Alice Lenshina's gospel home with them. The rule that widows were free to remarry whom they chose, or not to marry at all, may have made the church especially attractive to women. Churches were built, newsletters were circulated, subscriptions and church dues collected, and all of this activity reaching from the far northeast of the country to the towns of the Copperbelt and beyond took place on an essentially nontribal basis.

There are also some important ways in which an interethnic outlook is encouraged both symbolically and in a practical form in "orthodox" circles. These include the example set by African nuns. In addition to those who belong to the European missionary orders and to the contemplative monasteries, there are over 100 such groups with close on 9,000 full-fledged members in all.[29]

The girls who respond to the call of religious life give up the extended family and enter into a wider family by joining what is called a Congregation of Sisters. Although a few of these associations are recruited from one ethnic group alone—the single "tribe"—it is the aim to create a Christian community. Consequently, such a community may consist of women from several tribes, all different in their customs and taboos. A Congregation in Kenya, for example, is composed of Kikuyu, Luo, Kamba, Teso, and other peoples. This means that women of different tribes have learned and are learning to live and work together in harmony, often under the supervision of a woman tribally alien to most of them. It also means that the network of relationships established among the Sisters themselves helps to spread a wide, universal brotherhood among the local population.[30]

In addition, any ethnic tendency the local church may have is partly offset by its wider organization. Quite often, this involves branches in several towns, and there is a highly ramified system of "mother fellowships" centered on the church's headquarters in the capital. Since this brings together local representatives and delegates, there is a continuous intermingling at this level of leadership of women as well as men from different ethnic and tribal groups. Naturally, this also applies even more clearly in the case of internationally organized associations such as the Red Cross and the Young Women's Christian Association, whose purpose in particular is to bring women and girls into a worldwide fellowship. In Ghana, for example, the parent body is located at Accra, but in the local branches nearly every member has directly or indirectly some kind of responsibility that contributes to her importance in the group. This also means that each local officer has the prospect of serving at the national level, and so, if elected to Accra, a woman might find herself representing the national chapter on the world level. Thus, not only does the YWCA help women to acquire prestige in a modern way, it also sets up an important network of personal contacts that cross national as well as regional lines.[31] Nor should it be forgotten that some associations of a more indigenous character have a specifically philanthropic function. In Kumasi, for example, there are a number of such societies, and their motto, "everything will be left behind," symbolizes the charitable principles of their activities. In these organizations, membership is open to adults of both sexes, irrespective of ethnic group as well as religion.[32]

CONCLUSIONS

We have thus far endeavored to indicate some social implications of African urban women's associational and individual activities. We have argued that irrespective of these women's own aims, the effect is to confront ethnicity with a countervailing influence analogous to class consciousness. This does not necessarily mean that a very large number of women have ceased to belong to ethnic associations or that they no longer engage in ethnically oriented activities. It does mean that as a result of this countervailing influence, there are probably more opportunities for urban dwellers in general to adopt a nonethnic standpoint. This could be sociologically important because, as we have explained, the urban process is a matter of social differentiation as well as segmentation. Consequently, although the individual may be constrained by norms to identify himself or herself with a particular ethnic group, these constraints do not necessarily govern every social situation. Thus, although there are occasions on which, say, a Luo or a Yoruba may feel obliged to conform with expected Luo or Yoruba standards of behavior, this does not always happen. Rather, when fresh social or economic circumstances arise, there are also changes in the kind of alignment formed.

This is the principal reason why the countervailing influence spoken of may be sociologically significant. Admittedly, influences as indirect and as diffuse as these are incapable of altering attitudes wholesale. Even to imply that they did would be inadmissible. What we do suggest, however, is that the women's activities described have a cumulative effect upon social institutions in general. In particular, by decreasing the importance of ethnic roles, they open the way to alternative channels of social action.

We further suggest that there are two main reasons for the latter restructuring. The first is the female social solidarity initiated by militant leadership. True, there is nothing new about women's cohesiveness as such because a similar sense of fellowship strongly exists among rural women as well. The significant difference is that, unlike their rural counterparts, the urban associations engender esprit de corps by telling their members that women in general are their friends. They deliberately refuse to differentiate between village group or tribe. There is a greater emphasis, too, on both what women as persons want and what women in the mass can do for the wider community. That the former is the case in politics as much as elsewhere has already been explained, but the Guinean experience is especially striking. Thus, when Sekou Touré sought fully to integrate women within the party, it was they and not the men who objected. The women preferred their own separate organization on lines parallel to the way in which male members of the PDG were grouped.[33]

The second reason, we consider, derives from the way in which urban women in general construe contemporary events. We suggest this because, although many of the changes taking place in Africa today continue to be ethnically oriented, the outlook of the women is less "traditional" in some respects than that of the men. They (the women) have little or nothing to gain from maintaining the status quo. Consequently, whereas the older men in particular may be disposed still to operate through ethnic channels, the women's attitude tends to be more pragmatic and empirical. Thus, as the contrasting circumstances of Guinea and Kenya have demonstrated, the women were apparently as ready to "go it alone" as they were to regard the opposite sex as allies. True, in the former case it was a matter of the men wooing the women, but the fact that the latter gave as much as they got fitted in with traditional ideas of reciprocity. It was probably regarded by the women, therefore, merely as a practical way of discharging their side of the bargain.

NOTES

1. J. C. Mitchell, "Perceptions of Ethnicity and Ethnic Behaviour: An Empirical Exploration," in *Urban Ethnicity*, ed. Abner Cohen (London: Tavistock Publications, 1974), p. 25.

2. Ibid., p. 1.

3. Kenneth Little, *Urbanization as a Social Process: An Essay on Movement and Change in Contemporary Africa* (London: Routledge, & Kegan Paul, 1974).

4. Abner Cohen, "Introduction," in Cohen, *Urban Ethnicity* (London: Tavistock Publications, 1974).

5. See M. P. Banton, *West African City* (London: Oxford University Press, 1957); and Enid Schildkrout, "Ethnicity and Generational Differences among Urban Immigrants in Ghana," in Cohen, *Urban Ethnicity*.

6. See Kenneth Little, *African Women in Towns* (Cambridge: Cambridge University Press, 1975); and Philip Mayer, *Townsmen or Tribesmen: Conservation and the Process of Urbanization in a South African City* (London: Oxford University Press, 1961).

7. Kenneth Little, "A Question of Matrimonial Strategy? A Comparison of Attitudes between Ghanaian and British University Students," *Journal of Comparative Family Studies* 7, no. 1 (Spring, 1976). See also Kenneth Little, "Some Women's Strategies in Modern West African Marriage." Unpublished manuscript, n. d.

8. *The Ghanaian Woman* 1, no. 3 (1974).

9. I have stated this position in the past tense because a forthcoming paper by Dr. Wipper will argue that *Maendeleo* is in a period of decline (personal communication). See Audrey Wipper, "The Politics of Sex: Some Strategies Employed by the Kenyan Power Elite to Handle a Normative-Existential Discrepency," *African Studies Review* 14, no. 3 (December 1971).

10. An example is the eloquence of Ruth Habwe, *Maendeleo's* executive secretary. It is reported that she declared in one of her speeches: "For a long time, men have advocated equality; they loathed discrimination in any form in human society. But when it comes to practising what they profess, because of their own self-centred motives, they contradict themselves. . . . But time and again many of our leaders have expressed favour for the women—although when it comes to give [sic] a woman something they have always dialled O. . . . I am aware of the difficulties which men face when they consider the possiblities of women gaining positions of influence. They harbour the inevitable fear that men being superior to women, if women reached the same level, they would fall from the exalted stature they have exploited for so long." Quoted in ibid, p. 465.

11. Hilda Bernstein, *For their Triumphs and for their Tears* (London: International Defence and Aid Fund, 1975).

12. This whole assembly converged on the government offices. Then, after the women had stood in complete silence for thirty minutes, they sang a number of anthems, including a new freedom song. Its refrain was: "Now that you have touched the women, you have struck a rock, you have dislodged a boulder, you will be crushed!"

13. Ibid., pp. 46-47.

14. Also relevant in this context was the formation by white women of the Black Sash Organization. Although not directed against apartheid as such, this movement identified itself with the struggle of non-European women, and the latter's protests against discriminatory treatment have gained them many supporters. When, for example, the African women workers decided to boycott public transport in Johannesburg, a large number of middle-class white women assisted by giving lifts in their cars. Bernstein, *Triumphs and Tears*, pp. 41, 53, and passim.

15. Margarita Dobert, "Liberation and the Women of Guinea," *Africa Report* 15, no. 7 (October 1970); and Claude Rivière, "La promotion de la femme guinéenne," *Cahiers d'Études Africaines* 8, no. 3 (1968), passim.

16. As indicated, the PDG in Guinea is not unique in its claim that women have the same rights as men within the Party; sexual equality is emphasized by most African mass parties. As a result, the women have a distinct and parallel organization that is considered to be consonant with the traditional social pattern rather than a deliberate segregation of the sexes. This division provides a convenient basis for the organization of specific Party functions, which may include the holding of rallies, dances, and picnics. These are also run sometimes by associations and clubs, which, although not a component of the Party itself, are affiliated to it. For example, a Malian association of young women organizes dances to mark and honor meetings of the Union Soudainaise-Rassemblement Démocratique Africain (US-RDA). Its ostensible aim is to participate in parades and help at weddings, baptisms, or circumcisions within the members' families; but nearly all its members are the wives of relatively well-to-do people, including civil servants and business men who belong to this party. Claude Meillassoux, *Urbanization of an African Community* (Seattle and London: University of Washington Press, 1968), pp. 110-112.

17. See Lars Rudebeck, *Guinea-Bissau: A Study of Political Mobilization* (Uppsala: Scandinavian Institute of African Studies, 1974); and David Apter, *The Gold Coast in Transition* (New York: Atheneum, 1963).

18. Filomina Steady, *Female Power in African Politics: The National Congress of Sierra Leone* (Los Angeles: Munger Africana Library Notes, California Institute of Technology, 1975).

19. At the APC's annual convention in Freetown, the NCSLW delegates march according to their various branches and wear APC colors and badges. Also, at elections, each NCSLW member takes up the responsibility of ensuring that the people in her household and compound vote for the Party. When the APC government put to death some army officers held responsible for assassination attempts on the President, the NCLSW planned and staged a massive demonstration in support of the executions. It also helps the APC government's tactics. This is done by the women pushing forward imminent government policies by agitating and demonstrating in favor of these policies before they are introduced by the government. Then, the policies are often phrased as "demands" by the women, demands that have to be met so that the women will be placated. For example, at a rally, the NCSLW leader said that if the President did not declare a republic, the women of Sierra Leone would do it for him. Ibid.

20. Ibid. Emphasis added.

21. Also, although women merchants are not nearly so numerous as male buyers of produce and shopkeepers, some of them do conduct business on an extensive scale. In Nigeria, for example, these women trade chiefly in textiles, bought wholesale from overseas firms, a thousand pounds worth at a time, and sold retail through their own employees in bush markets as well as in town. Others trade in fish or palm oil, own lorries, build themselves modern houses, and have their sons educated abroad. Most of these "merchant princesses" are illiterate or semiliterate, but Bauer's reference to three Onitsha women who traded in partnership for twenty-five years and who had an annual turnover of £100,000 indicates that at least in individual cases women are commercially more successful than the great majority of their male counterparts. Some Ghanaian women, too, are said to have credits of thousands of pounds sterling with overseas firms and to act as intermediaries to women retailers. Little, *African Women in Towns*.

22. Swahili is, of course, the lingua franca of East Africa.

23. Communicated by Nici Nelson, to whom I am also indebted for permission to draw on the above passage from her unpublished paper, "Informal Sector Economic Activity in a Squatter Neighbourhood." This was presented at the African Urban Culture Seminar, London School of Oriental and African Studies, February 11, 1976.

24. Kenneth Little, *West African Urbanization: a Study of Voluntary Associations in Social Change* (Cambridge: Cambridge University Press, 1965).

25. *Ideal Woman* (Lagos), May 1974.

26. Georges Balandier, *Sociologie des Brazzavilles Noires* (Paris: Colin, 1955), pp. 145-148.

27. If active support of nationalism is any indication of an interethnic orientation, Abidjan prostitutes have also taken positive steps in the latter direction. Thus, when the women of Abidjan turned out in force in public demonstrations against the French, the women of the streets played as prominent a part as their "respectable" sisters. Jean Rouch and E. Bernus, "Note sure les prostituées 'toutou' de Treichville et d'Adjamé," *Études Éburnéennes* 6, (1959): 231-242.

28. Steady, *Female Power*, p. 25. By the same token, women traders vote for politicians sympathetic to their business interests. Hodgkin quotes the case of the Ghana woman who was asked what she was going to vote. She replied, *meko vote apatre* ("I am going to vote fish") as a Lancashire woman might "vote cotton." Thomas Hodgkin, *African Political Parties* (London: Penguin, 1961).

29. Sister Ann Gregson, "The African Sisterhoods," (Paper for School of Oriental and African Studies Seminar "Christianity in Independent Africa," London University, 1975).

30. Sister Ancilli, "African Sisters' Congregations," (paper for School of Oriental and African Studies Seminar, "Christianity in Independent Africa," London University, 1975). In respect of this and the preceding reference, I am indebted to Sister Gregson and Sister Ancilli for permission to draw upon unpublished material.

31. D. K. Fiawoo, "The Influence of Contemporary Social Changes on the Magico-Religious Concepts and Organization of the Southern Ewe," (Ph. D. diss., University of Edinburgh, 1959).

32. A. A. Bandoh, Unpublished manuscript, n. d.

33. Rivière, "La promotion de la femme guinéenne."

10

Voluntary Associations and Ethnic Competition in Ouagadougou

Elliott P. Skinner

Ethnically based voluntary associations in contemporary Africa function as much to give their members competitive advantage in the struggle for access to the resources of their societies as to provide the framework for social activities. Few voluntary associations in precolonial African societies were ethnically based. Among the acephalous Yako[1] and Igbo[2] in what is now Nigeria, membership in the often all-powerful voluntary associations was based on achievable qualities such as aggressiveness, hard work, and wealth. Such ascribed statuses as age, sex, and lineage connections often made it easier to become a member of a voluntary association, but ethnicity was never important in the structure or function of these organizations. Therefore, when these groups competed with each other for power and prestige in their respective societies, the ethnic factor was absent.

Even in the more complex and often multiethnic African states, ethnicity was seldom if ever a factor in the formation of voluntary associations. The *ogboni* society among the Yoruba was composed of ethnic Yoruba who had achieved distinction in their communities.[3] The *dopkwe* associations in Dahomey, which were organized previously for economic purposes but were often politically and socially important, had no ethnic basis.[4] The Poro and the Sandwe associations in the societies of Liberia, Sierra Leone, and the Ivory Coast, although they were politically powerful in their respective societies and had interterritorial links, were ethnically specific to their communities.[5] Here again ethnicity was not a factor in the struggle of the associations and their members for preeminence.

The castelike occupational groups that were found in the western Sudan were the nearest analogues to the ethnically based voluntary associations of contemporary Africa. Each composed of persons from the same ethnic group scattered throughout different societies, these associations were, nevertheless, as much voluntary as they were ascriptive. Although usually only persons from specific ethnic groups could ply certain trades or professions, not all members of those groups were obliged to do so. Moreover, persons belonging to a different ethnic group but who wanted to engage in a "caste"-specific occupation could and did assimilate into the group whose occupation they desired.[6]

While some of the more economically complex conquest states in Africa did develop ethnic systems, most of them did not. The relative absence of ethnically based societies in precolonial Africa was a function of the nature of these societies. The really small-scale societies of tropical Africa cannot be said to have had ethnic groups describable as social entities that, while inhabiting the same state, country, or economic area, consider themselves biologically, culturally, linguistically, or socially distinct from each other, and most often view their relations in actual or potentially antagonistic terms.[7] Some "Pygmy" and "Bushman" groups lived in symbiotic relationship with their more powerful neighbors, but significantly these hunters and gatherers exploited resources other than those exploited by their agricultural and pastoral neighbors. They normally had only intermittent relations with the people of these more complex societies, and, despite myths that provided the charters for these relations, the hunters and gatherers were constantly being displaced or absorbed.[8] The tendency of small-scale groups that came into prolonged contact with people who exploited the same areas or resources was to merge with them or to move away. They did not create ethnic systems, described by Cox, as power relations or a system of superordination or subordination between groups, with a hierarchy of access to their societies' resources.[9] And in an ethnic system, competing groups, to enhance their power, use their differences so as to delineate their groups' boundaries and give them cohesion in order to compete against their neighbors.[10] Thus the small-scale pastoral and agricultural societies seldom if ever formed ethnic systems as the hunters and gatherers did. When they came into contact with other groups, they either merged to form a large-scale statelike society, or one of them moved away.[11]

The tendency of most African states was to assimilate the conquered populations biologically, culturally, and socially. Thus, the Swazi of South Africa carried out a policy of intermarriage with conquered groups. They "established a complex network of economic and social obligations, gifts and services in 'in-law' relationship" that merged the diverse elements and "enriched the nation with new ideas, words, rituals and tools."[12] Similarly, the Ambomu, who had migrated into and conquered portions of the southeastern Sudan, amalgamated about "twenty foreign peoples to form the Zande people" as the people of the Azande state became known. It was still possible at the time of European conquest to find foreign elements "at all stages ranging from political absorption but cultural automony to total assimilation, both political and cultural."[13]

Apparently there were limits to the ability of African states to assimilate all groups who were brought into their orbit. Many states in Central Africa and the Sudanic fringe region formed ethnic systems, with the conquerors or local ruling groups in superordinate positions. Some subordinate groups in the Sudanese states and in Hima-ruled states in Ruanda and Burundi were incorporated in "castelike" positions with quite limited access to the total resources of these societies. Many of the subordinate groups exploited specific ecological zones or niches, or practiced specific occupations. And while myths provided the charters for structuring the "premise of inequality" in these societies, the potential, if not the fact, of physical coercion was ever present.[14]

A number of African states absorbed some conquered populations and migrants, incorporated others as subordinate groups, and assigned to still others the status of strangers. The invading Dagomba from Ghana quite easily absorbed and assimilated the Ninisi in the Upper Volta region, with whom they shared major sociocultural characteristics except for political organization.[15] The result was the Mossi states. The Mossi did not absorb but only incorporated the Fulani pastoralists who, according to tradition, drifted into their country and "submitted voluntarily to the authority of the Mogho Naba."[16] The Fulani remained socially different from the Mossi, even though in time they learned the Mossi language and adopted many aspects of Mossi culture. The Mossi were refugee traders who settled in the wards of the capital at Ouagadougou. These two groups had their own chiefs and headmen, and as part of the cortege of the Baloum Naba, the Chief Steward of the Palace and "Mayor" of Ouagadougou, they paid homage to the Mogho Naba. A very large percentage of them married Mossi women, including the daughters and sisters of former Mogho Nanamse, and of course their sisters and daughters married Mossi men, including the Mogho Nanamse. Nevertheless, the Yarse and Hausa populations remained distinguishable from the Mossi population by language (all are polyglot), religion (Islam), and especially by occupation (since they remained traders). They were not wholly absorbed, partly because they had the important role of conducting trade with the people of their original homelands. They opened their homes to countrymen visiting in Ouagadougou and represented their guests at court if and when there were problems to be resolved. It was, therefore, to the interest of the Mossi political hierarchy that the resident trader populations retain their linguistic and other skills necessary for dealing with foreign traders.

The ethnic system that developed within Mossi society had its analogues in other parts of the Sudan. Migrating pastoral peoples were partially incorporated into these states, foreign traders were incorporated as strangers, and comparable agricultural peoples were assimilated. The strangers in Sudanese towns lived in special wards (sometimes called *Zongos*) under the control of their own chiefs or headmen. Es Sadi described this pattern for Timbuktu as early as 1352: "Timbuktu at that time [of the Battuta's visit] was inhabited by people of Mimah and Tawarek (Molathemin), especially Masufa, who had a headman of their own, while the Melle governor was Farba Musa."[17] In some societies, these chiefs and headmen represented their followers at the local courts; in other areas, the local ruler appointed one of his officials to deal with the strangers and their chiefs. Comparable data from East Africa suggest that the twelfth-century ruler of Sofala welcomed traders from Kilwa to settle permanently in his realm. To gain their friendship so that they could act as local representatives for their visiting countrymen, he also offered them wives.[18] With few exceptions, the strangers were under the political control of the local African political authorities and stayed on only at the sufferance of their hosts.

European contact with and conquest of African societies created many more ethnic systems than had heretofore existed on the continent. The commercial activities of the early Europeans stimulated a vast movement of populations; and subsequent European conquest incorporated different societies within the same colonial territory. During the heyday of colonialism, many traditional African polities, societies, and linguistic groups were broken up and incorporated into various colonial territories. The result was that the very structures of many African societies were changed.[19] In some cases, the Europeans coopted the local kings and rulers into a system of indirect rule, or created formal leaders for acephalous populations. In other cases, the Europeans attempted to rule the societies directly. But whether the Europeans used indirect or direct rule, the result was that the African societies lost their independence and were brought into a worldwide economic system that subordinated their own economic institutions to that of the dominant Europeans. This occurred whether the Africans were forced or induced to work in mines or on farms and plantations, to produce cash crops, or to raise money for taxes by selling either their animals or their subsistence crops. In any case the Africans lost their economic independence, and thus politically as well as economically they became part of a colonial system dominated by the Europeans.

The major arena in which the colonized Africans encountered each other, the non-African middlemen stranger traders or functionaries, and their European colonizers, was the urban center. These administrative, commercial, or mining areas became the mecca for millions of Africans drawn or pushed there by the necessity of earning a new type of living. Many of these African migrants found shelter with autochthonous chiefs and people and later with kinsmen. When there were traditional noblemen among the migrants, some of them emerged as leaders or chiefs.[20] In the absence of such individuals or when the migrants came from acephalous societies, they reorganized commoners as headmen, elders, or *vieux*. Through these chiefs, leaders, or elders, the first generations of migrants utilized "primordial ties" as mechanisms for adaptation to urban life.[21] The

African groups that had traditional voluntary associations at home often introduced these entities into the towns. [22] The Yoruba peoples, whose "towns" provided the basis of a new type of urbanization, used their traditional voluntary associations such as *esusu* and others to good effect. Thus, during the early part of this century, some African groups living in the emerging towns had no voluntary associations, and their lives were regulated by their chiefs or headmen; others had both chiefs and voluntary associations. [23] However, regardless of their social arrangements, the Africans were still inward-looking and seldom actively competed either among themselves or with the Europeans for either economic or political power within the town. In fact, Mitchell argues that when such groups as the Kalela dancers arose, they began to compete with each other for prestige and used the educated Africans as a reference group. They did not challenge the economic status of each other, of the emerging African elite, or of the dominant Europeans. [24]

The emergence of a group of educated Africans gave rise to a new type of voluntary association, whose form and function came to influence those of the traditional organizations. The history of the new type of urban voluntary associations suggests that their raison d'etre was primarily political, even though they had some social and cultural functions. In contrast to the other voluntary associations, whose "integrative function" has been so stressed as to have been distorted, [25] these associations were not created by people new to the urban centers and thereby seeking new forms of social and cultural integration. Their founders were persons who had largely adopted European values and who were concerned with the problems posed by European economic, political, and social control. For example, the first voluntary associations to emerge among the elite in the Gold Coast were more concerned with politics, with the "colonial situation," than with traditional, social, or personal relations. [26] Kimble reports that by the mid-1800s,

> educated Africans, in particular, felt the need for integration with those who shared their *new interests* and *outlook*, and their command of English helped to break down tribal barriers. During the latter half of the nineteenth century they seem to have attached great importance to building up all kinds of clubs and societies to foster social contacts and common interests. At first these were nearly all modeled on European lines, with some attractive variations. [27]

Significantly enough, the Europeans in the Gold Coast, recognizing the threat to their power from these associations, disapproved of them, thereby incurring the anger of the educated Africans. For example, in 1864 the Philanthropic Society of Cape Coast "blamed the Government and 'white men generally' " for maligning its members as "half-educated and semi-civilized," and accused the whites of making it difficult for Africans to raise public funds. The Africans piqued by the alleged insult, announced their intention of appearing on behalf of the Gold Coast Protectorate before the 1865 Select Committee in London; but they did not succeed in doing so. For the next few years the energies of the educated handful were directed into political channels, especially while they were cooperating with the Chiefs in the Fanti Confederation.[28]

Had the British not rebuffed the educated Africans, they might not have sided with the representatives of the traditional system but instead have attacked both the chiefs and the colonial system. As it was, they joined forces with the only class recognized by the Europeans as politically significant.

Comparable events and processes took place in other parts of Africa. For example, on the Copper Belt (Zambia), where the British had imported rural traditional chiefs to administer to the needs of the growing urban industrial population, a number of these politically motivated voluntary organizations arose. About fifty of the "most economically and educationally favoured Africans in Ndola formed a local Native Welfare Organization similar to those at Lusaka, Broken Hill, Livingstone and the Roan Antelope mine." The leaders of the Africans in three of these centers indicated that their purpose "was frankly to safeguard the rights and interests of the natives of the Northern Rhodesia."[29] One district commissioner, recognizing in this organization a "natural creation" of "thinking and politically-minded Africans," sought to use it as a government advisory body for such purposes as tax assessment. However, most officials on the Copper Belt considered these organizations dangerous to their economic and political interests. They had no objections if such associations became debating societies (where their members could display their newly acquired English), but decided that they should not be permitted to discuss "political matters." This restriction was imposed ostensibly because the European officials felt that the educated Africans were not representative of the African population as a whole.[30] Later events were to prove the futility of this course. The traditional leaders whom the mineowners on the Copper Belt had imported to help administer and to "integrate" the urban African mine workers failed miserably in their "administrative" tasks but succeeded in the "social" ones. This left the voluntary associations, especially the mine unions, to aggregate the interests of the Africans and to compete among themselves and against the Asians and Europeans for control of the polity of Northern Rhodesia (Zambia).[31]

The history of voluntary associations in Ouagadougou, the capital of the colony and now of the Republic of Upper Volta, throws significant light on the multifaceted and evolutionary nature of voluntary associations in African urban centers. Departing from their usual policy of direct rule, the French permitted the defeated Mossi king, the Mogho Naba, and his chiefs to act as political leaders of their erstwhile subjects.[32] Thereafter, that is, from 1896 onward, the rural Mossi who migrated to town came under the supervision of the Ouagadougou-based provincial ministers who controlled their home districts. When the post-conquest migrant strangers such as the Dioula (Malian), Hausa, Songhay, and others arrived in town, they too were placed under the control of their own "chiefs" or "headmen" by the Baloum Naba, the chamberlain of the Mogho Naba's court. These migrant elders were held responsible for the social as well as administrative affairs of their charges.

The Mossi and stranger groups in Ouagadougou apparently had little need for many of the voluntary associations said to be characteristic of the populations of many emerging African towns.[33] And the Mossi from outside the Ouagadougou kingdom and the various stranger groups were apparently quite satisfied with the function of their *vieux*, as their elders were called. Thus, few associations developed, and those that did were designed to serve purposes other than those normally associated with voluntary associations in African towns.

One voluntary association among the Mossi that has had a long history in the town is called the *nam*. For the traditional Mossi, the concept *nam* means "that force of God that permits a mere man to rule over other men." Thus it represents such attributes as sovereignty, authority, and legitimacy. The term is associated with *naba*, the word for chief, the "father *ba* of the *nam*." Within Ouagadougou, an organization called the *nam* arose among young Mossi men; it was maintained for the sole purpose of doing charitable work. If, for example, an old woman or an old man needed a hut repaired, a field cultivated, or grain harvested in the outskirts of the town, the young men of the *nam* got together and did these tasks. Nevertheless the *nam* is not a *sisoga* (a cooperative work group), since its members do not help each other with chores of any kind.

The history of *nam* in the Bilibanbili ward of Ouagadougou is lost in time. It may once have been an agricultural association of Mossi youth, but today its functions are purely philanthropic and recreational. The head of the *nam* is called the Yung'Naba (literally "Night Chief"), a title that is equivalent to Mogho Naba; but since that title is forbidden to any man other than the king, the title Yung'Naba is used instead. The person named Yung'Naba must be kind, wise, considerate, jovial, and helpful. He is usually elected from among members of his generation and retires when his generation moves into the more serious affairs of life. The Yung'Naba is assisted by a number of young men (thirty-three according to informants, but this is doubtful since multiples of three are not used for any number associated with Mossi Men), who take any vacant chiefly title they prefer. Thus there is the Manga Naba, the Tema Naba, the Tengkodogo Naba, the Baloum Naba, the Nobere Naba, the Kam Naba, and so forth. There is apparently no relationship between the provenance of a young man and the title of the *naba* he chooses, and none of the functions is assigned to specific *nanamse* in the association. Any person can be asked by the Yung'Naba to perform a specific task for the organization or can volunteer to do it. However, as *chef de propagande* ("spokesman") in the organization, the Kam Naba has the duty to summon his fellow chiefs whenever there is a meeting of the group, either for recreational purposes or for voluntary work.

There are few requirements for membership in the *nam* except being an ethnic Mossi. Any young man from the ward who wishes to join can do so provided he is able to buy enough millet beer to entertain the other members on the day of his elevation to membership. An inquiry into the status of *nam* members showed them to be both Ouagadougou-born Mossi and immigrant Mossi. They are primarily illiterate tradesmen and laborers; a few are cultivators, a number of whom consider themselves to be unemployed because they have no noncultivating occupations.

The members of the *nam* have little contact with either the traditional or modern elites in the town, but, like most persons forming an organization, they had to have an official sponsor. They succeeded in being sponsored by the Municipal Councilor of their quarter, and they elected him their honorary treasurer. However, this official took almost no interest in the *nam* and placed little value on his association with that organization. This was perhaps beneficial to the *nam*, because it continued to be viewed as a nonpolitical organization. Its members either were not interested in the political struggle that took place after 1945 or thought their organization should not be involved in politics. At any rate, like other organizations that existed before the efflorescence and decline of associations through 1958-1960, the *nam* continued to function even after the revolution in 1966. The problem for the *nam*, however, is that it may not be able to adapt to rapidly changing Ouagadougou. One wonders whether future migrants from the rural areas will participate in a recreational institution modeled on a traditional institution, the chieftainship, which has difficulty modernizing itself.

The *gumbe* is an association that appeared in Ouagadougou during the early years of World War II. Either it came from the Ivory Coast via the French Soudan (Mali), or, as claimed by some of the present members, it was brought to Ouagadougou by young Mossi labor migrants who had been in the Ivory Coast.[34] The *gumbe* apparently reached its climax in the town around 1957, when there were six *gumbe* groups in various wards. Besides Koulouba, where it developed, *gumbe* groups existed in Tiedpalogho, Bilibambili, Paspanga, Dapoya II, and Ouidi. The groups were autonomous, but those in Bilibambili and Koulouba had close relations and often planned activities together. After 1958 the *gumbe* disappeared from Ouagadougou, as part of a phenomenon I shall discuss later, only to reappear in November 1964 in the Koulouba ward. Although the author attended several meetings of the new *gumbe*, most of the data were gathered from interviews with past officers and members of the organization.

As described by its members, the *gumbe* is a youth organization formed for the purposes of recreation and mutual aid and for educating the youth in such matters as public service and obedience to parents and authority. *Gumbe* members pride themselves on the multiethnic nature of their organization. And though the majority of the members of the revived *gumbe* are Mossi, there are members from all the other groups in the town, especially the Dioula. Another characteristic of the *gumbe* is that although it is "multi-confessional," that is, it has members from all the various religious groups in the town, it is decidedly Muslim in orientation. The *gumbe* does not meet during the holy month of Ramadan, nor does it permit any alcoholic beverages to its members. Yet, the *gumbe* is more modernistic than the typical Muslim association and has always included a substantial number of girls among its members. During 1958, the *gumbe* in Koulouba quarter had 120 male and 69 female members. Moreover, the female members of the *gumbe* were actively included in most of the affairs of the group.

The *gumbe* has always had a well-organized structure, a table of organization, and clear-cut functions. Its officials include a president, vice-president, secretary general, secretary-general adjunct, commissioner of finances, commissioner of police, and police attached to the president. There are also a female president and vice-president, but no other female officials. Although there are no age restrictions on membership in the *gumbe*, most of the members are between the ages of twelve and thirty, since younger persons are unable to pay the required dues and older people are too busy with affairs of their own. A prospective member of the organization is usually proposed by a member and is invited to attend a Monday night meeting, where he or she is either admitted or rejected. Persons are rejected only for the reason of bad character, and since such persons normally are known to the youth of the ward, they are not likely to be proposed for membership in the first place.

Each member of the *gumbe*, except the president, pays monthly dues based on rank within the association. The feeling is that the service of the president to the organization is enough. The vice-president pays 500 francs ($2.00 US), the secretary general 300 francs, and other officers, such as the commissioner of police, 200 francs, and ordinary members 100 francs. However, owing to "unemployment" and other problems causing financial embarrassment among the youth, the association does not insist that its members pay their dues every month.

As one of the aims of the *gumbe* is to instill obedience in the young people, discipline within the organization is quite strict. Members call each other "comrade," and no one is permitted to call the officers by name; they are addressed by their titles. Anyone violating the rules of the *gumbe* or who is guilty of misbehavior is summoned to the Monday meeting of the association, where he is judged and, if found guilty, fined. An important stimulus for discipline among the youths of the *gumbe* is their fear that adults and other authorities might misunderstand the aims and activities of the organization. The older people in the town, especially the Mossi, mistrust the interaction of young males and females in any organization and believe that they associate together for immoral purposes. This fear is due to the traditional Mossi prohibition of close public relationships between young males and young females. Furthermore, the older people fear that the modernistic tendencies of the organization can only lead to a breakdown in the traditional moral system.

The Catholic missionaries are especially hostile to the *gumbe* because most of its members are Muslim and because they sponsor mixed dancing without adult control. So strong is mission opposition to the *gumbe* that when the young Christians of St. Leon ward organized a *gumbe* and had their first public dance, a Catholic priest went to the meeting, took off his belt, and chased the participants away, exclaiming that "the devil is routed!" Because of such attitudes among many townspeople, the *gumbe* members are especially careful about the relationships between the males and females in the association. The president and his officials watchfully chaperone the young girls who attend the meetings and dances of the *gumbe* and escort them home when these affairs are over.

The most publicized and best-known activities of the *gumbe* are the open-air dances held on Saturday and Sunday afternoons and on the various holidays. For these performances the members gather in the street before the house of the president, where they place benches forming a square. The president, major officials, and the musicians are usually seated at the side of the square backing on the president's house, and the young people segregate themselves by sex on the other benches. The policemen of the *gumbe* ensure that order is maintained on the benches, they seat girls and visitors, and chase away the numerous young children who attempt to fill the benches and disturb the members.

All the members of the *gumbe* are recognizable by their dress or insignia. The president and the men wear shirts and long trousers, both of which have fringed sides resembling those worn by American cowboys. The men wear either a version of a western hat or a military-style overseas cap. When the latter are worn, badges indicating the rank of the wearer are pinned on them. The policemen normally do not wear this type of dress but use regular pants and shirts, overseas caps, bandoliers, and belts, the latter often decorated with red, white, and black, the colors of the republic. The girls most often wear what has emerged as "modern" African dress: wraparound skirts and blouses. The skirts are usually made out of imported cotton prints, and the blouses may be of the same cloth or of white cotton.

There is now a standard pattern of music, singing, and dancing associated with the *gumbe*. The musical instruments consist of a large drum called a "jazz," small square drums (really boxes covered with skin) called tambours, rattles, and gongs. The musicians beat a rhythmic pattern, changing the rhythm from time to time and stopping at intervals to tighten the skins on the drums by heating them over a fire. The singing is usually done by young women who sing through a "micro" in a high falsetto the numerous *gumbe* songs that have been transmitted from the Ivory Coast and Mali. The words to the songs are usually in Dioula, but occasionally the singers use More, the language of the Mossi, and even French. The dancing is stylized in the *gumbe* pattern. Young men and women dance solo and at times with each other, either in bodily contact or separately. When the youths dance separately, they usually compete with the musicians to see whether they or the musicians can dance or play rhythmic patterns that the other cannot follow. After each dance, or when a dancer has stopped dancing, he or she goes before the president and either salutes or curtsies in respect. The organization furnishes water for thirsty members and may even provide soft drinks when the treasury is full.

During the heyday of the *gumbe* in Ouagadougou, the young people participated in other social activities. Almost every Saturday morning the young men of the *gumbe* accompanied young girls to the pools at the northern part of the town and entertained them while they did their weekly washing. This activity ceased when the *gumbe* was deactivated and has not resumed, since there are now standpipes in the various quarters for washing clothes.

When the political activities began during the mid-1950s members of the *gumbe* performed both social and political functions for the politicians. During 1957-1958, they were engaged by individual politicians to drive around town in vans, singing praises for them and their parties. The *gumbe* was often used to sing songs of welcome for "high personalities" arriving in Ouagadougou either by train or by air. Indeed, some members of the *gumbe* considered this political function of "activating the masses" so important that one cannot help wondering whether it was not the impending elections in Upper Volta in 1965 that stimulated the president of the *gumbe* to reactivate the organization. True, he did not receive the presents, such as a small car, that certain politicians had promised him before and during the previous election, but he entertained hopes for the future. On New Year's Day 1965, he led his group in singing the praises of influential politicians and the mayor of the town. He said then that even if he did not get anything for himself, he at least hoped to receive some aid from the administrative authorities to buy instruments for the musicians and costumes for the members of the organization.

During its heyday, the *gumbe* provided material aid for those of its members who got married, either to other *gumbe* members or to strangers, and attempted to "organize joy" or festivities for the young couple. The *gumbe* also provided assistance to members who were ill or out of work and were especially helpful during funerals, a period of sudden and great expense. *Gumbe* members also sought out the members of foreign *gumbes* and gave them any help they could; for, as the president said, "It is important for strangers who visit Ouagadougou to be able to return to their homelands and declare how kind the youths of Ouagadougou are and how well they get along with each other. When this happens people in the foreign areas will not hesitate to welcome the young people from Ouagadougou to their lands."

Besides the *nam* and *gumbe*, there were very few voluntary associations in Ouagadougou before the end of World War II. Even the *nam* and the *gumbe* were very specialized types of associations, since they were primarily recreational and geared to the youth.

RISE AND FALL OF THE "NEW" VOLUNTARY ASSOCIATIONS

Politics was responsible for the rise and fall of most of the associations that appeared in Ouagadougou at the end of World War II. These associations were designed to compete for power and to control the state's resources, even though they masqueraded as mutual-aid societies. In 1946 a number of civil servants and clerks from the Koudougou region, residing in Ouagadougou and fearing that the indigenous people would capture power in the reemerging Upper Volta colony, formed an Association des Originaires de Ganzougou in an effort to protect their interests. Despite the obvious political purpose of this organization, its constitution stated that its goal was to provide social and material aid for its members. The organization did not last long enough to achieve its aim, since, when its key members were elected to political office in 1947, they lost interest in the association and it folded. Few attempts were made to form other associations until the passage of the *loi-cadre* in 1956 promised greater African participation in the government of Upper Volta.[35] As soon as this law went into effect, practically every possible group in Ouagadougou began to organize for its own advancement, and they all formed mutual-aid associations.

One of the first of the new voluntary associations to appear was organized by the major politicians of Ouagadougou. On December 12, 1957, these men formed the Association des Originaires de Zoundiweogho (Cercle de Ouagadougou), with its headquarters in the town of Ouagadougou. Characteristically, its declared objectives were: "(1) To unite in fraternal union all the people originally from Zoundiweogho; and (2) To provide mutual aid for them and defend their common interests." An explanatory note appended to this declaration stated that the association was "a fraternal union of persons organized for mutual aid and for the defense of the common interests of its members, relatives, and persons from a well-known region. All persons originally from Zoundiweogho, whether domiciled in that region or living outside it, regardless of religious or political affiliation, may join this present association." Nevertheless, this association, like its predecessor, never engaged in social welfare or mutual aid but actively mobilized the people of Zoundiweogho for political purposes.

Almost all the other associations that emerged in Ouagadougou during this period took the same form and had the same function as the association of Zoundiweogho. For example, an association known as Regroupemente des Originaires du Cercle de Kaya, founded in Ouagadougou, had as its objective "to provide mutual aid among ourselves and to defend our members." The political nature of this association was indicated by the nomination of a well-known politician from Kaya as its honorary president. The Association Fraternelle des Originaires du Cercle d'Orodara had for its objectives (1) to restore links of solidarity among all persons originally from the Cercle of Orodara and (2) to provide mutual aid among its members. The Union Fraternelle des Peulhs (Fulani) de la Ville de Ouagadougou was to provide mutual aid and to strengthen lines of friendship and brotherhood among its members. The Association des Originaires de la Subdivision de Hounde had the same objective as the others. The Association des Originaires du Gourma and the Association des Originaires de la Subdivision de Zabre gave slightly different goals as their raison d'etre. They both desired to organize persons from their respective areas "in a sentiment of solidarity," but besides safeguarding their interests they wished to ameliorate the material, moral, cultural, social, and economic conditions in their respective home areas. But, here again, the real aim of these associations was political mobilization.

A number of associations that appeared in Ouagadougou during 1957-1958 made little pretense about their political aims. For example, a group of young people from Bousse living in Ouagadougou and belonging to the political party Parti de Regroupement Africain formed the Jeunesse du P.R.A. de la subdivision de Bousse. They gave as their goals: (1) to unite all the youth of the P.R.A. originally from the Subdivision of Bousse and now living in the Cercle of Ouagadougou; (2) to pursue and defend the interests of the youthful adherents of the P.R.A. of the Subdivision of Bousse; (3) to develop and coordinate the cultural, artistic, and social activities of all adherents; and (4) to aid in rural construction. Almost simultaneously, three politicians, Gerard Kango Ouedraogo, Nazi Boni, and Joseph Conombo (then residing in Ouagadougou), sponsored an association known as the Mouvement de Regroupement Voltaique. It was billed as "an organization of Socialist workers, peasants, and all other manual or intellectual workers of the territory regardless of race [ethnic group], religion or [place of] origin."

Other groups in Ouagadougou did not wish to be left out in this drive toward formal association. Various religious, lay, and educational youth groups were formed. Among them was the Catholic Jeunesse Étudiante Chrétienne Feminine (J.E.C.F.) of Ouagadougou. This female group had as its aim "the spiritual, intellectual, and social development of its members, and giving service to the scholarly community within the bounds of scholarly rules." The Muslims of Ouagadougou, who felt discriminated against by the Catholic elite, organized the Communauté Musulmane de la Haute Volta: Siege Social—Ouagadougou. They indicated that:

> Our goal, pure and simple, is to inspire an Islamic conscience among Muslims. To unite all Muslims of both sexes, in order to have them to understand each other, and to help each other morally and religiously. To rid Islam of all corrupt and harmful influences and practices. To combat by all appropriate means, fanaticism, superstition, and the exploitation of the credulity of the Faithful. To organize trips to Mecca. To obtain official recognition of the Islamic Holy Days, and to follow the precepts of the Koran.[36]

The Protestants in Ouagadougou organized an Association des Éducateurs Protestants de Haute Volta in order to secure their rights and privileges. Not to be outdone, the secularists formed the Federation Voltaique des Oeuvres Laïques and gave as their objective: (1) to disseminate and to defend the ideal of secularism and (2) to contribute to the development and defense of the moral and material interests of secular educational and social institutions. These so-called secularists were primarily Muslim intellectuals alarmed over the growing identification of the government with the Catholic Church.

Sports groups such as the "Mimosas" were organized to "create and to maintain relations, especially sporting, among all of its members residing in Ouagadougou, and principally to practice football and eventually volleyball, basketball, etc." Mixed sports and cultural groups such as the Association Sportive et Culturelle de Ouagadougou dite "Amitie Africaine" came into existence for the purpose of "developing and broadening the personality of young girls and young women by physical, familial, social, cultural, and artistic education." Purely cultural groups such as the Troupe Théâtrale Arc-En-Ciel de Ouagadougou also organized an effort to perpetuate African customs and to enact plays dealing with timely topics. Occupational groups, such as the Syndicat des Petites et Moyennes Entreprises de Haute Volta, were founded to protect their interests against the large European companies and enterprises. Parents organized such groups as Association des Parents d'Élèves de l'Enseignement Libre de la Haute Volta to raise questions pertaining to the education of their children, among other things.

The stranger groups in Ouagadougou (that is, the Africans from other countries) also formed their own formal associations. But here the stated and real aim was self-protection against the growing political power of the local inhabitants. Strangers from the Ivory Coast organized the Association Fraternelle des Originaires de la Côte d'Ivoire in Ouagadougou; its objective was to give material and moral aid to its members in case of need and to entertain its members by organizing recreational, cultural, educational, and artistic affairs. The Soudanese (Malians) formed L'Amicale Soudanaise de Ouagadougou, and the Senegalese in Ouagadougou organized the Ressortissants du Fouta Senegalaise. The stranger groups that were most outspoken about the aims of their associations were the hard-pressed Dahomeans and Togolese, who faced the loss of top administrative positions now that indigenous politicians were taking over. The Togolese formed the Association de la Jeunesse Togolaise in Ouagadougou and gave as its goals:

1. To join in close feeling of union, solidarity, and fraternity, its members originating in Togo but residing in Upper Volta
2. To tighten the bonds of solidarity between the natives (of both British and French Togoland)
3. To secure, insofar as possible, the amelioration of the conditions of members and to come to their aid in the event of illness, hospitalization, or any lawsuit that would cast aspersions upon the dignity of the association. In a word, [to help in] all circumstances where the intervention of the group is necessary
4. To supervise the good conduct of each member in his public as well as private affairs
5. To use all available means to correct the morals of a member whose conduct risks sullying the dignity of his person as well as the honor of the ethnic group to which he belongs
6. To stimulate the total membership, whatever their social status, with the same desire to reaffirm the ties of sympathy among them or of greater fraternal ethnic solidarity, and to enhance their collective advancement
7. To share the pain and sadness of members who might lose their father or mother, and to aid them with a sum . . . francs. Regardless of which parent it is, the members will hold a wake of from three to four hours at the homestead of the bereaved. If the treasury is empty, the committee may decide to impose a contribution in excess of the fixed monthly dues in order to attain the sum indicated.[37]

Because most of the associations that emerged in Ouagadougou during 1956-1960 had a political basis and bias, the African politicians who formed the Conseil de Gouvernement of Upper Volta, fearing trouble, asked the town's young people to form Le Comité d'Entente de la Jeunesse à Ouagadougou, representing all the youth in town. The council warned that:

The time of divide and rule is not entirely past; this is why you are being asked to form a committee. Its job will be to make our ideas known, to discuss them

openly so that this discussion will shed the needed light on issues pertaining
to the development of the type of country we desire. We have the will and
courage to do this, but we lack the means. It is still evident that it is the
Government to which we must address ourselves. We are certain that the
Government will understand [38]

Although the politicians and many other persons in Ouagadougou
viewed the proliferation of associations with dismay, others felt that it was
good for the townspeople to create "true fraternity, or more exactly, groups
of fraternization, each one hoping to bring its brick to create the Africa of
tomorrow." Still others pointed out that the formation or "originaires"
from one place or another was counterproductive: people who normally
had little in common were banded together, and those with comparable
education and training formed different groups, often in opposition to one
another. One young Voltaic wrote from France that one night a phantom
came to him in a dream and said, "Soon you will see the creation of an
association having as its objective: to determine the lactation capacity of
Zulu grandmothers in South Africa."

Although the main reason for the rise of associations during 1956-1960
in Ouagadougou was that each interest group, corporate or not, sought to
mobilize its members to compete for economic and political power, the
problem was that by 1956 the French had already decided to yield political
power, and African political parties had appeared. These parties, led by the
high civil-servant elite, were better prepared than the voluntary
associations to fight for political power, and they exploited the
associations. Indeed, during 1958-1959, when the political parties fused,
split, or fell apart, the associations were left in limbo.

The political party that emerged victorious and formed the government
took a jaundiced view of the associations, considering them divisive
elements in the new nation-state. In 1961 the minister of interior and
security criticized and then banned the Union des Jeunes de Regions
Sindoises as a subversive organization.[39] Once this occurred, the other
associations folded and disappeared. There was no longer any need for
them since political mobilization was over. The dominant ruling party
procribed some minority parties, while others simply disbanded. In the
absence of important nonpolitical jobs, the defeated politicians abandoned
their erstwhile supporters in the associations and made peace with the
victors. By so doing, they attempted to acquire the junior, but still
important, positions abandoned by the French or seized from the ousted
Dahomean and Togolese strangers. Ouagadougou returned to a period of
relative political calm, and the informal groupings of strangers and
regionals, which had been superseded and overshadowed by the formal
associations, resumed their functions. They recommenced to provide the
mutual aid so important to new townsmen and to migrant groups coming
into town.

The oldest voluntary association among Europeans in Ouagadougou is the Amicale des Bretons. This group, composed of natives of Brittany, meets at least one Sunday a month for dinner and supports those of its members in need of material or spiritual help. However, the Bretons and their association are not taken seriously by the other Europeans in town, who accuse them of "regionalism." They declare that "wherever two or three Bretons are gathered together they found an *Amicale*."

The local branch of La Croix Rouge Francaise, founded in 1955, was the only voluntary association in colonial Ouagadougou that was jointly patronized by the important Africans and Europeans in the town. A report of June 1956 stated that there were 400 members of the group in the capital and that it had started to perform charitable work. During the Christmas season of 1956, Madame Bourges, wife of the governor of the colony, led a group of women in the distribution of kola nuts, cakes, and cigarettes to 340 ill persons in the hospital; and on January 12, 1957, the same group distributed 200 garments to small children at social centers around town. The Red Cross also inaugurated a number of new social service projects in the town. It organized festivities for children on Mother's Day; established a nursery for the children of working mothers, especially the children of African and European elite women; and, in 1958, began the first courses in lifesaving and first aid in the town.

With independence, the control of La Croix Rouge Voltaique passed into local hands, but Europeans continued to be members of it. The only problem then was that competition for the leadership of the organization arose between the wives of the traditional elite and those of the modern elite. The modern elites won, and the Red Cross did not suffer the fate of those organizations that had been created for political action. As a matter of fact, the Red Cross was declared a national utility and as such beyond the boundaries of partisan politics; at least this was the hope. By 1964, the Red Cross was still functioning in Ouagadougou and had established working relationships with foreign groups from which it received financial aid. For example, at a meeting of its national council on October 22, 1964, in Ouagadougou, the organization heard reports on the travels of its members to Europe and acknowledged correspondence and gifts from Red Cross groups in West Germany, Canada, Poland, Finland, the Netherlands, the United States, and Japan. The members also planned elections for a committee for the Junior Red Cross within the parent organization. Today it is still viable and performed valuable services for the Voltaiques during the recent Sahelian drought.

The Lions Club of Ouagadougou was founded after independence and, like the Red Cross, has the support of the government. In 1964 its twenty-one members included such influential Europeans as the manager of the Banque de l'Afrique Occidentale (B.A.O.) and a number of Africans, such as an important professor at the *Lycée*, the former secretary of defense of the African and Malagasy Union, and the mayor of the town. This organization, whose work includes giving charity to such institutions in the organization, whose work includes giving charity to such institutions in the town as the hospitals and social centers, has in the past been chaired by both Africans and Europeans. The Ouagadougou group has established contact with the parent organization in the United States, and Voltaique visitors to America have been received by the locals here. The only problem facing the local branch in Ouagadougou is a financial one, since many of the African members find it difficult to make as substantial contributions to charity as their European opposite numbers.

CONCLUSION

The character of the rise and fall of voluntary associations in Ouagadougou raises some interesting questions about their nature and evolution in urban Africa. In Ouagadougou, at least, most of these organizations did not emerge for social purposes. They functioned more as political mobilizers than as adaptive mechanisms for social and cultural change. True, the *nam* and the *gumbe* served traditional modernizing youths in Ouagadougou in the manner described by Little.[40] However, voluntary associations did not proliferate until elite Africans saw the opportunity of achieving ultimate political and economic power within the town. Once this had been achieved, the voluntary associations either withered away or were banned. Even the Red Cross and the Lions Club of Ouagadougou now operate under the aegis of the political elite.

The reasons the people of Ouagadougou used "sociological" types of voluntary association for political action rather than create strictly "political" clubs or parties to do so are largely historical *and* political. The French colonial administration, like its British, Belgium, and Portuguese counterparts, feared the creation of voluntary organizations and proscribed those not strictly limited to "harmless" or nonpolitical activities. In fact, if the educated Africans wished to form any kind of voluntary association, they had to submit a statement of its goals and a list of its officials to the colonial political affairs officer. If the association was approved, it sometimes received a subsidy; if it was not, its organizers were viewed with suspicion and penalized. Thus, everyone knew that the obligation to register voluntary associations was dangerous, since it gave the government valuable information about the aims and activities of their members.

When, in 1946, the first civil servants did organize a voluntary association, they chose the organizing formula least threatening to the French. They emphasized the overt "social" goals of their organization while covering their political objectives. One suspects that both the Africans and the Europeans chose to ignore the truism that in a colonial situation any type of organization is potentially subversive. The idea, however was to avoid trouble if possible, and therefore the "socio-logical" formula was used. The origin of the formula itself is obscure. It may have diffused to Ouagadougou from British territories through the intermediary of Togolese civil servants, many of whom had been educated in the Gold Coast. Once the formula had been used, it was retained long after there was any need for caution.

I suspect it was the presence of voluntary associations among the urban African elite class that stimulated their use among lesser-educated immigrants. The latter had initially been helped by their kinsmen and headmen, but in time they too desired to shed the cocoon of the traditional system. They had become wise in the ways of the gown and of the Europeans, and they needed vehicles for their new needs and aspirations. They adopted the model of the voluntary association, but having different needs from their educated brothers, initially at any rate, they used the associations for different functions. Some migrants certainly used the associations as adaptive mechanisms for coping with the exigencies of modern urban life, even when modern life involved morally questionable activities.[41] But as Banton shows in his data on the Temne in Freetown, Sierra Leone, migrants used the voluntary associations as much for sociological support as for intragroup and intergroup economic and political competition.[42] It is certainly the evolution of these latter and later functions that Hodgkin had in mind when he declared that the voluntary associations gave "an important minority valuable experience of modern forms of administration—the keeping of minutes and accounts and correspondence, the techniques of propaganda and diplomacy."[43] It may also be added that they enabled their members to compete more effectively in the multiethnic urban arenas. Here groups faced with competition from others used every vehicle at their disposal to gain an advantage. Ethnic groups used the voluntary associations because the ethnic systems developed during the colonial period had not yet given way to class alignments.[44] When, as in Ouagadougou, the emerging classes are finally established, the lions of the world might well unite against the sheep. Yet it is also possible that the emergence of different political economies might obviate the need for, or change the very structure and function of, voluntary associations.

NOTES

1. Daryll Forde, "The Governmental Roles of Associations Among the Yakö," *Africa* 31, no. 4 (October 1961).

2. Victor C. Uchendu, *Igbo of Southeast Nigeria* (New York: Holt and Rinehart, 1965).

3. Peter Morton-Williams, "The Yoruba Ogboni Cult in Oyo," *Africa* 30, no. 4 (October 1960): 362-374.

4. Melville J. Herskovits, *Dahomey: An Ancient West African Kingdom*, 2 vols. (New York: J. J. Augustin, 1938), 1: 72.

5. Kenneth Little, *The Mende of Sierra Leone* (London: Routledge and Kegan Paul, Ltd., 1951), p. 184; see also Michael Banton, *West African City* (London: Oxford University Press, 1957), p. 21.

6. Elliott P. Skinner, "West African Economic Systems," in *Peoples and Cultures of Africa*, ed. Elliott P. Skinner (New York: Doubleday, 1973), pp. 77-97.

7. Oliver C. Cox, *Caste, Class and Race, A Study in Social Dynamics* (1948; reprint ed. New York: Modern Readers Paperback, 1970), p. 317.

8. Colin M. Turnbull, *The Forest People: A Study of the Pygmies of the Congo* (New York: Simon Schuster, 1962); see also Elizabeth Marshall Thomas, *The Harmless People* (New York: Vintage Books, 1965).

9. Cox, *Caste, Class and Race*, pp. 317-318.

10. Fredrik Barth, ed., *Ethnic Groups and Boundaries: The Social Organization of Cultural Difference* (Boston: Little, Brown and Company, 1969).

11. M. G. Smith, "Institutional and Political Conditions of Pluralism," in *Pluralism in Africa*, ed. Leo Kuper and M. G. Smith (Berkeley: University of California Press, 1971), pp. 27-65.

12. H. Kuper, *An African Aristocracy* (London: Oxford University Press, 1947), p. 17.

13. E. E. Evans-Pritchard, "The Zande State," *Journal of the Royal Anthropological Institute*, vol. 93, pt. 1 (1963), p. 136.

14. Jacques Maquet, *The Premise of Inequality in Ruanda* (London: Oxford University Press, 1961); see also Jean Rouch, *Les Songhay* (Paris: Presses Universitaires de France, 1954).

15. Elliott P. Skinner, *The Mossi of the Upper Volta* (Stanford, Calif.: Stanford University Press, 1964).

16. A. A. Dim Delobson, *L'Empire du Mogho-Naba Coutumes des Mossi de la Haute-Volta* (Paris: Domat-Montchrestien, 1932), p. 90.

17. Henry Barth, *Travels and Discoveries in North Central Africa*, 3 vols. (New York: Harper and Bros., 1859), 3: 663.

18. G. S. P. Freeman-Grenvill, *The Medieval History of the Coast of Tanganyika* (London, 1962), p. 90.

19. Elliott P. Skinner, "Group Dynamics in the Politics of Changing Societies: The Problem of 'Tribal' Politics in Africa," *American Ethnological Society, Proceedings of 1967 Annual Spring Meeting* (Seattle: University of Washington Press, 1967).

20. Elliott P. Skinner, "Strangers in West African Societies," *Africa* 33, no. 4 (October 1963): 307-320.

21. Clifford Geertz, ed., *Old Societies and New States* (Glencoe, Ill.: Free Press, 1963), p. 105.

22. William A. Shack, "Notes on Voluntary Associations and Urbanization in Africa, with Special Reference to Addis Ababa, Ethiopia," in Niara Sudarkasa, ed. "Migrants and Strangers in Africa," *African Urban Notes*, ser. B., no. 1 (Winter 1974-75).

23. Michael Banton, "The Restructuring of Social Relationships," in *Social Change in Modern Africa*, ed. Aidan Southall (London: Oxford University Press, 1961) pp. 113-125.

24. J. Clyde Mitchell, *The Kalela Dance*, Rhodes-Livingstone paper 27

(Manchester: Manchester University Press, 1956).

25. James S. Colemen, "The Role of Tribal Associations in Nigeria," *Proceedings Annual Conference, West African Institute of Social Economic Research* (Ibadan, 1952. See also Kenneth L. Little, *West African Urbanization: A Study of Voluntary Associations in Social Change* (Cambridge: Cambridge University Press, 1965); and Shack, "Voluntary Associations and Urbanization."

26. The educated Africans who emerged in the developing urban centers of colonial Africa were quite different from their illiterate and often rural-born parents and kinsmen. They were educated by Europeans, frequently missionaries, and participated in modern social organization in their schools. Moreover, as "schoolboys" and the emerging African elite, they were outside the sociopolitical system of the chiefs and elders. They thus had the knowledge and opportunity to view colonial rule in a new and almost "European" light, and they were the ones to challenge foreign domination. In order to do so, however, they needed to organize.

27. David Kimble, *A Political History of Ghana* (Oxford: Clarendon Press, 1963), p. 146. Emphasis added.

28. Ibid., p. 147.

29. J. Merle-Davis, *Modern Industry and the African* (London: Frank Cass, 1967), pp. 86-87.

30. A. L. Epstein, *Politics in an Urban Community* (Manchester: Manchester University Press, 1958), p. 47.

31. Ibid., pp. 82-85.

32. Skinner, *The Mossi of the Upper Volta*, p. 155.

33. Elliott P. Skinner, *African Urban Life: The Transformation of Ouagadougou* (Princeton, N. J.: Princeton University Press, 1974).

34. Claude Meillassoux, *Urbanization of an African Community* (Seattle: University of Washington Press, 1968), pp. 116-130.

35. Edward Mortimer, *France and the Africans 1944-1960* (London: Faber and Faber, 1969), p. 258.

36. *Bulletin Quotidien d'Information* (Distribué par le Service de l'Information de la République de Haute Volta, June 8, 1963), p. 9.

37. Ibid., April 15, 1959, p. 10.

38. Ibid., July 12, 1957, p. 10.

39. Ibid., January 6, 1962, p. 10.

40. Little, *West African Urbanization.*

41. Thomas Hodgkin, *Nationalism in Colonial Africa* (New York: New York University Press, 1956), p. 90.

42. Banton, "The Restructuring of Social Relationships," pp. 113-125.

43. Hodgkin, *Nationalism in Colonial Africa*, p. 84.

44. Skinner, *The Mossi of the Upper Volta*, p. 194.

11
The Celebration of Ethnicity: A "Tribal Fight" in a Namibian Mine Compound

Robert Gordon

In this chapter, I seek to analyze a situation in which ethnic identity was apparently important, namely, during a "tribal fight" that occurred at a Namibian mine compound.[1] I was a somewhat reluctant participant in this fight, which was between the Kavango contract workers, who constituted the vast majority resident at the mine compound, and the Ovambo minority. It started one Saturday afternoon when most of the workers were off duty and drinking and continued until Monday morning, when the exasperated white officials were finally able to control the situation and the workers returned to work. During the fight, apart from assaulting members of the opposing ethnic group, the Kavango workers literally went on a rampage and destroyed, looted, or maliciously hid the possessions of their Ovambo coworkers.

This analysis is based on Gluckman's argument that "if we treat the mine and the tribe as parts of a single field, we see that within all the areas it operates, capitalist enterprise produces similar results. . . . What occurs in each area is affected by local variation, and the variant aspects also have to be studied."[2] It is hoped, then, that this analysis will contribute toward an understanding of the widely reported recent spate of "tribal fights" in southern African mining compounds.

These "tribal fights" raise a number of important questions and, apart from underlining Southall's recent assertion that ethnicity is the paramount problem facing contemporary Africa and Lewis's view that ethnocentrism is an inherent phenomenon,[3] would seriously appear to challenge Gluckman's classic statement that "the African newly arrived from his rural home to work in a mine, is first of all a miner (and possibly resembles miners everywhere). Secondarily he is a tribesman; and his adherence to tribalism has to be interpreted in an urban setting."[4]

Gluckman's stress on the preeminence of class structure over ethnicity seems to have been forgotten by his successors, who apparently prefer to relate the genesis of ethnic boundaries to factors affecting the competition

213

for environmental resources in a rational model.[5] Such an approach is tempting in this case, especially since a short while before the outburst, or "fight," some Kavango spokesmen approached the compound manager with the complaint that they were being pushed out of their jobs by the Ovambo workers. However, I find—and will show—that such an approach does not make the data at hand "intelligible." I will opt instead for a symbolic interactionist perspective.[6]

Following Gluckman, I want to show how, by focusing on the interface between workers (blacks) and management (whites), such an emphasis can assist us in understanding ethnic phenomena such as "tribal fights." Two theses are presented and argued. First, in terms of the individual and his awareness context, ethnicity achieves its collective importance only through negation. It is the denial of the moral worth of the other that is the raison d'être of ethnicity. Second, I will contend that the "tribal fight" was not a breach of normal and regular behavior; indeed, it was "expected" and wholly consistent with the context and somewhat peculiar class structure at the mine. In dialectic fashion, it served to increase worker consciousness. My use of dialectics in this case must be contrasted with the Marxian view, which would treat such outbursts as manifestations of "false consciousness" that hinder the emergence of worker consciousness by serving as a mechanism to channel off discontent. My line of argument is certainly not novel. Gluckman used it in 1955 to explain behavior in traditional society.[7] I am simply modifying and applying it to a different context. The implication of these theses is that I believe that the *ethnic* factor and its attendant problems have often been overstressed and too narrowly analyzed—all to the detriment of a deeper and more relevant analysis.

Organizationally, the mine was typical of those in white-ruled southern Africa. Since descriptions of such mines are reasonably common, this section will be brief and will focus specifically on how the migrants see the situation. The mine's labor force was divided into two distinct strata based on race. The upper stratum consisted of a small number of whites who were "permanent" and who filled the managerial and supervisory positions. The subordinate and larger stratum consisted of black short-term migrant contract workers who did all the manual, unskilled, and semiskilled work under close white supervision in what Gouldner has aptly termed a "punishment centered bureaucracy." Their numbers were large, since the mine was a labor-intensive operation. Built into every organization is a definition of how that organization views its participants. In the case of the mine, whites were unmistakably "superior" and blacks "inferior." Even off the job, blacks were forced to lead a tightly organized "round of life" in the compound, since whites and management believed that idle blacks only cause "trouble." Similarly, the mine provided for all the needs of the workers (except sex). Thus bureaucratically they were treated as anonymous "blocks of men," who by the nature of their station at the mine were easily replaceable.

This, in short, is the world the migrants have to contend with when they arrive at the mine in search of money. It is not surprising then, that one of the first tasks such migrants set themselves is to discover what the *Weta Zomina*—the "law of the mine"—is. This "law," or "recipe knowledge" as Schutz would call it, provides the migrant with information about the behavior of the whites, what the opportunities to make money are, and conditions in general. Discovering and formalizing this "law" is an ongoing process stretching right through the migrant's stay at the mine. It also provides the migrant with informal guidelines on how to stay out of "trouble" with whites and fellow workers and how to "make money" in the "informal sector." This "law" is an ideology that is a response to a problematic situation and a felt lack of needed information. Not only does it provide a map of "problematic social reality and motives for the creation of collective conscience,"[8] but the constantly evolving theory of the "law" also explains their situation to the migrants and provides them with moral justifications for their actions.

Reduced to its barest essentials, this "law" states that the blacks are oppressed by whites and that the whites are united in oppressing them. This oppression is possible because whites control the access to money, which everyone needs in order to survive. Whites are believed to be earning salaries in excess of R2,000 a month and are supposedly aware of the difficulties caused by the low wages blacks get. Oppression results from the fact that "whites think they will become poorer if they don't oppress blacks." It is based on fear, since "whites know blacks can do their jobs if given the opportunity."

As a result of their powerful position, whites treat blacks like "dirt": "They won't let us have women here because they think we are animals." "By calling me a 'kaffir' [a derogatory term], they mean I do not deserve respect and that they can behave any way they want with me." White superiority is expressed in the pattern of forced deference and the superior social facilities to which they have access. The power of the whites causes the blacks to view the whites as dishonest, ambitious, untrustworthy, fearsome, and above all unpredictable.

Promotion and demotion are seen to be based on the arbitrary whims of the whites, not on technical expertise or seniority. Few blacks have a comprehensive knowledge of the formal mine regulations pertaining to their behavior. In part this is because of the high labor turnover rate and because they are defined as "inferior" by the whites, who consequently regard it as unnecessary to explain the regulations comprehensively. There are also "informal house rules" that individual whites make and that vary from white to white; this further increases white unpredictability. The "inferior" status assigned to blacks also results in blacks often being blamed for errors their white supervisors make. The unpredictability of whites is emphasized through their use of coercive power. It is commonly accepted that if a white tells a black to do a task, the black must do it. It is immaterial whether the white is the black's supervisor or for that matter, an employee of the mine. Complaining to the white compound manager about such maltreatment does not help, since the migrants believe whites will not act against their white brothers. Whites will not fire other whites because they believe that it is easier to replace black employees than whites. In addition, should a black complain, he runs the risk of being labelled an "agitator" and being victimized. In a situation such as this, then, blacks have to be sensitive to the feelings, moods, and behavior of whites, for white behavior must be taken seriously even if it means the black must play a degrading and alienating role.

Indeterminancy is a characteristic of social life everywhere. In total institutions such as the mine, it is heightened by the distribution of power between black and white. Blacks see the mine's social order as dangerous, unpredictable, and almost chaotic. Consequently, they are forced to become "phenomenologists," not of their own daily life, but of their significant others—the whites and their brothers.[9]

SALVAGING SELF-ESTEEM

The black workers, by their own definition, are oppressed. The "law," however, also provides a framework within which they can react to the "demeaning" definition of themselves by attempting to reassert their self-esteem. This they do in a number of ways. For example, they deny the moral superiority of whites by emphasizing that whites are actually "barbarians," since powerful people do not act the way they do. They claim that while whites might know and control the material world, they are ignorant about the spiritual world; indeed, black diviners and herbalists are often claimed to have had successes in healing people where white expertise has failed. More especially, migrant workers see themselves as enduring this humiliating situation because they are *men* trying to care for their families. The importance of being a "family" man is stressed, and prestige is gained by maintaining and supporting one's family while at work through remittances. Yet paradoxically this leads to increased feelings of insecurity, because it leads the workers to worry about their homes. The epitome of manliness is, of course, standing up to the whites and "telling them like it is," but this involves a high risk of assault, and invariably the possibility of being fired is very real. A more common, but less prestigious, strategy is to simply resign from the mine and return home. This has the advantage of reducing the possibility of antagonizing the whites and keeping the door open for a possible return to the mine and a better white supervisor at a later date. A last strategy to keep one's self-esteem can be mentioned. It is being "clever" and entails the skills of exploiting the whites, of beating them at their own game and thus making fools of them, even while the whites are unaware that this is in fact happening.[10]

The blacks are caught in a merciless dialectic of being forced to earn money in a situation that demeans and humiliates them. In a situation such as this, skills at the social graces, not technical competence, determine how much money one will make. If a worker wants to make money "fast" and with a degree of security, the trick is to be labeled a "good boy" by the white supervisors and to establish ties of patronage with whites. White patrons are important, for they grant small favors, overlook irregularities, and will possibly support blacks' complaints or requests to the compound manager: "It was not enough to have a good case. ... One's European was expected to listen (and support his client). A European demanding justice for his African was not encouraging rebellion; he was upholding the generally accepted relationship."[11] But this strategy increases the tension in the dialectic, for it means that the black worker has to ingratiate himself with the white and thus humiliate himself even more. More importantly, a black who establishes patronage ties with whites runs the risk of being labeled a "sell-out," "informer," or "stooge" by his fellow blacks. To avoid such ostracism requires considerable maneuvering and social investment to demonstrate one's good faith. He has to demonstrate that he still subscribes to the ethic of brotherhood.

BROTHERHOOD: BLACK WORKER SOLIDARITY

It is a widely reported sociological finding that workers at the bottom of the organizational pyramid build up a self-defense organization to resist management. An English apprentice provides the following insight into this phenomenon:

As soon as [the supervisor] had gone the workers near me extended the unforgettable claustrophobic comradeship of the factory. It is a friendship generated of common experience, common income and common worktasks. Out of this shared pattern of experience grows a common culture of the work-place. And like other cultures it can never be fully understood by the outsider. ... On the first morning at work I began to learn all the expected patterns of response, all the rewards and reactions. ... I quickly learned the harsh language of aggressive friendship; the need to identify myself with the workgroup in opposition to all forms of authority from the chargehand up. Nothing must be allowed to threaten the cohesion of the workers, for only through this "sticking together" could we face the problems facing us.[12]

Comparative studies indicate that of all the industries, worker solidarity is most intense in mining because of the dangers, relative isolation, abnormal working hours, and uncertainties involved.[13] The extensive quotation serves as an analogy, because it is difficult to describe the intensity of the solidarity among the migrant workers at the mine. The additional facts that the blacks, to a far greater degree than elsewhere, do not accept the authority of their white supervisors as legitimate and that they have fewer rights than mine workers in other parts of the world, combine to provide an exceedingly tight worker cohesion irrespective of differences in origin or background. With so many migrants in an identical situation—being treated alike, having the same problems, and facing the same dangers—it is easy for them to identify sympathetically with one another. In such a situation, occupational and wealth differences are enveloped by brotherhood, since, as one migrant put it, "One can be rich one day and poor the next," and no matter how rich one is, ultimately one is "still a Kaffir." Survival in such an oppressive situation is seen in racial terms. Brotherhood transcends occupational groups, the rich and the poor, and ethnic differences, and unites all blacks at the mine. Worker consciousness, or brotherhood, is thus of a populist variety. Not only does brotherhood receive its main impetus as a reaction to oppression, but whites are also a direct stimulus. Blacks realize that "if you make trouble in the compound, and the compound manager finds out, you will be fired." Brotherhood is essential in the compound, where the lack of privacy and ready possibilities for theft and assault call for a strong moral order.

Brotherhood is needed for dealing not only with problems of "internal" order among blacks living in the compound; it is also an essential tool for confronting whites. According to Weber, restricting production is a natural part of the workers' unending struggle for a fair wage. Brotherhood provides the medium for the articulation of the group norms that regulate and restrict output. It prevents competition at the workplace, competition that would disturb established interpersonal relationships, and contributes to the stability of relations among blacks in the workgroup. It alleviates white pressure because it is believed that if one worker works hard, the whites will expect and pressure other workers to work harder.[14] Blacks at the mine have also long realized the importance of numbers. Whites are more impressed and are likely to investigate a complaint about a white supervisor's behavior if the whole gang walks off and lodges a complaint than if an individual complains. Group action also lessens the possibility of being labeled an "agitator." More important, as Alverson notes, the secret of the blacks' survival in the white industrial world lies in keeping the whites as ignorant as possible about the blacks.[15] Brotherhood delineates a distinct moral universe that specifically excludes all white and other nonbrothers. It provides the setting for a "private" culture in which blacks can be "themselves," masters of their "own" actions and responsible for them. The success of this screen is demonstrated by the mine security officer's continual stress on the need for cultivating black informers and his apparent lack of success in so doing.

In terms of ideal types, the etiquette of brotherhood is the antithesis of black-white etiquette. It emphasizes respect, trust, consultation, and dignity. Social order, certainty, and security are created by helping one's fellow migrant and thus deliberately creating a network of obligations through extensive borrowing and lending. In terms of social relationships, brotherhood is a series of interlocking networks and potential networks. Since deeds and not words are important, brotherhood is demonstrated through transactions involving material goods. The major form of purchasing material goods is by COD parcels from mail order houses; when the parcels arrive, the migrant invariably is forced to borrow cash from brothers, since COD parcels seem to have a propensity for arriving before payday. Money earned is "farmed out" in the form of loans or simply given to one's brothers for safekeeping, for to have large sums of money in the compound is to invite theft. On the other hand, many migrants distrust the white banks and, indeed, given their own isolation, find it inconvenient to place their earnings in a bank. As stated earlier, earnings are also often sent home with returning migrants. Such migrants obviously have to be trusted brothers, and this entails an extensive chain of reciprocity.

A migrant known to have a large sum of money has to be able to present an acceptable account for it. Large sums of money are attributable to "luck," which in turn is the result of witchcraft or "selling out" to the whites. Such people are ostracized, for not only does such behavior foster uncertainty (one never knows whom such a worker will inform on or what effects the witchcraft will have on others), but it is also a manifestation of the fact that the black is a bad brother; the money demonstrates that he was not using his wealth to help his fellow workers. Thus, in order to make "extra" money through patronage, pilfering, or other illicit activities, one needs the reputation of being a good brother. One has to be "clever." This is necessary not only for a post hoc account, but also to initiate and maintain these activities, for success invariably depends on the cooperation of one's fellow workers. Wealth is gained not as an individual but as a star of the team. Workers are forced to identify and demonstrate brotherhood, since noncompliance signals that the individual's loyalties do not lie with his fellow workers but rather with himself or the whites, and that he does not wish for, or need, their social and psychological support.

The most common ritual of brotherhood is beer drinking. Beer drinking is a communal affair that takes place at one of the illicit *shebeens.* The very fact that one is participating in an act of which whites disapprove and regard as illicit is a symbol of brotherhood. Drinking heavily is a means of demonstrating one's manliness, and drinking in company asserts brotherliness. I have already mentioned the importance of material transactions as manifestations of brotherhood. Buying beer and sharing it is one of the few, but most common, opportunities available for engaging in the material transactions that are so important to brotherhood. Drinking is also an important mechanism of conflict resolution. Where there is tension or disagreement between brothers, it is allowed to surface during drunken states in the form of insults and assaults. Being drunk means that such activities can be attributed to inebriation and not to the individual. Such drunken assaults also mean that other workers are not forced into the act of publicly taking sides and thus potentially dividing the workers into factions. Lastly, since drunken fights occur at communal drinking parties, other drinkers are able to stop the fight before serious injury results.

THE IMPORTANCE OF ETHNICITY

Having briefly examined the nature and bases of black worker solidarity, we must now examine the divisive potential of ethnicity. There is, of course, no such thing as an Ovambo, Kavango, and Angolan tribe. These terms refer to geographic entities, the boundaries of which were arbitrarily and artificially created by the white colonialists. Workers originally engaged by labor bureaus at Ondangwa were labeled "Ovambo," and the vast majority at the mine, who were engaged in Rundu, were labeled "Kavango." Within the Kavango area, there is a rich diversity of what are popularly known as "tribes," viz., the Kuangari, Hambukushu, Sambiu, Djiruku, and Mbunza; but the cultural and social distinctiveness of these "tribes" appears to be more mythical than real. For example, in the early 1960s, a survey of 364 household heads in the Sambiu tribal area showed eighteen different "ethnic" groups residing there: 98 Sambiu, 156 from the four other Kavango tribes, and 110 from thirteen "foreign" ethnic groups.[16] This diversity was also found at the mine, and I suggest that this very richness makes ethnicity insignificant. Indeed the lingua franca in the compound was *Kuanyama*, an Ovambo language that most workers had learned on previous contracts. Some "Kavango" workers openly stated that they were from Angola and that their tribal affiliation was Angolan Chokwe, Nyemba, Chimbundu, Mwuunkundi, and Kuanyama. In addition, the Ovambo at the mine were represented by workers from the Ongandjera, Kuanyama, Ondonga, and Ombalantu "tribes." It is difficult even to try and conceptualize a distinct cultural core for the Kavango and Ovambo groups. There is, for example, a demonstrated linguistic affinity between the largest Kavango "tribe," the Kuangari, and the largest Ovambo "tribe," the Kuanyama. The Kuangari are also said to draw their chiefs and headmen from an Ovambo family. Even the Odendaal Commission Report, which proposed the *bantustan* concept for the territory, noted the similarities and ties between Ovamboland and Kavangoland and suggested that at a later stage it would be advisable to merge these two areas into a single homeland. The evidence thus suggests that there was little foundation to the official South African police explanation that the unrest was due to traditional animosity between the Ovambo and the Kavango.

In maintaining social order in the compound, I have suggested that ethnicity is not important. The situation and its concomitant ideology of brotherhood militate against its becoming an important issue. Indeed, even the blacks have difficulty in physically differentiating among the ethnic groups. When individuals were encountered during the "tribal fight," they were asked first of all, *Watse nasie is jy* ("What nation are you?"). This is not to say that blacks are ignorant of the ethnic diversity in the compound. On the contrary, migrants are aware of the variety, but their ethnic stereotypes are not as uniform or consistent as they are with reference to the whites. Individual traits and variations are recognized and accepted to such an extent that single stereotypes of ethnic groups as a uniting force did not arise.

But if, as suggested, ethnic identity is of negligible importance, why did ethnicity become important enough to cause a "tribal" fight? To understand this one must examine the stereotypes the whites have of the blacks, for as Berger and Luckmann point out, "he who has the bigger stick has the better chance of imposing his definition of reality."[17] While the organizational ideology of the mine management does not cater to ethnic differentiation among blacks, the whites "supertribalized" the black workers into the following groups: republic "boys," "Ovambo boys," "Kavango boys," "Angolan boys," and "town boys." Mitchell's argument that supertribalism arises because people try to simplify a complex situation does not account for these ethnic labels, because whites are not aware of the ethnic diversity encompassed by the labels they impose upon the situation.[18]

Labels are not simply labels. Each one carries a specific moral content that varies from one to the other. Labels are evaluative: to label a person a "Padre" is not simply to specify an occupation, but to imply what sort of person he is, how he will behave, and why. Labels, including ethnic labels, help to explain why things are the way they are.

These ethnic labels are part of the white theory of black behavior and help to explain and justify why the "job" is not going as well as it should. The moral contents of these ethnic labels read more or less as follows: republic "boys" are the best. Allegedly they know how to work and if one beats them up, they will understand and be back at work the next morning without complaint. But unfortunately for the whites, there were no republic "boys" at the mine. Angolans, or "Porries" (Portuguese), are believed to be dumb but well-trained because the Portuguese are reputed to have been particularly brutal in their treatment of blacks. Because of their experience of Portuguese brutality, Angolans are regarded as being good boss boys because they are not afraid to beat up fellow workers. The mine had a handful of blacks who claimed to be Angolan. Ovambo were viewed as good workers and quick to learn, but extremely troublesome: "Just touch them and they go and lodge a complaint." The Ovambo, as was noted, were a small minority at the mine. The Kavango were reputedly the worst workers the whites had come across: they are lazy, stupid, unreliable drunkards. It goes without saying that workers from the Kavango formed the majority at the mine.

The relationship between the moral content of the ethnic identity of blacks and the number of blacks being supervised by whites should be noted. At another Namibian mine where there were more Ovambo than Kavango workers, the Ovambo were labeled lazy, unreliable, and troublesome; while the Kavango were believed to be good, reliable, and friendly, but inclined to drink too much. It is important to stress the *but* in the moral label applied to the smaller groups of workers. The *but* provides an escape hatch for whites to account for varied black behavior. Lastly, at the South African gold mines where the Kavango workers are a very small minority, the whites there believed that they were extremely good workers. Thus, large groups of blacks were "morally bad," smaller groups were "morally better," and the very small groups were the "best." The label thus did not depend on the blacks' ethnic identity. Many whites actively vocalized these moral contents of ethnic labels in their dealings with blacks in an effort to increase productivity. They felt that by insulting blacks in ethnic terms, they would stimulate "ethnic pride," which would in turn produce ethnic competition and thus stimulate productivity.

This is the perspective from which the Kavango complaint that the Ovambo were pushing them out of their jobs must be examined. In this regard, the compound manager's position needs to be examined more fully; after all, he is responsible for the hiring, firing, promotion, and demotion of the black workers. His views carry weight, and, more important, he did not articulate the moral contents of the ethnic labels in the same way as the other whites at the mine did.

The compound manager held an important position at the mine. This was recognized in the organizational hierarchy, where he was classified as a senior official with all the privileges and rewards attendant to this station. Of all the senior officials at the mine, his was the only position *not* based on technical qualifications. His position was based on subjectively defined "experience." Typically, compound managers are what Malinowski has aptly termed "practical men": they eschew paperwork and even reading. Traditionally, they gain their "experience" as ex-policemen, Department of Bantu Administration officials, or more uncommonly as youths who have grown up in the black areas and are fluent in one or two of the Bantu languages (which need not necessarily be applicable to the particular mine compound which they manage. This was the case at the mine.) As a rough rule, whites and the organization believe that the greater a white's fluency in *fanagalo*, the pidgin lingua franca used for black-white communication, the greater his expertise of the black workers. The compound manager went along with the sentiments and image of the black workers, and this influenced his policies. The last thing a compound manager wants is "troublesome" black workers. If there are "troublesome" workers, the compound manager's proclaimed expertise of "knowing the Bantu and how to deal with them" will be doubted by his white colleagues. He would thus lose the moral justification for his well-paid, powerful, and privileged position at the mine. The mine's labor force was thus recruited from the less troublesome Kavango area when the mine commenced operation, and the compound manager actively opposed the recruitment or promotion of the Ovambo workers. To do that would imply that he had made a "mistake" and would have brought his mystical expertise of "knowing the Bantu" into question.[19] But the compound manager was also under organizational pressure to have his labor complements (the planned number of blacks) up to strength, a situation that was aggravated by the high and unpredictable rate of labor turnover. Consequently, a small number of Ovambo workers who had finished or resigned their contracts in the vicinity of the mine and who had come looking for work were engaged (often illegally). Since they were hired as "stopgaps," they were placed in the job categories that had the highest labor turnover: i.e., the most unpopular jobs. In a few cases, Ovambo who had followed their white supervisors from other Namibian mines to the project were also engaged. In such cases, the compound manager was placed under white peer group pressure to engage these workers because "by their actions they had proved that they were 'good boys'."

Ovambo penetration of the Kavango enclave was minimal, as the following figures illustrate:

	Ovambo	Kavango	Others	Total
January	14	251	0	265
April	33	340	4	377
July	48	341	4	393
September	43	411	2	456

Source: compound manager's records

The Kavango claim that their jobs were being taken over by the Ovambo was thus more of a reaction to a white ideological claim made at the workplace than an action based on reality.

Since the whites take the matter of ethnic identity seriously, the black workers are forced to do likewise. As indicated, getting along successfully on a job does not require technical competence per se (especially since most of the work is manual or semiskilled); rather, it depends on whether the white supervisor defines the black as a "good boy" or a "bad boy." Since ethnic identity is important in this definition, it is relatively easy for Kavango to "pass" as Ovambo. Whites simply cannot differentiate among the ethnic groups at the mine, either physically, linguistically, or socially. The ethnic label that whites apply to blacks is based on what the blacks tell them. It is thus a simple matter for a black to switch his ethnic identity on the black-white interface; for the ethic of brotherhood prohibits fellow workers from informing on such identity switches. Even if confronted by a white, the black can simply claim that his father belonged to his new ethnic group.[20]

Having developed the analysis so far, it is now necessary to complete it by examining the outburst itself in greater detail, especially since, despite the various strong social controls such as brotherhood, fights between individuals of different ethnic groups were a frequent occurrence yet did not provoke "tribal" fights.

THE HOSTILE OUTBURST AS A TIME OUT

The "Trigger Event"

What activated the hostile belief that provided the moral justification for the outburst? Here it is necessary to look more closely at one of the black eyewitness reports.

After the initial fight between an "Ovambo" and a "Kavango," the Kavangos were surprised by the fighting and did not know the reason for it. When they saw how their friend had been injured, they decided to take revenge. But they had not as yet decided how to do it. They reported the case to the security officer, who, however, was otherwise occupied. They also summoned the part-time ambulance officer, who was irritated at having his Saturday afternoon sleep disturbed. The ambulance officer was also one of the most vocal advocates of "Ovambo virtues" and "Kavango badness" at the mine. He went to the compound to treat the injured Kavango, who was surrounded by concerned brothers. The eyewitness report continues: "He asked . . . what the problem was. I replied, 'There is fighting. The Ovambo want to fight against us and some of us have been injured.' He said, 'Show me where the injured people are. Where are the Ovambos who are fighting against you? This is nonsensical. You are a bunch of women complaining about nothing. I appreciated your abilities. I thought that you were really people. Take me to the Ovambos.' We refused. 'Furthermore you are cowards,' he continued. After this the Kavango began to retaliate."

As with other interethnic fights at the mines, insults had been traded between the "Kavangos" and the "Ovambos," but the ambulance officer's remarks served to twist the knife in the injured self-respect of the Kavango. Insults by whites must of necessity be tolerated, but the same insults by one's fellow blacks, brothers in the same situation, people from whom one expects compassion, sympathy, and understanding, simply could not be tolerated. The Kavango were simply trying to deal with an objection (which they believed was impossible to settle at an individual level) by employing the power of numbers.

Whenever the Kavango were asked what the fighting was about, the answers were always similar: the "Ovambo" had insulted them by calling them dumb and women. They felt that the Ovambo had done this to provoke them deliberately and that they had to react in order to "prove themselves." In the milieu of the compound, such insults are serious. The basic justification workers use for working in an "oppressive" situation such as the mine is that they are the epitome of men. To admit fear publicly or to ignore such insults is unmanly behavior. In oppressive and alienating circumstances, the ability to be "one's own man," which entails being capable of working the system illicitly, is an important feature. Thus, the insult of dumbness rankles; it implies that one cannot cope with oppression. The "Ovambo" justified their insults by claiming that the Kavango were dumb and womanlike because they were not used to money: after they had saved some money, they spent the rest on drinking. These explanations were quite similar to the white justification of their ethnic labels. The Kavango in turn insulted the "Ovambo" by calling them donkeys. They were donkeys because they worked so long and do not know when to stop and enjoy life. They were donkeys because they are also antisocial, for they are always found in pairs, never in groups and never caring about anybody but themselves.

In analytical terms, these explanations are essentially different strategies for resolving the difficulties implicit in the need to earn money in an alienating workplace. Since the relationship between power and ethnic conflict has been well established, it could be argued that one of these strategies would enable its practitioners to gain greater power. But power is not some monolithic force: it is socially defined, and definitions of power can vary. In this case, power can either be related to security and therefore derived from being a good brother, or it can be related to the ability to be independent from brothers. The contradictions between these two strategies can only be resolved at a mundane level: that of insults, the classical dilemma that inmates face in total institutions.

The Framework

The fight was preceded by heavy drinking. In a recent study of drunken comportment, MacAndrew and Edgerton have convincingly argued that individuals, when inebriated, are allowed to engage in behavior that is otherwise unacceptable. This behavior points out the inadequacies and failings of the social order and follows certain cultural "rules." This phenomenon they term "time out."[21] Similarly, I suggest, the "tribal" fight can be seen as a "time out," both as an extension of the drunken "time out" and as a "time out" in its own right. The behavior before, during, and after the outburst makes "sense" only when the actions labeled "ethnic" or "tribal" are interpreted and evaluated in terms of the brotherhood ethnic, not in terms of ethnicity. The etiquette of brotherhood created a social order and allows social interaction in the compound by blocking potentially disruptive information and behavior. It can become so effective that the truth, as perceived by the compound inmates, might not be allowed to surface.[22] The "tribal fight" allows an occasion to arise in which the truth could come out and be evaluated publicly. The "tribal fight" enabled the inmates to know which brothers they could "really" depend on and what the other brothers "really" thought of them.

The Black-White Interface

The "tribal fight," from the vantage point of the whites, was spontaneous. Not only was it an indication of the strength of the workers' private culture, but it was also a necessity: it had to surprise the whites in order for it to take place. The image of spontaneity and associated emotionality that it provoked gave the blacks license to ignore white requests or suggestions. It also gave whites license to accept such situations philosophically: "It just goes to prove what barbarians they really are." This nonreceptivity to suggestions was, however, highly selective. The whites were ignored, but workers listened to and obeyed their own leaders.

The outburst took place in areas away from the white residential area, which was seen as a neutral or safe zone where black combatants could rest or seek refuge. To take the action into this area would have provoked a white response, which would have forced the action to end and would have had serious repercussions for the blacks.

The outburst ended when both groups felt that if they continued, the opposition groups would have a justifiable case for engaging in socially dangerous acts such as witchcraft. Black leaders then approached the exasperated white officials and suggested that the South African police be called in to keep the peace.

Organizational Features of the Outburst

Although the outburst took the whites by surprise, blacks had been expecting "trouble" for quite a while. Many Ovambo interviewed afterward stated that they had been warned beforehand by Kavango to leave the compound over the weekend as there was to be "trouble": an act of true brotherhood. One Ovambo, when attacked by a group of irate Kavango, was protected by his Kavango roommates. Another Ovambo's possessions were saved by his roommates when a group of Kavango were on the rampage. Two Kavango were surrounded by a gang of Ovambo. The one was assaulted, while the other was left unharmed "because they saw who I was." An Ovambo was assaulted by other Ovambo because they claimed that he was siding with the Kavango. A Kavango had his possessions looted by Kavango allegedly for being in league with the Ovambo. The Ovambo who had adopted independence from the brotherhood strategy fled not to where their brothers might offer them protection and succor, but to their white patrons. Such workers were labeled as "sell-outs" or "lone jackals," and their possessions, even mine-issued blankets, were destroyed with a vengeance.

These examples can be multiplied many times over. They suggest that ethnicity was not so important. The issue was rather one of brotherhood. A worker with many brothers, or socially defined as a good brother, suffered little. On the other hand, even a common ethnic identity was not enough to save a bad brother from attack. Every worker assaulted by a gang, apart from those who received an anonymous shove or two, claimed to be able to identify positively the leader of the gang. Closer probing revealed that a grudge existed between them, a grudge that could be evaluated as a breach of brotherhood. Only single workers were assaulted. Workers in groups with their brothers (people on whom they could rely) were not assaulted. At worst a few stones were hurled at them.

A similar pattern can be discerned in the looting. Workers who had little money lost very little, if any at all. Those with large amounts were the hardest hit. Some Ovambo claimed that where the looters were identified, at least one of the looters had prior knowledge that the Ovambo had a large sum of money hidden away. Hoarding money is bad brotherliness. Looting and theft stress the importance of investing it the way a good brother should.

Resolving the Outburst: Brotherhood Supreme

The outburst, it will be recalled, was terminated by black spokesmen requesting that the South African police be called in. Two white policemen and their dog consequently made an appearance and warned: "There is to be no fighting. You must stop it. We have come to protect you. Everyone who has a kierrie or any other weapon must put it away. Anyone found carrying a weapon will be in trouble." Their effect was more symbolic than practical, for they served to remind the blacks where the ultimate power lay and to stress the blacks' common "oppression." If the police caught a black, irrespective of ethnic group, he could expect the same treatment.

The Ovambo workers only moved back to the compound the day after the "fight," but during the interim all the symbols justifying the outburst had been removed. During the night, people of unknown identity overturned and emptied the drums of beer at the *shebeen.* Injuries sustained during the outburst, which were worn like badges of moral indignation, were now treated and bandaged. Signs of overt ethnic identity, for example, crude notices in their rooms proclaiming "Kavango Room," were removed. Marks of the outburst (overturned chairs, scattered clothes, broken bottles) were removed, and, where possible, the possessions of Ovambo workers were collected and given to a close brother of the Ovambo for safekeeping and return to the rightful owner.

The next day, the Ovambo returned and discovered their losses, which they proclaimed publicly. Many of the stolen goods, but little of the money, was "rediscovered." Where the Ovambo was a good brother, the Kavango workers tried to help him. Some Kavango looters were informed upon in the interests of brotherhood. Where the looter was a good brother, his account that he had stolen in the heat of the moment was deemed acceptable; where the looter was a bad brother, his account was unacceptable. His acts were "proof" that he was unreliable and could not be trusted. Such workers could not expect support from their brothers if the Ovambo sought revenge. Most of the workers who resigned from the mine in the days following the outburst were not those who had suffered loss or injuries, but those who had failed to measure up to the standards of good brotherhood.

The ethic of brotherhood was reinforced and strengthened in other ways as well. Some Kavango opened a collection list for donations to replace unaccounted for or unacceptable Ovambo losses (by standards of brotherhood). Workers on and off duty went about their jobs and leisure activities in groups to prevent assaults on individuals. These groups were composed of both Ovambo and Kavango. Brotherhood ties were single-mindedly stressed and manipulated for security, and ethnic group identity was deliberately minimized. The Bantu Affairs Commissioner who visited the compound shortly after the outburst told the compound manager to house the Ovambo workers separately from the Kavango because of their alleged traditional animosity. The Ovambo workers refused to accept this and deliberately housed themselves in groups of two or three in rooms with a Kavango majority.

Whites at the mine saw the outburst as proof that the black workers were children and that they would kill themselves were it not for white intervention. Some whites used the outburst as an opportunity to have "agitators" and other "bad boys" fired. Management also felt that the only way to prevent such outbursts from occurring was to implement its rules and regulations on black behavior more stringently. Black drinking was to be more strictly controlled, and the workers were to be given a stern lecture on what exactly their jobs were. Searches for illicit liquor and weapons were to be more frequent and thorough in the future, and informers were to be actively encouraged, since it was felt that they were essential for the smooth operation of the mine. Such actions increased the oppressiveness of the situation and underlined the importance of brotherhood and worker solidarity.

CONCLUSION

In this chapter, I argued that the "tribal fight" was not a chaotic, "normless" event—as is often portrayed in the literature and press—but that it had its own rules of conduct. Although the behavior was "rational," the new "action theory" approach to ethnic behavior (which treats the participants deductively, as puppets who react to objective and observable inequality in a preordained and predictable way[23]) does not enable us to understand the "tribal fight." What is crucial is how the workers who participated interpreted these "objective" conditions and events. In this regard, the vital feature for the participants was indeterminacy. Such an approach also allows one to specify the unintended consequences of their actions, which is something social scientists of the neo-rational school persistently overlook.

In their efforts to cope with this uncertain environment or turbulent situation, the black workers have consciously created and emphasized brotherhood. But they have doubts about the efficacy of brotherhood, and this emphasizes the dialectic inherent in the situation: brotherhood alleviates and promotes anxiety at the same time. Doubt results in continual stress on the importance of brotherhood, which again emphasizes the opposite.[24] So the vicious spiral continues, to be resolved only by leaving the mine or through collective action such as a "tribal fight" or riot.

Within the cultural milieu of the southern African mines, "tribal fights" form an important part of the folklore, not only of the blacks, but also of the whites.[25] The belief in the ever-present threat of conflict of this kind reaffirms the importance of brotherhood for the black workers; for the whites, it reaffirms their justification for maintaining the system. The outburst itself provided a "time out" in which conflict and disagreements could be dealt with in public. It allowed open expression of discontent and thus helped to solve conflicts and doubts and to strengthen the ethic of brotherhood. Brotherhood arose not only in a dialectical tension from an external threat but also from an internal threat, a point overlooked by many students of ethnicity and worker solidarity.

It is also important to note, as Simmel pointed out long ago, that conflict, whether ethnic or worker, is not necessarily destructive or abnormal. Only withdrawal from society is abnormal. To withdraw from compound society is to be a bad brother, and this is what the outburst was about. In a sense, then, the "tribal fight" was a purification and rejuvenation ceremony of brotherhood. Bad brothers and those who doubted the morality of brotherhood left the mine. Those who stayed were those whose faith in brotherhood was validated and confirmed. But at the same time conflict served to strengthen brotherhood, it also made the workers more aware of the inconsistencies and uncertainty of their situation. Hence, it sowed the seeds for further conflict.

Ethnicity was a surface feature, not a fundamental cause of the "tribal fight." People have many identities, some of which are more important than others. Contradictions involved in one set of identities can only be resolved through articulation and recourse to other, "lesser" identities. The fact that the outburst was conducted in "tribal" terms was more a product of the whites' definition of the situation. As Goffman has pointed out, to do otherwise would challenge the whites' definition of the situation and make further "illicit" or private activity more hazardous, if not impossible. Essentially, then, the "tribal fight" was not an expression of ethnicity but rather of antiethnicity.

NOTES

1. Data for this paper were collected during a thirteen-month period from November 1973 to December 1974, while the author was employed as a personnel officer catering to the needs of the white employees. The outburst occurred in September 1974. A somewhat revised version of this analysis of the "tribal fight" emphasizing the ritual aspects is to be found in Brett Williams and Robert Gordon, eds., *Anthropological Studies of Total Institutions* (Champaign: Stipes, 1977). For a general description of compounds within southern Africa, see Francis Wilson, *Migrant Labour in South Africa* (Johannesburg: Ravan Press, 1972).

2. Max Gluckman, "Malinowski's Funcational Analysis of Social Change," in *Social Change: The Colonial Situation*, ed. Immanuel Wallerstein (New York: John Wiley, 1966), p. 223.

3. Aidan Southall, "Forms of Ethnic Linkage," in *Town and Country in Central and East Africa*, ed. David Parkin (London: Oxford University Press, 1976); and Ioan Lewis, *Perspectives in Social Anthropology* (Harmondsworth: Penguin, 1976).

4. Max Gluckman, "Anthropological Problems arising from the African Industrial Revolution" in *Social Change in Modern Africa*, ed. Aidan Southall (London: Oxford University Press, 1961) pp. 68-69.

5. See, for example, several papers and the introduction in Leo Despres, ed., *Ethnicity and Resource Competition in Plural Societies* (Chicago: Aldine, 1976).

6. There are numerous "subschools" within the symbolic interactionist perspective. See e.g., B. Meltzer, J. Petras, and L. Reynolds, *Symbolic Interactionsm* (Boston: Routledge Kegan & Paul, 1975) for a recent statement in this regard. My own analysis leans so heavily on Erving Goffman, *Asylums* (New York: Anchor, 1961) and his *Relations in Public* (Harmondsworth: Penguin, 1972) that rather than undertake the almost impossible task of continually citing his work, I would simply like to acknowledge my debt and inspiration to him, particularly to his classic *Asylums*.

7. Max Gluckman, *Custom and Conflict in Africa* (Oxford: Blackwells, 1955).

8. Clifford Geertz, "Ideology as a Cultural System" in *Ideology and Discontent* ed. D. Apter (London: Free Press, 1964) pp. 64-70.

9. Goffman, *Relations in Public*.

10. For a detailed exposition of this point, see Nils Braroe, *Indians and White* (Stanford, Calif.: Stanford University Press, 1975).

11. Elizabeth Colson, "Competence and Incompetence in the Context of Independence," *Current Anthropology* 8, nos. 1-2 (February-April 1969): 94.

12. Quoted in P. Warr and T. Wall, *Work and Well-Being* (Harmondsworth: Penguin, 1975) p. 69.

13. For a fruitful discussion on the nature of this cohesion and the structure of mining communities in general, see M. Bulmer, "Sociological Models of the Mining Community," *The Sociological Review* 23, no. 1 (1975).

14. Studies on "informal work regulation" are plentiful. For a representative bibliography, see Warr and Wall, *Work and Well-Being*.

15. Hoyt Alverson, "Labor Migrants in South African Industry" in *Migration and Anthropology*, ed. R. F. Spencer (Seattle: American Ethnological Societyy, 1970).

16. J. P. Bruwer, *Die Matrilinere order Van Die Kavango* (unpublished manuscript, 1966).

17. Peter Berger and Thomas Luckmann, *The Social Construction of Reality* (New York: Anchor, 1967).

18. J. Clyde Mitchell, "Race, Class and Status in South Central Africa," in *Social Stratification,* ed. A. Tuden and L. Plotnicov (New York: Free Press, 1970).

19. As we know from exchange theory, once a choice has been made, in this case with regard to the source of labor, that choice imposes constraints of future choices. By the time of the outburst, the compound manager had established friendly relations with the (white) official on the Kavango in charge of labor recruiting there and believed that because of these friendly ties he was receiving "superior" workers.

20. R. J. Gordon, "Informal Labor Organization in Namibia," *South African Labour Bulletin* (in press) provides an indication of how widespread this phenomenon is.

21. C. MacAndrew and R. Edgerton, *Drunken Comportment* (Chicago: Aldine, 1970).

22. This part owes much to an article by P. Dennis, "The Role of the Drunk in a Oaxaca Village," *American Anthropologist* 77, no. 4 (December 1975). The very clandestinity of brotherhood generates uncertainty and suspicion. Goffman captures this nicely: "By the very nature of the criminal enterprise, the participants must doubt the apparent discretion of those properly in the know [i.e., brothers] and the apparent unsuspectingness of those who should not know [i.e., whites]; and yet, he must thrust himself into assuming that those who appear to be properly going about their business really are." Goffman, *Relations in Public*, p. 330.

23. Despres, *Ethnicity and Resource Competition.*

24. Best illustrated by the fact that interestingly enough, while all weapons are illegal in the compound, during the outburst practically every worker had a concealed or obvious weapon, to use official phraseology.

25. In the South African mines, the older "practical" compound managers are vehemently opposed to the introduction of young university graduates into their jobs, precisely because they lack "experience" in handling "tribal fights" and riots in the compounds—evidence that they are not as infrequent as one would suspect.

PART 3
Color, Racism, or Ethnicity

One component we mentioned earlier as frequently being involved in ethnic group identification is that of physical type. In its more harmless context, this may involve differences in stature, shades of pigmentation, or similar phenotypic variations used as temporary markers between different groups. But these boundaries are more fluid and can be ignored much more easily than those that are used in certain cases to mark social classes or castes.

When a society is divided on the basis of color alone, such racism can only change as color is deemphasized and as the color caste system evolves into a class system. In an ideal situation, such ethnicity then assures freedom, equitability, and dignity for all. That is the positive, if idealistic, note on which this collection of essays concludes.

12
Formative Factors in the Origins and Growth of Afrikaner Ethnicity

J. H. Coetzee

Let me begin by saying that I largely agree with Glazer and Moynihan that "ethnic identity has become more salient, ethnic-assertion stronger, ethnic conflict more marked, in the past twenty years."[1] Whatever the reasons, in this quest for research and dissemination of knowledge on the subject of ethnicity, the Afrikaner "ethnos" has surged into the center of attention, too. Recently, several approaches to an analysis, description, and understanding of the Afrikaner, Afrikaner nationalism, and Afrikaner ethnicity have been attempted. These efforts undoubtedly produced valuable material, although I believe that the results tend to remain fragmentary and to reveal lack of depth and insight.

A glance at the contributors to *Ethnicity* shows that they hail from the fields of political science, sociology, and history.[3] The absence of anthropologists, both social and cultural, from this academic company is striking. The same applies to the authors of collected papers on Afrikaner ethnicity and related fields. Philosophers, theologians, and journalists have joined the ranks of the above mentioned writers in their efforts to come to a clearer understanding of Afrikaner ethnicity. But any criticism about what has been published both in the general field of ethnicity or on the specific theme of Afrikaner ethnic consciousness and features should rather be laid at the doorstep of the anthropologists. Political scientists, sociologists, and historians cannot be taken to task for not approaching the subject from an ethnological angle. But even those anthropologists who have ventured in this direction (such as Barth)[4] are still subscribing to the traditional textbook definition and connotation of ethnology as a purely comparative ethnography[5] or as a history of a people.[6]

This is not the appropriate place to argue for a new or different approach to ethnology as a discipline distinct from social and cultural anthropology. Hence, I merely state that in broad outlines my views on the specific subject rather correspond to the views of Shirokogoroff[7] and Mühlmann.[8] The concept of ethnos as a process; centripetal and centrifugal forces and movements; aspects of internal ethnical equilibrium; influences of the interethnic milieu; and the leading ethnos as a mechanism of adaptation and cultural remodelling and as a factor of changes—all serve as a frame of reference for a hypothesis on the origins and growth of Afrikaner ethnicity and ethnic features.[9] The concepts of interethnic pressure and adaptation to local conditions far removed and different from the previous habitat and ethnic milieu, both in space and time,[10] and systems of selection (*Siebungssysteme*) regarding acceptance or rejection of "foreigners"—all appear to have an important bearing on the Afrikaner situation.[11]

The topic for this discussion is expressly limited to one aspect of Afrikaner ethnicity, viz., the comprehensive web of processes, attitudes, conditions, and characteristics intertwined in the origin and growth of this ethnicity. This requires major attention to the historic background and, of course, sounds a warning against the pitfalls of the "kulturhistorische" school of thought and methodology of Schmidt, Koppers, Graebner, and Ratzel. Bastian's effort to explain the diversity of cultures and culture groups through his idea of "der Mensch in der Geschichte" is not wholly devoid of merit.[12] Boas's "rigorous demonstration of historical relation-ships and the new light he shed on the processes involved in such (i.e., regional cultures) contacts" are the more remarkable when evaluated against his training as a geographer.[13] Lewis provides an acceptable conclusion to his discussion on history and social anthropology: "More generally, historical data are not merely relevant, but are quite decisive in evaluating a given society's own view of its past. People's views of time, and their ethnocentric 'history' are very much part of the picture which even the most particularistic anthropologist seeks to delineate."[14] No ethnicity can be comprehended or explained in isolation from the history of the group of people who identify themselves as such. The recognition of the historical background as an illuminating factor in the existence of ethnic units and of ethnic relations, for instance, pervades the sociological analysis of ethnic dynamics by Hunt and Walker.[15] The same trend is visible in the handling of pluralism in Africa by Kuper, Smith, et al., and it is explicitly summarized in Thompson's contribution.[16]

This approach should, however, neither discount nor overemphasize the factors of time and space in the events of men; man himself is naturally bound to both. The important point to keep in mind in all cases is to avoid any trace of determinism and to be able to evaluate the relative effects of the whole gamut of contributing forces in each specific case. Thinking in terms of "laws" rather than of general trends is fatal to any study of ethnicity.

Before proceeding to the empirical details of the study, some general lines of treatment have to be suggested. My basic point of approach includes the presence of two apparently conflicting forces or movements;

both, however, are in effect active in the production of the end result. These can be designated as processes of fission and fusion,[17] processes of centripetal and centrifugal forces,[18] processes of ethnogenesis and schismogenesis,[19] or integrating and disintegrating forces. Attaining a consciousness of ethnic identity usually includes the effective presence of both kinds of forces. My explanation and tracing of the origin of Afrikaner ethnicity is done within this framework.

Another aspect to keep in mind is the relative effect of the various formative factors within and dependent on the diverse sets of conditions or situations within which they operate: stronger and more prominent in one set of circumstances, less in another. Two examples may illustrate this point. The space factor (the territorial distance or separation from the original ethnic group) is much the same for Afrikaners of Dutch, German, and French descent and for South African English of British stock. The time factor (the stretch of time signalling the physical separation from the parent body or bodies) differs almost by half. The main body of progenitors of the Afrikaners has come to South Africa since the second half of the seventeenth century; the ancestors of the English South Africans have come more or less since the first quarter of the nineteenth century. The coloureds and the South African Indians (insofar as the former represent a deeper rootedness as an African ethnic group(s), while the Indians are still keeping up a stronger and clearer profile of "separate" or "foreign" ethnic and cultural consciousness) seem to accentuate a preliminary conclusion corresponding with the difference between Afrikaners and South African English. They seem to point to the more potent influence of time as a schismogenetic and ethnogenetic factor. Even so it remains to be ascertained whether time was the only factor in the origin and growth of a separate or new ethnic consciousness. That the time factor, as opposed the spacial factor, is not the sole determining factor, however, is suggested by the fact that other immigrants, arriving much later than or simultaneously with the Britishers of 1795 and 1820 and afterwards, passed the barrier to the "laager" of Afrikaner ethnicity, while the English-speaking South Africans are largely still keeping out of the fold.

The aim of this chapter, then, is to analyze and identify the factors that contributed to the Afrikaner ethnogenesis. This means identifying and explaining those conditions that promoted the fusion of certain groups and individuals and their assimilation into a new and distinct ethnic body and those conditions that stood in the way of this process, the respective roles of distance in time and space for immigrant groups or stratification into conquerors and conquered. This aspect of interethnic pressure can hardly be overemphasized in the South African situation.[20] In close combination with the influence of physical features, moreover, language, culture and civilizational niveau, religion, and ideology have to be given some attention.

HISTORICAL BACKGROUND

A summary view of the basic and relevant historical features of the ethnogenetic milieu of the Afrikaner appears to be the most logical point of departure. Giliomee's history of the development of the Afrikaners' self-concept is a valuable, concise, and explanatory framework.[21] However, his focus was directed toward the Afrikaner's political self-concept, and for this reason he relied mainly on pronouncements of political leaders who endeavored to articulate political self-consciousness.[22] The ethnic concept, however, precedes rather than follows the political consciousness, and, as Fortes and Evans-Pritchard have pointed out, the ethnic unit and the political unit do not always coincide.[23] The Afrikaner ethnos and ethnic consciousness toward the end of the nineteenth century spread over two republics and two colonies in the area of the present Republic of South Africa and also in Rhodesia and Angola. After the Anglo-Boer war, sections of the Afrikaner ethnic unit settled in Kenya and in Argentina. In both Angola and Argentina, a process of schismogenesis had a divisive effect, causing the lesser part to assimilate with the population of the respective areas and the rest in due course to return to South Africa.

In 1652 Jan van Riebeeck and a party of officials, artisans, soldiers, and sailors established a refueling station at the Cape of Good Hope. This was done in the name of the East India Company and for the benefit of its fleets passing the Cape on their errands to and from the East. No notion of a permanent sojourn or settlement, let alone a "volksplanting" (literally, the planting of a new people) entered the minds of those responsible for this pragmatic venture. But that is exactly what happened. A new group consciousness did appear, and the Afrikaner stepped onto the stage of history.

Van Riebeeck's company came mainly from the Netherlands and had a sprinkling of German and other nationalities. Within four decades, they were reinforced by a number of French Huguenot families. Efforts to determine how much each national group contributed differ: 50-53 percent Dutch, 27-28 percent German (and even more), 17-25 percent French, and 4-6 percent from unnamed nationalities.[24] The respective numerical strength of the different components was, however, neutralized by other factors, among which political bonds and authority were extremely important.

Fusion of the various European groups was the result of two apparently opposite processes: one of dissociation from their respective ethnic parentages, the other a complementary process of mutual identification. Both Dutch and German were quite voluntarily, although unconsciously, borne along with this outgoing tide to enter a new harbor of ethnic existence and security. They had no explicit reasons or rational intention of falling "victims" to schismogenesis. The French were in a rather different situation: fleeing the wrath and violence of their Roman Catholic countrymen, they were forced from their country and eventually settled in South Africa as a substitute fatherland. At the Cape of Good Hope, the temporary officials, the garrison, and artisans on the Company's payroll were joined by a new class, viz., "free burghers" tending their own land. A further social and economic development was the gradual movement of farmers into husbandry and extensive pastoral practices and trekking into the hinterland. The French especially contributed to the cultivation of vineyards. These developments brought conflict both with the administration and with the aboriginals—first with the pastoral and seasonal migratory Hottentots and the hunting and gathering Bushmen. As the farmers moved further east, conflicts arose between them and the Xhosa tribes, whose mixed subsistence economy included hunting, primitive agricultural pursuits, and extensive pastoralism.

This part of the population to a considerable extent grew apart from those around Cape Town and developed a kind of social and personal independence not always appreciated by the authorities of the Cape. The "conquest" of the Cape by the English in 1795 and the short interim regime of the Batavian Republic (1803-1806) were the first of a series of military and political clashes between the British and the "Boers." These clashes continued to the end of the nineteenth century and, as a result of the greater military power of the former, always ended with the subjugation of the latter.

It was clear, however, that the descendants of the Dutch, Germans, and French at the tip of Africa were changing in life-style, attitude, and loyalties. Time and space tended to sever their bonds with the ethnic groups in Europe from which they had originated. Restricted communication could not neutralize this "drift." The new and very different physical and social milieu, to which they had to adapt and which presented its own challenges, set afoot a process of "growing away" from the traditions of the old countries. The differences between local conditions and those in France, the Netherlands, and Germany brought different adjustments, different interests, and different approaches.

Two short notes should be added. Time was needed to bring about a dissociation between the South African or Cape Dutch and the Dutch of Europe. Even so, the French Huguenots did not immediately reject or neglect their French affinities and ethnicity, notwithstanding the fact that they had been rejected. Their gradual identification with a "new" ethnic—not a "Dutch"—group was promoted partly by the fact of having to stay in Africa permanently and partly by the policy of the powers-that-were, i.e., a policy that spread them out over the settlement among those of Dutch descent. Even under these circumstances, they clung to their French heritage in economic activities, church, school, and language for quite some time.

The second remark concerns the first German components of the population at the Cape. They almost totally assimilated with the rest of the population without even leaving any easily observable trace of their German culture except works of art and their names. Their imprint on the new ethnic unit and its culture is far less explicit than the French legacy. This poses a new question: why were the first German settlers totally submerged in the originating ethnic group, while those who arrived early in the second half of the nineteenth century and later on tended to maintain their cultural identity as distinct subgroups?

As the early Dutch, German, and French components clustered around a new nucleus and as they increasingly dissociated themselves from their forebears, the use of the terms *ethnogenesis* and *schismogenesis* is justifiable.

At the same time, another process began. Coming to South Africa without any racial ideology, the earliest representatives of the European peoples both officially and spontaneously took up contact (mainly for the sake of trading) with the native inhabitants. History is quite clear about their efforts to accept these people in social relations, education, and religion. "King" Harry's sister, who was christened Eva, baptized, and unofficially adopted by and cared for in the home of the commander, was eventually married to the second-in-command at the station, while the bill was footed by the Company.

Within a decade or three, however, relations began to change, and the process of "integration" was limited. This was to become one of the most characteristic traits of Afrikaner ethnicity, a trait deeply ingrained in and accepted by Afrikaners and a trait utterly unacceptable to apparently everyone else since the nineteenth century.

Dissociation with the European ethnic roots and the substitution of new ethnic boundaries went hand in hand with the gradual nonacceptance of the indigenous peoples as ethnic and social partners. This latter process cannot be called schismogenetic. Various reasons for it have been advanced, but most are peripheral and tend to oversimplification. This not uncommon development confronts the student of Afrikaner ethnogenesis with a further question.

A previous paragraph hinted at the rather uncommon phenomenon that, as opposed to the original European immigrants, groups arriving at a later stage do not seem to have met with the same fate. This is especially true of Jews, Germans, English, and, during the last decades, southern Europeans, including Portuguese. It is not uncommon to hear the reproach that Afrikaners have closed their ranks to immigrants or that Afrikaners have lost their power of assimilation. This, of course, causes great concern about English-Afrikaner relations and is of special political significance.

Why did the original tendency of ethnic assimilation change, and why did a new indigenous ethnic unity consisting of components derived from Europe, Asia, and Africa begin to grow?

THE FACTORS INVOLVED

The ethnogenetic process does not imply a one-way process. Seemingly opposing movements may in the overall situation prove to be complementary. The willingness to accept and be accepted are both involved. The leading ethnos or ethnic nucleus may be willing to assimilate newcomers, but the latter may refuse to be assimilated.[25] Or the contrary may be the case. Both patterns are observed in South Africa. This willingness naturally differs under specific conditions and seems to depend on various circumstances. It is axiomatic that ethnicity is in flux, that it is dynamic rather than static.

Interethnic Pressure: "Afrikaner" and Britishers

Shirokogoroff propounded the thesis of interethnic pressure. Consciousness of one's own ethnic existence is stimulated and intensified by meeting "other" groups.[26] Hence the nature of the contact has an effect on the reaction to the possibilities of acceptance or assimilation. The *Fremdphänomen* has a bearing on this issue.[27] Introduction to foreign cultures is accompanied by a kind of culture shock, which tends to result in enmity, jealousy, and a feeling of being threatened.

Within this frame of reference, it appears that the change of attitude in the Dutch-German-French group toward newcomers from overseas, especially toward the British, can be at least partly explained. Originally the Dutch, Germans, French, and whatever other Europeans lived at the Cape enjoyed the same social and political rights and privileges. The Dutch were the leading ethnos: they formed the numerical majority, and the Cape of Good Hope was administered by a Dutch company. With a system in which there was no franchise, any political differences were not accentuated. The Dutch and the Germans seems to have shown an easy mutual acceptance very soon. They were culturally not far apart, and the shortage of women especially affected the German section of the population.[28] Through the early years of their settlement at the Cape, the French Huguenots naturally longed for their country and countrymen and may have cherished ideas of returning home someday (their contracts of settlement kept that option open). But there was little goodwill between Dutch and French. As a matter of fact, the board of the Company decided that families who were either Dutch citizens or subjects of a German state would be encouraged to emigrate to the settlement but that no French were to be encouraged.[29] The cultural and ethnic distance between the two was, however, decreased by other factors: the adaptation to a similar milieu, the same religious confession, and, above all perhaps, a common struggle against the corruption of the administration.

In 1795 and again in 1806, the British arrived as conquerors and rulers. The Cape Dutch culture was superseded by the British culture of the dominant group; Dutch authority was replaced by British power. The real effect of this change was felt from 1815, when the new rulers "made attempts to anglicise the colony linguistically and to recast political and social institutions in a British mould."[30] Giliomee's statement that power also determined social stratification applies to the Cape Dutch–English relationships as well.[31]

At this stage, the Dutch, Germans, and French had gone far toward becoming a distinct ethnic unit. Whatever the young Hendrik Bibault meant when he referred to himself as an Afrikaner as early as the year 1707, when he was arrested and had to appear before the *landdrost*, is not clear. The significance of this name lies in the fact that it indicates a new term of reference, one that differs from the older French or Dutch or German and points to the consciousness of a different group identity. This trend was growing, and terms differentiating them from the European countries and nationalities increased in popularity. This included the names *Africaan, Kaapenaar,* and others. Giliomee rightly concludes that Afrikaner consciousness was enhanced on the eastern frontier by the peculiar problems and circumstances of the colonists. "Having to be largely self-reliant for their settlement and maintenance, they forgot the old country and came to realize that they were of Africa—their only true home."[32]

The mutual assimilation of French and Dutch, for example, was facilitated, on the one hand, when the French were forced to accept the fact that they were not destined to return to France and that they had to adapt to the same milieu as the Dutch. The conflict with the authorities, especially during the regime of the younger Van der Stel, made an important contribution toward the process of fusion.[33]

The English introduced a new factor, one that had been largely absent during the previous period. Their presence roused the attitudes and feelings that accompany interethnic pressure. (At this stage, I am only referring to relations among those of European extraction.) The French element did experience this position, being a minority compared to the Dutch. The factors enumerated above seem to have weakened their opposition to fusion. The Dutch formed the majority, or leading (dominant) ethnos, and their position was not seriously challenged by the French or Germans. Important in this respect was the fact, too, that the ties binding them to the Netherlands were wearing thin. A discontinuity arose. The effects of limited communication—the more limited the further they moved away from the peninsula—were enhanced by the almost total lack of new immigrants from the "old country." As a matter of fact, it appears that their attitude toward East India Company officials and, subsequently, toward the Batavian Republic reveals that they experienced these as a foreign power and a foreign people. Ethnic identities were being differentiated, the essence of embryonic ethnic pressure was thereby appearing.

When they arrived, the English displaced the Dutch as the leading ethnos and ruling component. The latter became a subject group, thereby for the first time sensing the threat of a dominant ethnic minority. This problem became all the more complicated as it entered various aspects of life: cultural differences, especially in language and religion, differences in attitude toward the indigenous peoples, and, as a result, differences in policy toward these people, a policy unsatisfactory to the emerging Afrikaner people.

The violent subjugation of the Afrikaners to British authority and the rise of British culture to dominance continued for the duration of a century. Contact with the British differed from the contact among Dutch, German, and French: it was one of conflict. This must have had a bearing on the readiness (or lack of it) of the Afrikaners to accept the British as part of their ethnic unity. This does not mean that no Britisher ever entered the fold of Afrikanerdom; the British who did were exceptions rather than the general rule. Those who had no scruples or choice in throwing their lot in with Afrikaners were accepted. In the Orange Free State Republic, which after 1854 more or less escaped the attention of British imperialism, this process appears to have reached proportions not attained in, say, the Transvaal. In the latter, however, Dutch immigrants were accepted and welcomed and taken in as part of the Afrikaner people. This is true of the German immigrants during the latter half of the nineteenth century as well, but to a lesser extent; coming as families and settling as cultural groups, the Germans greatly associated with the Afrikaners but at the same time maintained their German identity.

An attitude of exclusiveness toward Britishers is one feature of Afrikaner ethnicity during its formative stages. This eventually extended toward almost all *uitlanders* ("foreigners"). The irrational side of all this—so characteristic of every aspect of the ethnic phenomenon—was brought to light during the second war of independence, when volunteer units of Dutch, Scandinavians, and other Europeans under leadership of Irish, French, and Dutch fought and died bravely for the cause of Afrikaner independence against the British aggressor.

An additional factor, one that accentuates the attitude of exclusion of the British, is the difference in general life-style, culture, and religion. In a certain sense, this may be secondary. In the absence of strong interethnic pressure in the form of political conflict, these differences would probably not have had the repellent power they actually had. With political strife and threats to independent Afrikaner nationhood, the differences in language, culture, and religion were emphasized; they exemplified the threatening foreigner and served as a demarcation of ethnic identity. In the Afrikaner's attitude toward and ethnic evaluation of the blacks, physical differences were generalized as the primary mark of otherness. But as for Afrikaner-Britisher relations, physical differences scarcely existed. An interesting feature, however, strikes the observer, viz., the tendency on both sides to express their mutual disdain in terms of physical identification: *Rooinek* ("red neck") and "hairy backs." An indication of the secondary, or at least less important, place taken by cultural differences between Afrikaner and English is the vast impact of English culture and language on Afrikaner culture and language, notwithstanding several movements against this peaceful cultural penetration.

Against this background, language and religion were the most important demarcating aspects. Additional items were probably dress, dietary customs, eating and drinking habits, and economic pursuits (there was a time when the Afrikaners were called *Boers*, i. e., "farmers," both as an honorific and as a contemptuous appellation). Language differences naturally hampered communication, but they were primarily seen as the most direct expression of everything British and everything anti-Afrikaner. One of the main reasons for this reaction to the English language was the attempt by British administrators to anglicize the Cape Colony. Another was the arrogant attitude taken in both official and popular circles toward the new Afrikaner language; even today it is referred to merely as the *Taal* ("language"), and the name *Afrikaans* is avoided. In its earlier stages of development, it was also spoken of as kitchen Dutch—by both English and true Dutch. The *taalbeweging* (movement for the recognition of Afrikaans as the actual spoken language of the Afrikaners officially and unofficially), which had its inception in 1875, was bound up with a national consciousness indicative of an ethnic unit reaching its maturity. An insult to its language was an insult to the people using that language. Language was to become the clear manifestation of ethnic boundaries. Gupta sees language as one of the important marks of ethnicity in the history of political evolution and political situations.[34] It provides a bond of unity among its speakers and defines a line of separation marking off one speech community from another. In this he subscribes to the findings of Shibutani and Kwan, although they limit their observations primarily to other aspects.[35] In the history of Afrikaner ethnicity, this was undoubtedly the case. On the one hand, it was a centripetal force for Afrikaners, but on the other it was a barrier against English.

"In some communities religion may serve as a useful index of ethnic identity. This is especially true where there are sharp differences in overt behavior—in ritual and dietary practices."[36] In the case of South Africa, the differences in religious beliefs and practices cannot be bypassed. The reactions of the groups concerned appear to be inconsequent. The first Dutch officials arrived at the Cape about three decades after the important Synod of Dordrecht, which is viewed as a "reformation" of the Protestant churches of the Lowlands. Van Riebeeck, his crew, and those who came after them—all brought with them some of the reformed Christian spirit of Dordrecht. But it was almost a generation after Dordrecht; even the prayer read by the commander on his arrival contains an element of the formalism that had begun to infiltrate the reformed churches at this stage.

The arrival of the French Huguenots was of prime importance. They were fugitives for the sake of their religious convictions and their Christian confession. As far as dogma and confession were concerned, the two groups were essentially similar. I get the impression, however, that the French exercised a healthy influence on the Dutch and that the French reformed faith resulted in a characteristic brand of Afrikaner religious belief and life. This was the Christianity that accompanied the new Afrikaner ethnos that was moving inland and that eventually settled in the northern parts.

Another remark seems to be necessary. Toward the end of the eighteenth century and very definitely since the nineteenth century, new trends in European Christianity—rationalism, humanism (and later also Pietism and Methodism)—left their marks on those who lived closer to the original settlement. The frontiersmen, because of their greater isolation, escaped these influences to a large extent.[37] This differentiation of religious ideas and practices among Protestant Afrikaners still exists, although it is largely manifested in the diversification of reformed churches. The use (abuse?) of the church in the process of Anglicization—even more than the disparity in confession and religious practices between Afrikaner and British Protestantism—brought about the cleavages and accentuated the overall cultural differences. Lutheran Germans and Scandinavians were more acceptable to the Afrikaners ethnos that British Protestants.

This combination of general cultural differences—sharply profiled in language and religion—gave birth to the Christian-national concept so expressive of Afrikaner ethnicity. This concept was especially introduced and implemented in the field of education. Its intent was to bring up younger generations in the spirit of the reformed faith as confessed and practiced by the Afrikaner so as to pervade their whole life; its intent was to tie education to a South African loyalty and to do it through the medium of the national language, Afrikaans. The concept of Christian-national, translated to Reformed-Afrikaans in the specific situation, was viewed as the means to safeguard the future existence of the Afrikaner people. The answers to the questions posed by interethnic pressure from those of British descent and orientation was part of the ethnogenetic process of Afrikanerdom.

Interethnic Pressures: The Blacks

A "second ethnic front," as it were, was contact with blacks. This soon became contact-in-conflict, and it posed questions that had to be met in a different way. Historically, this must be assigned the first and foremost place in the origin of Afrikaner ethnicity.

With regard to the European settlers' attitude to the indigenous peoples, two features have to be pointed out. The first is that the earliest settlers originally had no scruples about accepting the aboriginals as equals,

however exotically different they were. They arrived with no known race
ideology. The only condition for acceptance was acceptance of
Christianity. Even this was rather widely interpreted, as seen in the custom
and the orders of the authorities that children of heathen parents, especially
slaves, were to be baptized. Entries in van Riebeeck's diary give the
impression that he was aware of the state of civilization of the inhabitants
but that he cherished hopes of bridging the gulf by way of contact and
education.[38] That was before he left the Netherlands for the Cape. Once at
the Cape, however, his optimism appears to have faded. When he left after
a decade, the situation could be summarized as follows. Acquaintance had
been made with a number of Hottentot tribes, especially with the
Watermen (Watermans), an impoverished group under their leader,
"King" Harry, who acted as interpreter. Contact with Bushmen was
sporadic. The first slaves had arrived. Relations with the Hottentots were
defined by trade, and relations with the slaves were defined by labor. All
attempts to live peacefully with the Hottentot tribes had come to an end;
the latter persisted in rustling the Company's cattle and killing the
herdsmen. Further efforts for peaceful coexistence went unheeded and
came to a climax with the first Hottentot war in 1659. Distrust and
antipathy are clearly observable during these years. The Hottentots
justified their behavior by pointing out that the immigrants were settling on
their land with no intention of moving away. Socially, however, the colored
inhabitants were treated without any serious discrimination. The first
"mixed" marriage was solemnized in 1656, and the second on record was in
1666. Within the first four decades, more or less sixty children of "mixed"
blood were born at the Cape—and an unknown number of these came from
extramarital relationships.[39]

   Gradually changes set in. The roots of this change did not feed in an
ideological substratum. What changes occurred were due to experience.
Although color was no restriction, the difference in level of civilization—a
very deep and broad gulf, indeed—had an inescapable divisive effect.

   Van Riebeeck himself and his successors lost much of their optimism
regarding the rehabilitation of these people.[40] As fate would have it, those
coming from Europe were in fact destined to make their first contacts and
acquaintances with the lowliest tribes on the scale of civilization. These
included the beach scavengers, impoverished Hottentots who lived on what
the sea offered without the slightest exertion on their part. The inland
Hottentot tribes' nomadic ways, their hygiene, their aversion to regular
work, their standards of honesty and sobriety—all were essentially different
from that cherished by the Europeans. The gulf widened when the
Bushmen entered the scene; the result was an unpreventable clash between
a pastoral-agricultural culture and a hunting and gathering way of life.
Simple bias and prejudice do not really account for the situation that
resulted from these early contacts, whether in South Africa, elsewhere in

Africa, the Americas, or in Australia. To the contrary, the attitudes of the Europeans came to be based on their experience of differences between themselves and the aboriginal peoples; as opposed to their original concepts of equality, these attitudes were pragmatic and empirical. They became biased and prejudiced when the historical (and hence temporary) differences in the respective stages of civilization came to be seen as inherent and unchangeable. This usually develops by a process of generalization and projection, a process in which color (as one instance and not necessarily the only) is evaluated as intrinsically and genetically bound up with inferior cultural characteristics.

The contact of the Company's representatives at the Cape with the aboriginal peoples and with slaves (the first batch were from Angola and Guinea) gradually changed their original attitudes and intentions.[41] It lay the foundations for a social, economic, and political structure that was to be characteristic of a later epoch.

It is remarkable that linguistic differences were not a potent divisive force. The really meaningful factors were civilization and religion. It was not a question of sophisticated theological or dogmatic differences; it was a clear demarcation between Christianity and heathenism. And Christianity was the sine qua non for acceptance into European society.

The religious differences ran parallel to the gulf between the European and indigenous civilizations. The latter included the whole life spectrum— from hygiene, physical cleanliness, and ownership and usage of land to feasting and government. Apart from linguistic differences and the resulting lack of communication, a common language could itself hardly have bridged the gulf between civilizations. Conflict had to ensue, not so much as the bitter fruit of bad faith but rather as the result of total misapprehension on both sides.

The numerical relation also contributed to conflict. A small white community, although superior in techniques and skills for the control of nature, found itself face to face with a majority of aboriginals. The numbers, the tremendous cultural differences, and especially the manifest enmity of the inhabitants transformed the *Fremdphänomen* to a confrontation with enemies. The British were unacceptable as a dominant group; the colored ethne were to be endured as subjects and inferiors, both because of their less developed culture and because of their numerical power. Social stratification was based on color, and color was viewed as the manifestation of both ethnic and qualitative differences.

All the differentiating characteristics of Khoi, San, slaves, and Negroids reduced to one common feature: color. It seems remarkable, however, that the appellation *nonwhites* dates from a later period. The older names were more discriminating: Bushmen, Hottentots, and Kaffirs. The latter was even applied as an ethnic name to that Negroid part of the population now known as Xhosa peoples. At a later stage, the term *native* was used to indicate the aboriginal peoples. In any case, the difficult-to-formulate complex of differences was expressed by a common symbol, viz., skin color and other physical features more or less characteristic of the general racial appearance. All the differences in civilization were projected onto color. At this stage, bias and prejudice entered the scene: difference of color was equated with difference of religion, civilization, aptitude, and intellectual capacity. White meant the superior qualities of Christianity, the civilized way of life, education, economic superiority, and privilege and political power. Black came to mean heathens and barbarians, the unskilled worker, the lower social classes, and lower intellectual capacities. But, unlike the usual class system, this white-black structure was rigid and lacked vertical mobility. As elsewhere, even where difference of color is absent, i.e., Hima and Iru,[42] the hierarchical order cannot be escaped.

At the same time, this concept found another expression, one that might even be regarded as the first phase of a racial ideology. Peculiarly, this led back to religion. Differences, racial and otherwise, tended to be explained in Old Testament idiom. The people of color were taken to be the kindred of Ham and his son Canaan, who were destined to be hewers of wood and drawers of water. A purely historic situation of temporary nature was thus evaluated and explained in terms of a fixed destiny. Thus embedded in a peculiar interpretation of Old Testament teaching, this development became a further feature of Afrikaner ethnicity: the nonacceptability of colored people.

This development has two aspects, that of whiteness and that of being a chosen people. Afrikaner ethnicity maintained and still maintains its European biological and racial heritage. As in all aspects of ethnicity, there were exceptions to the general trend. Mixed marriages were solemnized; as for extramarital miscegenation, no further evidence is needed. What happened is analogous to what happened in monoethnic, but multiclass, societies such as Russia, France, Germany, and Great Britain. Men from the higher classes abused girls from the lower classes, literally leaving the girl with the baby. This biological or sexual departure from the norm happened in South Africa, too. The difference is, however, that in this case those of the "lower class" also belonged to the "other ethnic" group with a different color. It is difficult to gauge the intensity of disapproval when comparing, for instance, the British and South African reactions to such situations. However, it is possible that stricter censure on the part of the Afrikaners explains both the nonacceptability of the (unfortunate) fruits of these misadventures and the social stigma that fell on the miscreants. Censure of the culprits and nonadoption of children of mixed blood safeguarded the white ethnos, but they inadvertantly helped create a variant, but culturally and biologically related, ethnic group.

The idea of a chosen people is related to the first racial ideological approach as a justification for the stratified ethno-social structure. This question should be treated with the utmost care. I use the indefinite article, *a* chosen people. In South Africa, it might have amounted to the same in practice. Surrounded by heathen tribal clusters, the Christian settlers almost naturally observed the analogy between their situation and that of Israel. Being aware of their precarious situation, not theoretically as much as from physical experience, they placed their hope and faith in God Almighty. Since there was a strong stress on the Old Testament (the children of Ham) and since the Afrikaners were a pastoral people with a strong patriarchal social structure, the analogies even increased in the Afrikaner ethnic consciousness. On the one hand, this doubtlessly culminated, in the minds of an isolated people leading a simple life and being quite Biblically minded, in the idea of a people chosen and called by God to bring civilization to this wilderness and the light of the gospel to the heathen. On the other hand, most were sufficiently versed in Biblical teaching not to identify themselves with the true chosen people of God, viz., Israel. This is not unique. The Pilgrim Fathers and their successors nursed the same notion. The Britishers and French, in a more secular way, justified their imperial expansion as the divine calling of a civilizing mission. Dr. Livingstone was one of the typical examples of the British ethnic ideal in this regard.

Whatever the relative weight and significance of being *a* or *the* chosen people, this aspect, derived from religion, was woven into the web of Afrikaner ethnicity during its infancy.

Impact of "Foreign" Ideologies

In dealing with the contribution of religious belief to the ethnic sense of the Afrikaner, it is necessary to touch upon a related aspect. What other trends of thought influenced Afrikaner ethnicity during its formative phases? On the purely religious niveau, passing notice has been given to the humanistic and rationalistic influence of the Enlightenment and the ideas of the French Revolution. Viewed in toto, this was limited to some, not all, of the Afrikaner people. The frontier farmers in the isolated northeastern parts of the Cape Colony almost totally escaped the early nineteenth-century theological innovations and philosophical trends. These were largely restricted to the Cape and the more settled environment. Humanism was later followed by various other religious trends, including Revivalism, Pietism, and Methodism. The more conservative elements of the population stayed in the faraway parts and after the Great Trek settled in the northern republics. The result was a duality in Afrikaner religiosity and a partial crystallization in the establishment of three separate reformed churches. To a large extent, the variations in Afrikaner ethnicity found expression in the various religious institutions.

Like religious trends and influences, ideology may have made contributions to Afrikaner ethnicity. There was no ideology of racial superiority when the first Dutch officials established the refueling station. Experience revealed considerable differences between themselves and the aboriginal inhabitants, differences they interpreted and evaluated erroneously. This led to a race attitide that was part and parcel of the ethnic composite, one that tended to evolve into an ideology. The opposing attitudes on race—the Ham-theory of the Afrikaner vs. the so-called Negrophile attitude preached and practiced by English missionaries, upheld by British governors, and professed by the leaders of the English community—came to be one of the most pregnant differences between Afrikaner and English and further accentuated the mutual demarcation and opposition.

Some historians are of the opinion that the French Revolution had a direct bearing on Afrikaner political identity, especially on the establishment of the late eighteenth-century republics (Graaff-Reinet and Swellendam). This is doubtful. Rumors of what happened in France surely reached some of the frontier farmers, but they could only have been faint, sporadic, and dubious. Thus, French revolutionism probably met the same fate as humanism and other European trends of thought. What reached the frontiersmen may well have stimulated the growing desire for independence and self-government, a desire that originated from their needs for economic and physical safety and the inability of the Cape authorities to provide that security. Signs of a sediment of French revolutionary ideology are lacking in the eighteenth- and nineteenth-century phases of Afrikaner ethnicity. The Afrikaners' almost pathological desire for political independence was an earlier harvest reaped from South African soil. If the French revolution did have an effect on the formation of Afrikaner ethnicity, it was clearly limited to one of the three ideas, viz., liberty. Equality and fraternity never entered the picture, at least in their logical and overall application. On these points the differences between Afrikaner and British policy toward the blacks culminated.

The striving for independence and democracy was born in the conflicts with the East India Company authorities and was strengthened in the fight against British domination. What happened is comparable to the events preceding the American war of independence. This clash was the essential manifestation of a people signaling the end of schismogenesis from their ethnic source and proclaiming the birth of a new and unique ethnicity. This is a manifestation of the natural craving of every ethnic unit to wield power.

Events and trends, economical, political, and social, during the twentieth century, affected the Afrikaner ethnos significantly in more than one respect. This, however, demands a separate study.

NOTES

1. Nathan Glazer and Daniel P. Moynihan, eds., *Ethnicity: Theory and Practice* (Cambridge, Mass.: Harvard University Press, 1957), p. 25.

2. A good example is Hendrik W. van der Merwe, ed., *Looking at the Afrikaner Today* (Cape Town: Tafelberg, 1975).

3. Glazer and Moynihan, *Ethnicity.*

4. Fredrik Barth, ed., *Ethnic Groups and Boundaries* (London: Allen and Unwin, 1969).

5. The following are typical examples. David Bidney, "The Concept of Value in Modern Anthropology," in *Anthropology Today*, ed. A. L. Kroeber (Chicago: University of Chicago Press, 1965), p. 684; Charles Winick, *Dictionary of Anthropology* (Paterson, N.J.: Littlefield, Adams and Co., 1964), p. 193; Victor Barnouw, *An Introduction to Anthropology*, 2 vols. (Homewood, Illinois: Dorsey Press, 1971), 2:20.

6. A. R. Radcliff-Brown, "The present position of anthropological studies" in *Report of the centenary meeting of the British Association for the Advancement of Science* (London: Report of the British Association for the Advancement of Science, 1932).

7. S. M. Shirokogoroff, *The Psychomental Complex of the Tungus* (London: Kegan Paul, 1935), pp. vi, x, xi. Also Shirokogoroff, "Die Grundzüge der Theorie von Ethnos," in *Kultur*, ed. C. A. Schmitz (Frankfurt: Akademische Verlaganstalt, 1963), p. 254.

8. Wilhelm Mühlmann, *Methodik der Völkerkunde* (Stuttgart: Enke, 1938), p. 1. See also idem, *Rassen, Etnien, Kulturen* (Berlin: Hermann/Luchterhand Verlag, 1964), p. 47.

9. Shirokogoroff, *The Psychomental Complex of the Tungus*, p. 14.

10. Ibid., p. 17. See also Mühlmann, *Methodik*, p. 175.

11. Mühlmann,, *Methodik*, p. 156.

12. Adolf Bastian, "Der Völkergedanke," in Schmitz, *Kultur*, p. 54.

13. Robert H. Lowie, *The History of Ethnological Theory* (New York: Holt, Rinehart and Winston, 1960), p. 147.

14. I. M. Lewis, ed., *History and Social Anthropology* (London: Tavistock Publications, 1968), p. xvii.

15. Chester L. Hunt and Lewis Walker, eds., *Ethnic Dynamics* (Illinois: The Dorsey Press, 1974).

16. Leonard Thompson, "Historical Perspectives of Pluralism in Africa," in *Pluralism in Africa*, ed. Leo Kuper and M. G. Smith (Berkeley: University of California Press, 1971), p. 351.

17. E. E. Evans-Pritchard, "The Nuer of the Southern Sudan," in *African Political Systems*, ed. M. Fortes and E. E. Evans-Pritchard. (London: Oxford University Press, 1950), p. 284.

18. Shirokogoroff, *The Psychomental Complex of the Tungus*, p. 14.

19. Mühlmann, *Methodik*, pp. 183, 231.

20. Shirokogoroff, *The Psychomental Complex of the Tungus*, p. 17. See also Mühlmann, *Methodik*, p. 229 and idem, *Rassen, Ethnien, Kulturen*, p. 59.

21. Hermann Giliomee, "The Development of the Afrikaner's Self-Concept," in Van der Merwe, *Looking at the Afrikaner Today*, p. 1.

22. Ibid., p. 2.

23. Fortes and Evans-Pritchard, *Africal Political Systems*, pp. 22.

24. F. C. L. Bosman, "Die Franse stamverwantskap en kulturele bydrae tot die Afrikaanse volk," in *Kultuurgeskiedenis van die Afrikaner*, ed. C. M. van den Heever and P. de V. Pienaar, vol. 1 (Cape Town: Nasionale Pers, 1945), p. 188.

25. Shirokogoroff, "Die Grundzüge der Theorie von Ethnos," in Schmitz, *Kultur*, p. 280.

26. Shirokogoroff, *The Psychomental Complex of the Tungus*.

27. Wilhelm Mühlmann, *Geschichte der Anthropologie* (Frankfurt: Athenaum Verlag, 1968), p. 14.

28. George McCall Theal, *History of South Africa*, vol. 3 (Cape Town: Struik, 1964), p. 370.

29. Ibid., pp. 388, 392.

30. Giliomee, "The Development of the Afrikaner's Self-Concept," in Van der Merwe, *Looking at the Afrikaner Today*, p. 10.

31. Ibid., p. 4.

32. Ibid., p. 9.

33. Ibid., p. 5. See also Theal, *History of South Africa*, 3:454.

34. Jyotirindra Das Gupta, "Ethnicity, Language Demands, and National Development in India," in Glazer and Moynihan, *Ethnicity*, p. 466.

35. Tamotsu Shibutani and Kian M. Kwan, *Ethnic Stratification* (London: MacMillan, 1970), p. 76.

36. Bouke Spoelstra, *Die Doppers in Suid-Afrika, 1760-1899* (Cape Town: Juta and Co., 1963).

37. J. H. Coetzee, "Rasseverhoudinge in Suid-Afrika, 1652-1952," in *Koers* 19, no. 4 (1952).

38. Giliomee, "The Development of the Afrikaner's Self-Concept," in Van der Merwe, *Looking at the Afrikaner Today*, p. 2.

39. Theal, *History of South Africa*, 3:157.

40. Ibid., p. 77.

41. A. M. K. Oberg, "The Kingdon of Ankole in Uganda," in Fortes and Evans-Pritchard, *African Political Systems*, p. 121.

42. M. Fortes and E. E. Evans-Pritchard, eds., *African Political Systems* (London: Oxford-University Press, 1950), pp. 121*ff.*

13

The "Coloureds" of South Africa

R. C. Adams

In order to approach the subject of ethnicity among the South African "coloured" population, both in relation to other ethnic groups and among themselves, it is necessary to discuss the racial classifications used. South Africa is legally divided into four primary racial categories: Africans (the Bantu-speaking peoples), white, Asian, and "coloured." Each of these groups is further subdivided according to traditions rather than law.[1] The African peoples are segmented tribally. The white population is divided linguistically between English-speaking and Afrikaans-speaking groups. The Asians are mainly divided on religious grounds between Muslims and Hindus. The "coloured" population is subdivided into a half-dozen smaller groups owing to differences in religion and ethnic origin.

The Africans, or Bantu-speaking peoples, of South Africa comprise the largest segment of the total population, namely, 68 percent. They are followed by the whites at 19 percent of the total. The "coloureds," at slightly less than 10 percent, form the third largest group. Of the coloureds in the republic, according to O'Toole, 25 percent live in greater Cape Town.[2] At 54 percent of its population, they are the largest ethnic group in the city.

The South African government has taken upon itself to label people according to their ethnic origins. Officially, we are never told what a coloured man is, but rather what he is not. According to the definition given in the Pensions Act of 1928, the coloureds are defined as a residue, a catchall race:

A Colored person means any person who is neither white nor
    a) a Turk or a member of a race or tribe in Asia; nor
    b) a member of an aboriginal race or tribe of Africa; nor
    c) a Hottentot, Bushman or Koranna; nor
    d) a person who is residing in a native location . . . under the same conditions as a native; nor
    e) an American Negro.

But to come to a definition of an "ethnic group"—Abner Cohen writes that "an ethnic group can be operationally defined as a collectivity of people who (a) share some patterns of normative behavior and (b) form a part of a larger population, interacting with people from other collectivities within the framework of a social system. The term *ethnicity* refers to the degree of conformity by members of the collectivity to these shared norms in the course of social interaction."[3] As for (a), the so-called coloured people do not have normative behavior patterns that distinguish them from white South Africans. Therefore, one cannot speak of a coloured culture in the same sense as one would speak of Xhosa or Zulu culture. To assume, as whites and some academics do, the existence of a specific coloured culture is clearly unfounded. If one assumes there is a coloured culture, one must prove it.

In his *Culture and Poverty*, Charles Valentine writes that anthropologists and others have abused the concept of culture. They have extended it to anything that seems on the surface to be a culture.[4] Valentine is undoubtedly correct. The fact that anthropologists and others have abused the concept is evidence of uncritical observation and an inability to attain at least a modicum of objectivity. Expertise on the concept of culture gives no one the right or any reason whatsoever to abuse the concept. It is useless to put forward the analogy between the fact that every family in a certain culture has a specific subculture that distinguishes it from other families in the same culture, and the assumption that different ethnic groups in the same culture have different cultures. Such an analogy is too simplistic, in regard to the coloured people of South Africa.

It is possible that the phrase *coloured culture* points to an inherent racialistic attitude. Lucy Mair maintains that there are anthopologists who, merely by choosing anthropology as a profession, became racialistic, so that they believe in segregation or differentiation, which is often accompanied by discrimination: "but it must never be forgotten that there are anthropologists whose professional studies have led them to support apartheid."[5]

If, for instance, the phrase *coloured culture* is not applicable, can we then speak of a "culture of poverty"? According to Valentine, similar phrases are "lower-class culture," "low-income life-styles," "slum culture," and even "dregs-culture."[6] He states that indeed "these labels are part of what amounts to an intellectual fad of attributing a 'culture' or sub-'culture' to almost any social category. . . . None of these phrases refers to an idea of seminal importance like the concept of culture itself. They all present attempts to extend the application of that concept. Moreover, they are all misapplications of the original concept."

A report in *the Argus* entitled "Poverty 'Alarming'" relates the following:

Mr. D. J. de Villiers (Nat, Johannesburg West) said in the Assembly this week that poverty among the "coloured" people, especially those in the lower income groups, had reached alarming proportions. . . . It was estimated that more than half of the Coloured population were victims of the sub-culture of poverty, in which social deviations like drunkenness, violence and workshyness were accepted as normal. Coloured children in this shadow-world perpetuated the sub-culture. An intensive programme of social reform was needed in which the Coloured people themselves would have to play a leading role.[7]

The views in this report are consistent with Oscar Lewis's belief thaᴧ poverty creates its own culture and that the culture of poverty is self-generating. "Once it [the culture of poverty] comes into existence it tends to perpetuate itself from generation to generation because of its effects on children. By the time slum children are aged six or seven, they have usually absorbed the basic values and attitudes of their sub-culture and are not psychologically geared to take advantage of changing conditions or increased opportunites which may occur in their lifetime."[8] Lewis goes on the suggest that the culture of poverty has the following characteristics: a relatively high death rate and low life expectancy; low levels of education; low participation in organizations such as labor unions or political parties; no participation in medical care or other welfare programs; little utilization of city facilities, such as stores, museums, or banks; low wages and little employment security; low skill levels; lack of savings or access to credit; and no food reserves in the house.

Other characteristics include lack of privacy, frequent violence and child-beating, consensual marriage, and frequent child and wife abandonment. Thus, families are often centered around the mother, and authoritarianism in the family is marked. According to Beals and Hoijer, other common factors are a sense of resignation or fatalism and the machismo (hypermasculinity) complex among men and the martyr complex among women.[9]

One may ask, "What is the meaning of poverty?" Valentine answers that "the primary meaning of poverty is a condition of being in want of something that is needed, desired, or generally recognizd as having value . . . , the condition of being poor does have a central significance: the essence of poverty is inequality."[10] But such poverty is of course relative. The deprivation the poor feel or experience is always relative to the affluent, the opulent, or those who live in comfort. Poverty is frequently related to poor education, political powerlessness, job discrimination, and the like.

Thomas Gladwin comes to the logical and most acceptable conclusion that poor people, in whatever ethnic group they may be, have basically the same values. According to Beals and Hoijer, "Gladwin's most challenging assertion is that the poor do not have different goals, values and attitudes from those of the major culture, rather they are frustrated and prevented from realizing these goals. Consequently they seek satisfying alternatives which may not be understood by the majority culture." [11] Since the poor frequently cannot organize effective strategies for achieving community power, they tend to capitulate and fall into a negative evaluation of their own abilities. "They are psychologically crippled by negative identity feelings," "lack of motivation to strive," and "inability to delay gratification or to plan ahead." [12]

When Gladwin came to the conclusion that poor people of whatever ethnic group have basically the same values and attitudes as the majority culture, he was speaking about social strata in different ethnic groups in the same society, e. g., the United States of America, which is not a pluralistic society. But the Republic of South Africa is a pluralistic society, as Radcliffe-Brown observed long before it became the fad to speak about cultural pluralism. [13] According to Clyde Mitchell, Furnivall first introduced the term *pluralistic* and found three outstanding characteristics of plural societies. [14] First, as he puts it, "there is a medley of peoples." Each group "holds by its own religion, its own culture and language, its own ideas and ways." Radcliffe-Brown also picks out this point as the main characteristic of what he calls "composite societies." Redfield's "heteronomous societies" are much the same. A second characteristic that Furnivall isolates is that in a plural society, the different sections of the community live side by side, but separately, in the same political unit. Here then, is the "plurality" and the "society." But the significant anthropological problem is the effect each has on the other. Here Furnivall raises his third characteristic, on which he places considerable emphasis. In plural societies, he maintains, the "members of the different sections meet as individuals but only in the market place in buying and selling." It follows from Gladwin's statement that whatever the ethnic group under consideration may be, the values of the people within the social strata of a specific society depend largely upon their environment—social, economic, and political. Is the theory of a coloured culture still watertight? The terms *coloured* or *brown* or, for that matter, a *coloured* or a *brown culture* point to pigmentation. They have nothing to do with culture.

Jan van Riebeeck arrived at the Cape in 1652 to set up a halfway station, and it is sometimes jokingly said that the first coloured people were born nine months after his arrival. The so-called coloured population of the Republic of South Africa has for more than three hundred years developed along Western cultural patterns. Why, then, do whites speak of a coloured culture? They probably want to maintain the racial status quo.

The so-called coloured population speaks English and Afrikaans. Among them we find the same churches as we find in the white community, e.g., the various Anglican churches, the Roman Catholic church, the various Protestant churches, as well as Muslims and various other sects. The so-called coloured population has the same type of social organization as we usually encounter in Western industrialized society and the same kinship system, that is, the same nuclear family of father, mother, and dependent offspring. Naturally, there are variations on the theme. We do encounter matrifocal families, especially among the lower working class, but then matrifocality is exclusive neither to the lower working class, nor to the coloured people as an ethnic group. Generally speaking, the coloured people's kinship system is cognatic, and the children take the surname of the father if the parents are legally married. The coloured population's educational structure has developed and is still developing on the Western pattern. Coloured children attend primary school and high school, and some attend university. Although they share Western culture with the politically dominant group, they are separated from them by a color bar.

The coloured people's social status is determined in part by inferior social services that cement their rank in the racial hierarchy. They pay taxes at the same rate as whites, but they do not receive the same social services in return. "For example, the rate at which welfare grants are dispensed is differentiated racially. All pensions, disability grants, and child mainten-ance grants in South Africa pay about twice as much for white individuals."[15] Petty apartheid, such as segregation in public places, is also a useful tool: it effectively destroys the dignity of the coloured.

O'Toole maintains that the whites in South Africa want to maintain a strict correlation between status and color and that they have therefore imposed apartheid laws.[16] Through these apartheid laws, the Nationalist Party and most whites (because it is through them that the Nationalist Party maintains its rule) want to reduce the coloureds and the whole black population to permanent inferiors by "proving" to them that they are uncivilized. Laws are designed to limit blacks' and coloureds' opportunities for advancement.

Over the years several major statutes have drawn even finer the line between coloureds and whites. The Nationalist Party, which came to power in 1948, has implemented these statutes:

1949—Mixed Marriages Act: barred "Coloureds" from marrying Whites.

1950—Immorality Act: made sexual intercourse illegal between Whites and non-Whites.

1950—Population Registration Act: provided every South African be classified according to his race and that he carry a card identi-fying his race.

1950—Group Areas Act: segregated residential areas.

1956—South African Act: provided for a separate voter's role for 'Coloureds.'

1963—Coloured Persons Education Act: put coloured education under the Department of Coloured Affairs.

1968—Representation of Coloured Persons: segregated the Coloured group politically by removing its representatives from Parliament.[17]

The so-called coloured people are oppressed socially, economically, and politically. In this respect, they are no different from the entire black population of South Africa. Both are the victims of the white man's exploitation and his economic and political opportunism.

It is generally recognized that large numbers of coloureds, whites, and blacks share the same cultural symbols and that many coloureds are indistinguishable from whites. To assure differentiation, whites repeatedly point at the negative aspects of coloured community life and identify coloureds with laziness, dishonesty, violence, immorality, and particularly drunkenness.[18]

Because there are definite indications of an ever-increasing middle and upper class among coloureds, the phrase *poverty culture* (or *culture of poverty*) is not applicable to all coloureds, even though most are still members of the lower, poor class. A coloured university professor still remains a coloured. On the other hand, even though a minority of whites live in a culture of poverty, a poor white still remains a white. Poor whites, who are mainly illiterate and unskilled, treat wealthy, educated coloureds as their inferiors. According to O'Toole, the apartheid system and the superiority complex of whites give rise to the rule that "the lowest-ranked white man must have higher status than the highest-ranked 'Coloured' man."[19] Immigrants from Greece or Portugal who can neither speak English nor Afrikaans are immediately given higher status in the society than any coloured man. This practice is resented, because it essentially means that coloured (the whole black population as such) have a very low status in the country of their birth and that newcomers who come to exploit them are given higher status.

The differences among whites, coloureds, and blacks as broad ethnic categories are due to political, social, and economic conditions in South Africa. These conditions come from the much hated (at least among the black population) apartheid system. Consequently, the black population of South Africa is chronically threatened. This has unfortunately left its mark.[20]

The rise of black consciousness during the past five years is of considerable significance. Unfortunately, only a few academics and students have been active in advocating black consciousness, notably student organizations such as the South African Students Organization (SASO). However, they and others are constantly harassed by the police. Blacks in South Africa are, according to tradition and law, discriminated against politically, socially, and economically, purely on the grounds of color. To be black has nothing to do with pigmentation; rather, it reflects a trend of thought and awareness. To regard oneself as black is to put oneself on the road to emancipation—not only social, political, and economic emancipation, but also psychological emancipation.[21]

Most coloured people do not want to be labeled black, either because they do not know what black consciousness is all about or because they feel that they will thereby lower their status in the racial hierarchy. Therefore, they will sooner strive to identify with the whites, who in any case reject them, than threated their own position by identifying with the Bantu-speaking peoples of South Africa.

A great controversy over this issue has erupted. A glance at the letters to the editors of the various South African newspapers is enough to see that the coloured people are divided on whether they should remain coloured in order to secure their position, that is, a step higher than the Bantu-speaking peoples, or whether they should be black. Actually, in the latter case, they would not threaten their own position in the racial hierarchy—they would merely show the white man that they are proud of what they are and that they can unite with other black peoples to form a solid front against a common oppressor. However, opinions always differ, and it will take time to reach a consensus on this matter.

GENERAL THEORETICAL ISSUES

What theoretical models and concepts can one use to describe ethnicity as a dividing force in such a complex and heterogeneous society as South African society? One of the most appropriate models is that of Mitchell, namely the stuctural-categorical-personal model, with a special emphasis on the categorical order. According to Mitchell,

There appear, in fact, to be three different orders of social relationships which are characteristic of large-scale societies—possibly of all societies—but particularly of urban social systems. These are:
   a) the *structural order* by means of which the behaviour of people is interpreted in terms of action appropriate to the position they occupy in an ordered set of positions, such as in a factory, a family, a mine, a voluntary association, a trade union, political party or similar organization;
   b) the *categorical order* by means of which the behaviour of people in unstructured situations may be interpreted in terms of social stereotypes such as class, race, ethnicity, "Red" and "School" among the Xhosas of East London;
   c) the *personal order* by means of which the behaviour of people in either structured or unstructured situations may be interpreted in terms of the personal links individuals have with a set of people and

the links these people in turn have among themselves and with others, such as the social networks of the families, in Bott's study. These are not three different types of actual behaviour: they are rather three different ways of making abstractions from the same actual behaviour to achieve different types of understanding and explanation.[22]

According to Pendleton, these three types of social relations frequently overlap. Thus a person might form part of the personal friendship network while simultaneously being subject to some form of social categorization. In the last analysis, the three types of social relationship have as a common focus the interaction between people.[23]

ETHNICITY IN SOUTH AFRICA

Ethnicity Within the Coloured Group

Barth has pointed out that an ethnic group is assumed to meet the following qualifications:

1. is largely biologically self-perpetuating
2. shares fundamental cultural values, realized in overt unity in cultural forms
3. makes up a field of communication and interaction
4. has a membership which identifies itself, and is identified by others, as constituting a category distinguishable from other categories of the same order[24]

The second qualification does not bear on the coloured people, because they have always developed on Western cultural patterns, that is, the dominant culture pattern in South Africa. As already mentioned, one cannot speak of a coloured culture in the same sense as one would speak of Xhosa or Zulu culture. If the term *ethnicity* is extended "to denote cultural differences among isolated societies, autonomous regions, or independent stocks of populations such as nations within their own national boundaries, it will be of little use." The differences among the Chinese and the Indians in China and India, respectively, are national, not ethnic, differences. "But when groups of Chinese and Indian immigrants interact in a foreign land as Chinese and Indians, they can then be referred to as ethnic groups. Ethnicity is essentially a form of interaction between groups of different descent operating within a common social context."[25] For Glazer, the term *ethnic* refers to a particular descent group, even though the group may share some common cultural aspects. For him, *ethnic* is also a part of a family of terms of similar or related meaning, such as *minority group, race,* and *nation.*[26]

The term *race* of course refers to a group of people that is defined by "common descent" and that has some typical physical characteristics. When one speaks of race, one somehow implies racial purity. However, racial purity no longer exists, because there has been interbreeding on a tremendous scale over the centuries. Racial purity has thus been eliminated. Consequently, the term *race* is an anachronism.

Race refers to the biological aspect of group differences and the similarities of phsyical characteristics within groups; *ethnic* refers to a combination of the cultural aspect and a putative biological element (the assumption of common descent). Ethnic groups tend to be of great significance in individual and self-identification.

Recently Mafeje said that ethnicity (he uses the phrase *urban tribalism*) does not pervade African social life and that it exists more in the mind of the social scientist than in African social life itself.[27] This may well be so, but on the whole, ethnicity in South Africa is a real dividing force. Several external determinants, notably the apartheid system implemented by the Nationalist government under the leadership of John Vorster, ensure that this dividing force remains. This has made the black people, especially the coloureds, even more color-conscious than their white counterparts.

Ethnicity also has to do with a set of meanings that actors attribute to certain symbols, signs, or cues by means of which they are able to categorize other persons—this is a way of distinguishing outgroups from ingroups. As Mitchell states:

> The symbols, signs and cues may be costume, hairstyle, language, facial scarification, diet, or any combination of similar diacritica by means of which the actor is able to "label" or categorize the person who presents these cues. Once the cues have been recognized and interpreted and the ethnic category thereby established then the actor has available to him a set of expectations of the person's behaviour towards him. Note that the perception of the ethnic category, although fundamentally a psychological process, from the point of view of social analysis is significant for its *consequences* for behaviour and how people construe that behaviour.[28]

There is little ethnicity within the so-called coloured group. However, there is a feeling of hostility among the lower working class, the middle class, and the upper class. Of course, this division among lower working, middle, and upper is too simplistic, since there are many variations within these social strata. For example, individuals who are financially well off but who lack the educational, social, and cultural background of a higher class may not clearly fit any of them.

But what criteria can we employ in order to establish to which class a person belongs? A few criteria are education, occupation, behavioral norms, and income. Some of these criteria flow directly from others—if a person is academically or otherwise highly qualified, he will probably have a corresponding occupation and therefore a high income. He will have many ways to accelerate his social mobility and attain more prestige and status, for example, a luxury house, a car, or a television set. One who has only a primary-school education will probably not have a highly paid job, and he will therefore not be able to own such status symbols unless he obtains them through illicit means. His social mobility will be restricted, and it is doubtful whether he will ever reach the upper class. As defined by Glazer, "social mobility refers specifically to movements between the strata of society, from one occupation, income level, education standard, to another."[29]

Social classes cannot be equated with ethnic groups. One is not born an unskilled worker, a clerk, or professional. One can be born into a family whose head may hold such an occupation. Social theories that insist that birth determines one's occupation and income are wrong. If they were right, we would not be so interested in "social mobility." *Social mobility*, however, can never mean a change from one ethnic group into another, for example, as in South Africa, where white-skinned coloureds with European features frequently pass themselves off as white. Just as *social mobility* applies specifically to the class phenomena of society, so do we need specific terms or phrases that apply to the movement between ethnic groups. Here we may use the terms *assimilation, acculturation,* or *conversion.*[30]

Few academicians would argue with Glazer's view that class (and caste) results in horizontal divisions in society and that ethnicity leads to vertical divisions.[31] These divisions are very clearly expressed among the peoples discussed here.

As already noted, there is little ethnicity between members of the coloured group. In fact, a dynamic class struggle exists between the social strata within the coloured group. Language and religion are significant factors in this struggle. As far as language is concerned, there is a hostility between the Afrikaans- and English-speaking groups within the coloured group. Upper-class professional people now tend to speak English at home (not just in business), to send their children to English middle schools, and to reject the Afrikaans language because it is the language of the oppressor—the white South African. However, there is a faction, mainly academics, who consider Afrikaans as much the language of the coloured people as that of the white South African—which is correct. Most coloureds are bilingual; it depends entirely on which language a person regards as his home language. The hostility among the social classes can be regarded as snobbism, and snobbism is no synonym for ethnicity. The coloured people can be regarded as an ethnic group, and their feelings of ethnicity are usually directed toward other ethnic groups in their sociocultural environment. These ethnic groups are mainly white South Africans, the various Bantu-speaking groups (which comprise the vast majority of the South African population), Muslims, and Indians. Muslims and Indians are also popularly regarded as coloured.*

---

*This statement is not legally correct, since Indians are actually defined as Asians. See also the introductory paragraph of this chapter—Editor.

One type of ethnicity that can be perceived among the coloured people is religious, that is, between Christian coloureds on the one hand and Muslims and Indians who adhere to Islam on the other. Hostile feelings, whether latent or manifest, actually come from both sides. Christian coloureds may not acknowledge it, but they have hostile feelings toward Muslims and Indians. In their ordinary speech, Christian coloureds also have derogatory names for Muslims and Indians. These hostile feelings may be attributed to the fact that Indians and Muslims always were and still are the shopowners and the traders among the coloured people and that the Christian coloureds have the notion (not always groundless) that these shopowners and traders usually cheat their customers. The Christian coloureds see the Muslims as exploiters, for, in effect, the latter let their own people suffer in order to get rich themselves. A report in the *Weekend Argus* is a striking example of the hostile feelings between the two religious groups. It is entitled "Race Row over Port Elizabeth's Mixed Marriage."

Port Elizabeth—A young Indian man who last year married the girl he loves has been forced to leave his mother's house and has been virtually ostracised by the Indian community because his wife is Coloured. The marriage and its aftermath has led to a race row here. The young man, Mr Mahendra Daya, 24, has accused Indian leader and chairman of the South African Organization for Desegregation, Mr Raman Bhana, of doing his best to prevent the marriage.

Mr Bhana, who is a leader of the Gujerati Temple to which Mr Daya's family belong and the chairman of the temple's welfare section—as well as leader of the S. A. O. D. and one of the city's leading anti-segregationists—has denied the accusations.

In spite of the opposition from his widowed mother—he is the sole breadwinner—and his family and the temple elders Mr Daya decided in September last year to go ahead with his marriage to Miss Faith Scholtz, the daughter of a leading Coloured family. Her father, Mr P. Scholtz is a commissioner of oaths.

The wedding took place in the St. James Catholic Church in Schauderville. No relatives or friends of Mr Daya's attended the ceremony.

Mr Daya said this week that the day before the wedding an attorney's letter was delivered to him on his mother's behalf ordering him to remove all his belongings from the house. He was also warned that he would no longer be recognised as a member of the Indian community.

Mr Daya is adamant that the action taken by his mother and the Indian leaders would never have stopped him marrying his wife, they were living happily in Schauderville with an aunt of his wife and he is attending her church.

"We are in love and this is all that matters. The fact that my relatives and friends stayed away from our wedding does not bother me," he said.

Mr Daya has hit out at Mr Bhana for the part he played in the affairs. "I see Mr Bhana as fighting against apartheid and racial dicrimination but not practising what he preaches."

Mr Bhana this week denied he had played a major part in separating Mr Daya from his family and having him ostracised by the Indian community.

"It is true that 10 elders of the temple, including myself, went to see the family at the request of Mrs Daya before the marriage took place."

"However, at no time was Mr Daya threatened. We merely pointed out to him the religious and other problems he and his wife would have to face if he married a Coloured girl. We also told him how difficult it would be for their children."

He dismissed the statements that Mr Daya would be thrown out of the temple as "nonsense."

"The doors of a temple like those of a church are open to all," he added.[32]

Reports such as this emphasize that Indians and Muslims also have hostile feelings toward Christian coloureds—purely on religious grounds. Each group wants to maintain its religious exclusiveness. Muslims and Indians seldom convert to Christianity and enter a Christian church when they marry Christian coloureds. In the article cited above, Mr. Daya says that he now attends his wife's church, but this is not usually done. In most of such cases, Christian coloureds convert to Islam so that their children will not suffer the traumatic experience of having parents of different religions.

It is tragic that even color is a divisive force within the coloured community. Apartheid, which is the official governmental policy in South Africa and which discriminates against blacks, has made the coloured people all the more color-conscious, even more so than their white counterparts. With no reason at all, ordinary lower-class coloureds usually rank people with white skins and European features higher than those who have brown or black skins and African features. Light-skinned coloureds with straight hair have derogatory names for darker-skinned coloureds with frizzy hair.

In the coloured community, Muslims and Indians (who are generally the shopkeepers and traders) also have a reputation for recruiting teen-aged girls from the rural areas surrounding the Cape. These girls are then usually brought to Cape Town under the pretext that they will work for a certain Muslim family and that the girls will receive schooling and pocket money. Their actual salary will be sent to their parents in the rural area.

Coloureds who live in rural areas on the farms are often poverty-stricken. They receive meager weekly wages because they live rent-free on the white man's farm. The white man usually buys their groceries for them at a general trading store, and on Friday evenings he deducts these amounts from their weekly wage. Sometimes farmers do not even pay their laborers in cash, but on Friday evenings they distribute wine among their workers as payment. This is the celebrated and much criticized "tot-system." This distribution of cheap wine usually leads to alcoholism, because these coloureds have many financial problems and no recreational facilities. The tot-system has been much criticized, but few farmers have done away with it.

When a family then gets the chance to get rid of one mouth to feed (that is, when a wealthy Muslim family comes to recruit a teen-aged girl who will work as a domestic servant for them), the parents are all too glad to do it. However, they feel sorry afterward. In most cases the Muslim family never sends the girl's salary home, and the family loses track of its daughter. Muslim families often maltreat the girls. They never keep their promise to send the girls to school, and they reduce the girls to mere slaves. Many such cases have been reported to welfare organizations. One recent case, which made the headlines in the newspaper, concerned a Muslim couple who threw paraffin over their servant girl and put her on fire just because she did not do what the woman told her to do. The girl was badly burned. A court case followed in which the Muslim couple was convicted. Such unfortunate events widen the gap between Christian coloureds on the one hand, and Muslims and Indians on the other hand. This does not diminish ethnicity, but rather augments it.

ETHNICITY AMONG THE COLOURED PEOPLE
AS IT IS DIRECTED TOWARD OTHER ETHNIC GROUPS

When I did field research during 1975 in a coloured government housing project sprawling on the Cape sand flats, one of my objectives was to find out where the so-called coloured people place themselves in the racial hierarchy of South Africa. I found out that they definitely place the whites at the apex of the hierarchy, then themselves, and, at the bottom, the various Bantu-speaking peoples of South Africa.

This may be one of the reasons why most coloureds, who live in great poverty, do not want to identify with the black peoples of South Africa; they think that by doing so they will lower their own status and position. Therefore, perhaps in order to heighten their status and position, they identify themselves with white South Africans, who reject them in any case. The coloured people, then, are in a terrible dilemma. They do not want to lower their position and do not feel that they really belong culturally to the indigenous Bantu-speaking groups. They feel closer to the whites, culturally and otherwise, but when they want to improve their position, they are rejected. Consequently, they do not know which way to turn.

Black consciousness, a pride in being black, has attracted more students and academics over the past five years, and one can only hope that this continues. However, the ordinary man in the street does not know what black consciousness is all about. His political knowledge and activity are vitually nil, and he insists that he was born a coloured and will die a coloured.

On the other hand, whites see coloureds as a threat to their own position. Similarly, coloureds see the Bantu-speaking peoples as a threat. The blacks in South Africa do not really realize their potential—if they only stand together, they can achieve much in a short span of time.

Members of these ethnic groups (the so-called coloureds, the whites, and
the Bantu-speaking peoples) have stereotyped perceptions of one another.
These stereotypes, however inaccurate or accurate they may be, are
enforced by the political ideology of apartheid, or the euphemism of
separate but parallel development—which in any case is not parallel.

Just as Americans teach their children the myth of the "American
Dream," that is, to strive to get to the top of the social ladder to be as
affluent as possible, South African children are taught the myth of racial
superiority.[33]

Rules on the separation of whites and blacks in South Africa are strictly
and precisely defined. Whites and blacks may not marry, but there are no
restrictions against a coloured's marrying an Indian or a member of one of
the Bantu-speaking groups. "Exclusiveness" in the last three groups is more
social than legal.

However, the apartheid ideology has proved to be ineffective and
downright racist. It is hated and rejected by the black population of South
Africa, and in this sense the blacks are united against a common enemy: the
white minority government, the Nationalist party, and the whites who keep
this party in power. Apartheid differentiates among people of different ethnic origins and does so purely on the grounds of physical features and
pigmentation (skin color). Blacks (the Bantu-speaking peoples and the so-
called coloureds) are discriminated against politically, socially, and
economically. First, they are not accepted as full citizens of the country in
which they were born. Recent immigrants from European countries who
cannot even speak either of the two official languages are given higher status than blacks who were born there. Blacks have no say in the running of
the government, and in order to calm a potentially explosive situation, the
white government has created a puppet parliament for the coloured people,
the Coloured People's Representative Council (CPRC), which in any case
has no power to change the white political superstructure. However, blacks
have seen through this ineffective camouflaging of political opportunism,
and they now demand direct representation in parliament. Another
method the white government uses to break down the dignity of the black
man is through petty apartheid, by placing white/nonwhite signs on every
corner, especially in government departments and agencies, such as post
offices, railway stations, and in all other public transport, such as buses and
taxis. Restaurants also display these signs, and some of them cater
exclusively to whites. Recently a few restaurants and hotels have been
"opened" to blacks, although the managers of such hotels still must ask the
minister of the interior whether he may accommodate coloureds, let them
have a drink in the bar or dance in the lounge. The government is also
inconsistent in applying its policy, because blacks from neighboring states
can use the hotels' facilities in the same manner as white South Africans,
but without having to apply for special permission. Blacks from neigh-

boring states may also dance with whites in these hotels, but black South African men are not allowed to dance with white women.

As previously noted, members of these ethnic groups have stereotyped perceptions of one another. Categorization according to these stereotypes is enforced by artificial separation. In other words, official policy in South Africa is that blacks should live segregated from whites, physically and otherwise. Blacks are not allowed to attend white schools or to mingle socially with whites. The whole economic system in South Africa is also discriminatory, because whites get much larger salaries than blacks, even when they are doing the same work or have the same qualifications. If the discriminatory system of apartheid is done away with, members of the different ethnic groups will mingle more freely with one another and ethnic feelings will probably diminish with the passage of time, although they will not cease to exist.

In South Africa each ethnic category is distinguished by stereotyped attributes of physical appearance, dress, language, role behavior, gestures, and other characteristics. As Pendleton states, "Usually it is a combination of attributes which signals identification. Physical appearance includes skin color, body type, hair type and hair color. Most of the Bantu-speaking peoples have a dark brown or black skin color, 'Coloureds' have brown or light brown skin color, and most Whites have a light or fair complexion. The Bantu-speaking peoples and many 'Coloureds' have dark, kinky hair while Whites and many 'Coloureds' have straight or wavy hair."[34] Whites and coloureds dress similarly, and urban Bantu-speaking people also wear Western fashions. Usually only the rural Bantu-speaking peoples in the tribal areas still wear traditional clothing, such as the red blanket among the traditionally oriented Xhosa-speaking peoples. Coloureds usually use the derogatory term *Boer* when they refer to Afrikaners. They also say that Afrikaners (South African whites) are culturally and otherwise unsophisticated and that they (the coloureds) are just as good as, if not better than, the Afrikaners. According to them, the Afrikaners are far behind the times, especially when they are compared with the Americans and Europeans, who are seen as much more enlightened and broad-minded than the Afrikaners.

One must remember, however, that the white population of South Africa is just as much internally stratified as the black population. Coloureds also have ethnic stereotypes of Jews, Portuguese, and Greeks. They say that Jews are stingy, money-hungry, and that they are always trying to con you into something. Portuguese and Greeks, they say, are dirty and unsavory.

Most of the coloured's ethnic hostility is directed toward the Afrikaners, whom they hold responsible for the apartheid system. They say that however liberal an Afrikaner may seem to be, he still remains a "boer" and therefore cannot be trusted. More sophisticated coloureds will usually speak English when addressing an Afrikaner, and afterward they will laugh at him for not being able to speak English correctly or for answering back in Afrikaans. The use of English carries some prestige, especially when a person is speaking to a South African "boer" (i. e., farmer). Coloureds know that rural South Africans generally dislike the English and therefore dislike the English language.

However, coloureds in the lower socioeconomic strata have the notion that whites should always be in the authoritative positions and that a person who has a white skin and European features should automatically rank higher than any black person. One will not encounter this attitude among the higher socioeconomic strata and educated people. Coloureds in the lower socioeconomic strata also use a term denoting respect when ad-addressing a white person, such as *baas* for a white man and *madam* or *miesies* for a white woman. These coloureds also think that whites are on the whole morally and religiously superior to coloureds, but of course this attitude is groundless. One can attribute this attitude to the fact that the mass media, especially the newspapers, emphasize and exaggerate crime among the coloureds. The whites are also economically and otherwise better off than coloureds, and therefore coloureds who take the whites at face value think that latter are more moral and religious.

The ordinary man in the street is not very interested in politics. He wants primarily to keep his belly full and to secure a roof over his head. He bluntly says that he was born a coloured and that he will die a coloured. He knows little or nothing at all about black consciousness and says that the government and politics are the white man's affair, not his.

As far as ethnic feelings directed toward the Bantu-speaking peoples of South Africa are concerned, coloured people on the whole stereotype what a person "really is" according to skin color, language, and behavior patterns. Coloureds also have a derogatory name for Bantu-speaking persons, namely, *Kaffir*, although they do not know the real meaning of the word. Coloureds and whites treat all Bantu-speaking persons alike. The white and coloured populations are not significantly aware of the elite status of many Bantu-speaking persons. With a few exceptions, it is not particularly important whether persons from these ethnic categories are from the elite or low socioeconomic class; their attitudes and behavior toward one another are essentially the same.

Bantu-speaking peoples complain that coloureds think they are better, and that this attitude is getting worse. Coloureds, they say with resentment, want to be treated like whites. They say that coloured people sometimes treat a Bantu-speaking person worse than whites do. In any case, the coloureds can only dream about the solidarity that exists among the Bantu-speaking peoples.

CONCLUSION

As shown in this chapter, ethnicity is a divisive force in South Africa, and it is reinforced by the white government's official policy of apartheid.

Against this background, categorical relationships take place between people whose interaction is patterned according to some social category that is mutually recognized. Coloureds, blacks, and whites may work together or have similar qualifications, but owing to the strength of categorical stereotypes, they maintain social distance.

NOTES

1. James O'Toole, *Watts and Woodstock: Identity and Culture* (New York: Holt, Rinehart and Winston, 1973), p. 13.

2. Ibid.

3. Abner Cohen, ed., *Urban Ethnicity* (London: Tavistock Publications, 1974), p. ix-x.

4. Charles Valentine, *Culture and Poverty: Critique and Counter-Proposals* (Chicago: University of Chicago Press, 1968), p. 15.

5. Lucy Mair, *An Introduction to Social Anthropology*, 2d ed. (Oxford: Clarendon Press, 1972), p. 290.

6. Valentine, *Culture and Poverty*, p. 15.

7. *The Argus*, 14 May 1975.

8. Valentine, *Culture and Poverty*, p. 69.

9. Ralph L. Beals and Harry Hoijer, *An Introduction to Anthropology*, 4th ed. (New York: The Macmillan Company, 1972), p. 652.

10. Valentine, *Culture and Poverty*, pp. 12-14.

11. Beals and Hoijer, *An Introduction to Social Anthropology*, pp. 652, 659.

12. Valentine, *Culture and Poverty*, p. 93.

13. Alfred R. Radcliffe-Brown, *Structure and Function In Primitive Society* (London: Cohen and West Ltd., 1952), p. 202.

14. J. Clyde Mitchell, "Tribalism and the Plural Society," in *Black Africa*, ed. John Middleton (London: The MacMillan Co., 1970), pp. 264-265.

15. O'Toole, *Watts and Woodstock*, p. 18.

16. Ibid., p. 20.

17. Ibid., p. 19.

18. Ibid., p. 28, 30.

19. Ibid., p. 30.

20. R. C. Adams, "Kinship, Family Structure and the Family Life of Five Families in Bishop Lavis," (Research Project Report, University of the Western Cape, 1975), pp. 1-11.

21. Ibid., p. 2.

22. J. Clyde Mitchell, ed., *Social Networks in Urban Situations Analysis of Personal Relationships in Central African Towns* (Manchester: Manchester University Press for the Institute of African Studies, University of Zambia, 1969), pp. 9-10.

23. Wade C. Pendleton, *Katatura: A Place Where We Do Not Stay* (San Diego: San Diego State University Press, 1974), p. 8.

24. Fredrik Barth, *Ethnic Groups & Boundaries* (Boston: Little, Brown, 1969), pp. 10-11.

25. Cohen, *Urban Ethnicity*, p. xi.

26. Nathan Glazer, "The Universalization of Ethnicity," *Encounter*, February 1975, p. 8.

27. Archie Mafeje, "The Ideology of Tribalism," *Journal of Modern African Studies* 9 (1971): 7.

28. Cohen, *Urban Ethnicity*, pp. 22-23.

29. Glazer, "The Universalization of Ethnicity," p. 11.

30. Ibid.

31. Ibid.

32. *Weekend Argus,* 14 February 1976, p. 15.

33. O'Toole, *Watts and Woodstock,* p. 3.

34. Pendleton, *Katatura: A Place Where We Do Not Stay,* p. 88.

14
South African Indians and Economic Hostility
Bridglal Pachai

By any criteria, in a country of rigorous standards and colossal denials experienced by black peoples, Indians, who today number a little more than 700,000 and constitute 2.8 percent of South Africa's population, have made a valuable contribution to the South African economy out of all proportion to their meager numbers. Yet a more despised and reviled community in South Africa could not be found. How have a successful people become so stigmatized as to occupy the lowest rating in a country in which the record points in the other direction? The main answer to this conundrum lies in public denunciations by white detractors and in government policies pursued in the nineteenth and twentieth centuries aimed at giving legislative support to the denunciations. Indians were in effect victims of the vice-grip of the mentality and politics of racial superiority of the Anglo-Saxons and other Europeans in the nineteenth and twentieth centuries (this is ably reviewed in Huttenback's most recent book).[1]

The assertion of European racist superiority manifested in attitudes and laws was aimed at keeping down the number of non-European immigrants so as to protect the present and the future interests of the European rulers. In all parts of white settlement in the British Empire (Natal, New South Wales, and British Columbia, for example), immigration laws were introduced to stem the tide of Indian (and other Asian) immigration so as to allay fears of being swamped be Asia's abundant population. In Natal, by 1893, thirty-three years after the first immigrants arrived from India, the Indian population exceeded that of the white population by 46,000 to 45,000. In view of this there seemed to be sufficient reason in the European rationale for the introduction of an immigration law in 1897 prescribing a test in a European language. This virtually closed the door to Asian immigration in the pre-Union period and became the formula for other countries of white settlement. To leave no doubt as to the government's determination to exclude Indian immigrants, the post-Union Immigration

Restriction Act of 1913 invested immigration authorities with the administrative power to bar persons whose economic standards and habits of life were deemed to be unacceptable to white South Africa. By administrative action, not by racial decree, the South African government set out to keep the numbers down, adopting a "closed-door" policy that the Indian government was prepared to go along with provided those already resident in the country would be accorded every opportunity to raise themselves, as permanent citizens, including the ultimate conferment of full civic rights.[2]

In the end, nothing of the kind emerged: the quid pro quo was a nonstarter; hostility lived on when population figures were not even remotely a threat to white predominance. In the South African context, interethnic relations are complicated by the fact that groups are labeled "South African" and "foreigners." Whatever the extent of the discrimination against indigenous blacks, they cannot be considered "foreigners"; similarly, the coloured population is as local to South Africa as the Dtakensberg Mountain. The whites, because of their dominant position, consider themselves South African in spite of the fact that they owe their numbers to immigrants who arrived over many centuries. Indians have been labeled foreigners for decades. Their social stability and economic success did little to remove this stigma. Above all, they originated in a subcontinent that was already heavily populated. South Africa would not welcome any of this abundant population even if their resettlement could be arranged. In short, interethnic divisions and relationships in South Africa are extremely complex. This chapter will deal with but one aspect of this complexity, namely, the economic hostility against Indians.

THE INDIAN POPULATION AND TRADE AND INDUSTRY

Indians made their first appearance in Natal—through a state-sponsored immigration scheme, which began in 1860 under Law 14/1859. This was terminated unilaterally by India in 1911 in protest against Natal's treatment of its free Indian population. This population was a natural development: immigrants stayed on legally to become permanent residents by choice and raised families who grew up as South Africans, knowing no other home. Traders came in on their own under Law 15/1859 primarily to meet the commercial needs of the laboring community and added to the free population what was numerically a drop in the ocean. Their very presence and their considerable commercial prowess raised a veritable tempest in white politics generally, but they, their supporters in India and elsewhere, and a number of influential whites in South Africa itself more appropriately regarded this not as a tempest but rather as a tempest in a teapot.

The beginnings of Indian trade were very modest. In 1870 the first two Indian traders appeared on the scene to serve an Indian population in

Natal of 6,000. In 1880 the number increased to 50 traders for an Indian population of 20,536. In 1885 there were 115 traders for an Indian population of 30,159, a European population of 36,701, and an African population of over 100,000.[3] Clearly, there was no question of overtrading. It was a matter of opening up small retail stores throughout the country, in urban and rural areas, and of taking essential goods to consumers of all races at competitive prices. While European traders were interested in settled town areas served by modern amenities, were fastidious about opening and closing hours, were impatient over bargaining, were inclined generally toward wholesale trade, viewed their trading centers as working places restricted to working hours, were disinclined to learn the home languages of their customers and to converse in the customers' non-European languages, the Indian traders were the exact opposite in all these matters.[4] The trader-consumer relationship was generally extremely fraternal, and the trading place was the Indian trader's kingdom on earth, to which he devoted all his and his family's energies, acumen, and time. Since he lived sparingly, had little inclination for luxury goods or luxurious living, and was more frugal and abstemious than his European counterpart, he could do with smaller margins of profit. These practices were described as unfair according to the rules defined by European traders. Whatever the one-sidedness of the rules, the European traders wielded considerable political pressures: they were the lawmakers. The issue of unfair trade competition was raised in the *Volksraads* of the Orange Free State and the Transvaal, the Legislative Councils and Assemblies of Natal and the Cape, and the Houses of Parliament of South Africa. Racial prejudice was the common bond holding the political pressures together, but trade rivalry or jealousy was the spark that set off the pressures.

One aspect of the prejudice against Indian traders was that they exported their profits and did not reinvest them for the economic benefit of the country in which they made their profits. Whatever figures are available do not suggest that huge sums of money were exported.[5] After all, profits were not all that large, and whole families lived off the proceeds. The export charge was more of a red herring than an actual problem, but it was nonetheless a good rallying point for whipping up emotions. What critics then and later overlooked was that the Indian trader found himself in the center of a violent and emotional controversy, the result of which was that he was torn between two loyalties: having moved away from the old, he was unwelcome in the new. Consequently, he kept his options open. If he did not invest as much as he ought to have done, this was due to both the dilemma of commitment and the lack of encouragement and opportunity.

The Orange Free State is an excellent example of the arguments against Indian traders. It is the one part of South Africa that succeeded in keeping them out completely. The first Indian traders entered the Free State about

1882, and by 1884 two or three stores were started. This aroused a fierce agitation, in spite of the fact that the actual number of Indians resident in the Free State was minimal: "at no stage were there more than 15 Asians in the Free State before 1900."[6] Yet in 1884 three petitions were handed to the *Volksraad* stating that ". . . many Asians are settling in this country and that the laws of our state were made without our lawmakers having thought that such a people would settle amongst us. These people occupy a country with the sole purpose of making money which is not spent in that country. It is generally known that in Natal where a large number of the people live, all their riches, with the exception of minor expenses for personal support, are sent to India."[7] The familiar charges of underselling and unfair competition were echoed by the Harrismith and Bloemfontein Chambers of Commerce, then under the control of English and German businessmen. Nor did it stop at that. The press picked up the issue and, as was generally the case, found itself divided. *The Friend* supported the traders, while *De Express* made it clear that "our wish is to prevent in a lawful way and by lawful means that an undersirable element should nestle in our midst, by such restrictive measures as the authorities will lay down."[8]

What *De Express* implied by "lawful means" was nothing less than bringing political pressures to bear on the lawmakers to use the law as an instrument to preserve white trading monopoly. Whatever support the Indian traders had, such as from *The Friend*, was not consistent or permanent. Indians were not constituents to be served, since they had no citizenship rights. Ordinance No. 29 of 1890 made sure that the Indian presence would be eliminated by the undercutting of its economic base: "No Asian would have the right to trade and/or farm. Evading the law was not possible any longer and infringement was punishable. Asians wishing to settle in the Orange Free State would have no other economic means than acting as labourers on the same level as the African population." Dr. van Aswegen goes on to say, "With this law the fate of the Asians in the Orange Free State was sealed. At this stage there were only 9 Asians in the country."[9] So much for the fear of numbers.

The other part of South Africa where Indians were in small numbers was the Cape, whose Indian population numbered under 3,500 in 1899, most of whom were in commerce in the urban centers of Port Elizabeth, Cape Town, Kimberly, Mafeking, and East London. Their plight was not an easy one: "On the whole and in spite of steady industry and frugal living standards, Indian traders made 'a very poor living at the best.' Their shops were generally in poor areas where they served a most important purpose, selling at a small profit margin to the poor of the community (including whites), among whom they lived. Consequently many did a turnover of no more than £30 per month and their insolvency rate was high." Between 1905-1907, for example, the Indian insolvency rate accounted for 79 percent of the total in the Cape Town area.[10] Insolvency for the sake of defrauding creditors—a baser practice—was sometimes resorted to.

Until Indians were indentified separately in the Cape in the 1891 census (1,453 persons, or under 1 percent of the population), there was no political agitation against them as a people. Thereafter the Indian community, especially its trading component, came under unfavorable mention. Mrs. Bradlow tentatively ascribes this behavior to "mass psychology" and "the irrational reactions of a developing xenophobia." The usual pattern of legislative enactments followed: the Immigration Restriction Act of 1902; amended in 1906 to tighten things up; bolstered by the General Dealers Act 35 of 1906, which "was meant finally to deliver the coup de grâce to economic competition."[11] There was nothing in the Cape comparable to the total exclusion in the Orange Free State or the total harassment in the Transvaal and Natal, both of which reveal the darkest sides of the economic hostility toward Indians in South Africa.

Although in the Orange Free State and in the Cape there was no need for contract or other forms of labor in large numbers, the plantations of Natal and the mines of the Transvaal suffered from chronic labor shortages. As long as labor could be guaranteed by contracts that would terminate in the country of the laborers' origin, there was no cause for alarm. But as this guarantee was not specially catered for in the immigration laws, laborers remained in the country, largely as free laborers and only in small numbers as petty hawkers, small retailers, and insignificant wholesalers.

In the 1880s there was an organized clamor for the exclusion of free Indians from Natal for a whole host of reasons, ranging from insanitary habits to the fact that they roamed about the country and competed in trade. Harry Escombe, who was later to become attorney general and then prime minister, spoke on these matters in the Natal legislature:

> The honourable member for Umvoti County summarised his objections to the Indian population in a very few words, and under two heads. He said he objected to the Indians being allowed to roam about the country and to compete in trade. Very many of the European population roam about the country, some on political errands, and some on cattle-stealing errands. . . . Another objection is that they compete in trade. I wish with all my heart that they could compete more, because competition is for the benefit of the country. What does it mean? It may mean an injury to the trader, but it means a profit to the consumer, and it is an extraordinary thing the Colony is so misled that it cannot see that the competition of the Indian trader is for the benefit of the whole Colony.[12]

The benefit factor argued for so long and so vehemently failed to divert the economic hostility. The Indian Immigrants' Commission under Mr. Justice Wragg reported in 1887 that Indian trade competition was legitimate in its nature "and that it would be unwise, if not unjust, to legislate against them." Yet legislation was heaped upon legislation aimed at crippling Indian traders. In the Transvaal, Law 3 of 1885 required Indian traders who settled in that republic for trading purposes to register at a cost of £25, later reduced to £3. Separate areas were set aside for both residence and trade, despite an understanding reached with the British government that traders would be exempted from the segregation law. A number of resolutions were passed in the Transvaal *Volksraad* limiting Indian trade, but these were rarely enforced since President Kruger himself held the view that Indian traders were of service to his people. In this he was supported by a number of petitions.[13]

Of all the places in South Africa before Union, the greatest economic hostility and the highest enforcement rate were evident in Natal, the most English part of South Africa. Here the Dealers' Licences Act of 1897 wreaked havoc on the Indian trading community, since it granted licensing officers discretionary powers to issue or to refuse trading licenses. Between 1895 and 1908, 11,765 licenses were issued to Indian traders, and 20,472 licenses were issued to non-Indians, mostly whites. There was nothing of this scale in any other part of the country, since about 80 percent of the Indian population resided in Natal. Consequently, when renewals were refused periodically, considerable agitation was generated on all sides. Around 1908 it was virtually impossible to obtain a new trading license. The two issues that became central were annual renewals as well as transfers of existing licenses.

In order to find a permanent solution to recurring hostility, the Natal government introduced two bills in 1908 along the lines of the exclusion policy of the Orange Free State: Bill No. 5 provided that no new trading licenses would be issued to Indians after 1908, and Bill No. 6 provided that all Indian trading would cease after December 31, 1918, on payment of compensation. The bills were passed in Natal but were vetoed by the British government. Natal was not prepared in 1908 to take the bull by the horns and ban all Indian immigration. Such an act would have meant economic ruin. In fact, a commission deliberating at this time on the consequences of the loss of Indian labor to Natal reported a year later that Natal could not survive without Indian labor. It reported that "evidence shows that demand is made in many quarters that another adequate supply of labour shall be guaranteed before Indian immigration is stopped." The commission reported that 6,149 Indians were employed in general farming, 7,006 on sugar estates, 3,239 in coal mines, 1,722 on tea estates, 2,371 on railways, 1,949 as domestic servants, 1,062 in corporations, 740 in brick yards, 606 in wattle plantations; and 422 were employed by landing and shipping agents.[14]

Rather than tackle immigration, which would be economically suicidal, the Natal government grappled with traders. The hypocrisy was clear. As the *Natal Mercury* noted, "It is the rankest hypocrisy on the part of the Colonial Secretary [of Natal] to say that the Bills [trading bills of 1908] are conceived in the interest of the native population. . . . Could there be anything more ludicrous than the practical shelving of a Bill to stop Asiatic immigration and the passing of two Bills to prevent Asiatics, and even Natal-born Indians, from trading in any shape or form?"[15]

The Natal government hesitated to take measures to stop Indian immigrants from entering the country in 1908 because of the labor needs of the colony. Once the Indian government terminated emigration in 1911, the South African government proceeded to bar Indians after 1913 through administrative action. A voluntary repatriation scheme was introduced in the Indian Relief Act of 1914, which provided for free return passages and bonuses for individuals and families who wanted to settle in India. Very few took advantage of the scheme, and a new approach was worked out in the Cape Town Agreement of 1927, which recognized that the term *repatriation* was a misnomer. The *assisted emigration* scheme that replaced it offered better facilities and greater cooperation between the governments of South Africa and India, but it was even less popular. In spite of its lack of popularity or success, the provisions remained on the statute book until they were finally repealed in 1975. One person took advantage of the scheme in 1970, and that was the last.

In addition to the repatriation and assisted emigration schemes, an abortive attempt was made in 1933-1934 to work out a colonization scheme to settle South African Indians in British North Borneo, British New Guinea, and British Guiana.[16]

What was apparent, but not publicly acknowledged by all the parties concerned, was that South Africa's Indian population would not voluntarily elect to remove itself from the country to which the pioneers had come by invitation and in which succeeding generations were born and reared. Though only in 1961 did the South African prime minister acknowledge that Indians in South Africa were a permanent part of the country, this was always so in actual legal terms. What white South Africa did to handle "the stranger that is within our gates" was to cripple him economically and socially.[17] The Indian trader, as we have noted, was the prime target in all parts of South Africa. Even after the Union of South Africa was established in 1910 and the immigration gates were closed, thus ensuring that the Indian population never reached 3 percent of the total population, there was no respite for the Indian trader. The Asiatics (Land and Trading) Amendment Act (Transvaal) of 1919; the Class Areas Bill of 1923; the Areas Reservation and Immigration and Registration (Further Provision) Bill of 1925; the Asiatics (Transvaal) Land and Trading Act of 1939; the Trading and Occupation of Land (Transvaal and Natal) Restriction Act of 1943; and the Group Areas Act of 1950—all aimed in one form or other at the economic strangulation of the Indian trading class in South Africa by severely restricting opportunities for both residence and trade.

If it were simply a matter of discriminatory legislation, racial antagonism was not necessary to achieve white monopoly of trade. In South Africa, racial antagonism was engendered first, and discriminatory legislation followed. In spite of the fact that a white superintendent of police could testify in the Natal of 1885 that "in the present condition of Natal I do not think it possible to substitute a White for an Indian population. . . . I could deal with 3000 Indians with the staff that I have, but if there were 3000 corresponding White British workmen, I could not,"[18] the most uncharitable things ever said about any community in South Africa were said about the Indians: "The man in the street hates him, curses him, spits upon him, and often pushes him off the footpath. The Press cannot find a sufficiently strong word in the best English dictionary to damn him with."[19]

This has been the milieu in which the South African Indian has lived, worked, and traded since 1860. The public image presented by his detractors has been that of an unworthy person. He is generally equated with trade, wealth, and dishonesty. But the facts belie this image. The following figures show the distribution of Indian workers in industry in Natal in 1970: agriculture: 4.8 percent; mining: 0.4 percent; manufacturing: 41.9 percent; construction: 6.8 percent; commerce: 24 percent; transport: 5.1 percent; services: 15.3 percent; and other: 1.5 percent.[20] Although over 58 percent of the total Indian labor force in South Africa are factory workers, only a little more than 21 percent are engaged in commerce.[21] As to the charge of wealth, the average monthly per capita income of the various groups in South Africa in 1973 show that the stereotype is misplaced: white R184, Asian R37, coloured R29, and African R10.[22]

This is not to say, however, that there are no rich Indians. There are wealthy individuals in all groups in South Africa, but they are not bracketed as a community as Indians are. The sins and successes of individual Indians are visited on the community as a hangover from the public image so viciously created in days gone by. Individual successes record another side of the picture: investment in manufacturing industries has shown a marked increase and is expected to reach R60 million in 1980; the number of Indian-owned factories has increased sixfold from 1972, from 110 to 712; Indians own some 400 garages and service stations today in Natal and 52 footwear factories; and the all-Indian New Republic Bank opened in 1971.[23] These are but selections from the economic scene of 1976. They underline the survival of Indian initiative and enterprise in trade and industry in spite of the long and painful chapters of economic hostility and denials in South African history.

The assessment of the South African Indian journal, *Indian Opinion*, in 1907 at the height of economic hostility, is generally true for a later period as well. Under the title "Business is Business," it noted:

Much of the antagonism to the Asiatics in South Africa rests upon a trade basis. . . . The fact of the matter is, of course, that, just as the South African does not farm, but looks on whilst the weeds grow among the mealies, so the South African trader does not trade, but expects that a benign Providence will make up for his lack of initiative and flagrant inefficiency. There are two classes of traders in South Africa who really understand their business—the Semitic immigrant from Eastern Europe and the Asiatic. The latter is the former's chief competitor. Hence the agitation largely initiated by him to drive the Asiatic from the country. The cry has caught on with those who have failed in trade by reason of their natural ineptitude. Hence, also, the charge of unfair competition levelled against the Asiatic. There is a great deal of ignorance in the whole affair and not a little hypocrisy. . . . When will they learn to look within for the reasons of their own failure, before they seek to lay the blame upon the shoulders of others?[24]

The significant facts are that South African Indians are not a predominantly trading community; they are not even a wealthy community; nor are they either an honest or dishonest community. They are composed of individuals as heterogeneous in their occupations, incomes, aspirations, and achievements as any other people. The economic hostility that has raged for so long and so fiercely has tended to lump a community into a single category with monolithic virtues and vices. This is as incorrect and as inapplicable to this commuity as, for example, it is to the Jewish community. Traders, like other professionals, are best considered as a class rather than a community tarred with a single brush.

NOTES

1. Robert A. Huttenback, *Racism and Empire: White Settlers and Colored Immigrants in the British Self-Governing Colonies, 1830-1910* (Ithaca, N. Y.: Cornell University Press, 1976).

2. Bridglal Pachai, *The International Aspects of the South African Indian Question, 1860-1971*, (Cape Town: Struik, 1971) pp. 59-68.

3. B. B. 1870-1885: N. and I. B. B. Vol. 66. Commission Report, Chapter 40, Natal Archives, Pietermaritzburg.

4. The testimony of a white consumer signing himself *Zululander* appeared in the *Natal Mercury* in about 1908 or 1909 and is reproduced in H. S. L. Polak, *The Indians of South Africa, Helots within the Empire and How they are treated* (Madras, 1910), p. 20. Part of this reads "In the first place, he [the Indian trader] makes storekeeping his especial study. He is not part storekeeper, part farmer, part money-lender. He is, above all, extremely obliging, and no trouble is too great so long as a customer can be pleased. In this he is a striking contrast to great numbers of European storekeepers. . . ."

5. The figures are available for the entire period of organized immigration to

South Africa, i. e., up to 1912, but the selection below covers a ten-year period to illustrate the paucity of the amounts remitted to India. The figures include remittances by laborers and traders.

Year	Amount Remitted		Year	Amount Remitted
1877	£97		1882	£440
1878	£183		1883	£754
1879	£272		1884	£901
1880	£280		1885	£589:3:3
1881	£270		1886	£516:7:1
			1887	£585:16:0

H. P. Chattopadhyaya, *Indians in South Africa. A Socio-Economic Study* (Calcutta, 1970), p. 103.

6. H. J. van Aswegen, "The Asians in the Orange Free State," in *South Africa's Indians: The Evolution of a Minority Community*, ed. Bridglal Pachai and Bala Pillay (London and New York: Dalhousie-Longman African Series, forthcoming) n. 4.

7. Ibid.

8. Ibid.

9. Ibid.

10. Edna Bradlow, "Indians at the Cape during the period of Responsible Government," in Pachai and Pillay, *South Africa's Indians.*

11. Ibid.

12. J. T. Henderson, ed., *Speeches of the Late Right Honourable Harry Escombe* (Pietermaritzburg, n.d.), p. 109.

13. Polak, *The Indians of South Africa,* pp. 71-74.

14. Bridglal Pachai, *The History of Indian Opinion* (Cape Town, 1963), p. 100.

15. *Natal Mercury,* August 5, 1908.

16. Bridglal Pachai, "Aliens in the Political Hierarchy," in Pachai and Pillay, *South Africa's Indians;* and Pachai, *International Aspects,* chaps. 2-4.

17. A caption in the Vancouver *World,* November 19, 1906, cited in Huttenback, *Racism and Empire,* p. 177. This refers to the immigration of Indian Sikhs to British Columbia. Between January 1 and the end of November 1906, 2,193 Indians arrived in Canada.

18. *The Collected Works of Mahatma Gandhi,* vol. 1 (Ahmedabad, 1958), p. 264.

19. Ibid., pp. 159-160. Open letter dated December 19, 1894, written by M. K. Gandhi and addressed to the members of the Natal legislative council and the legislative assembly.

20. Gavin Maasdorp and Nesen Pillay, "Indians in the Political Economy of South Africa," in Pachai and Pillay, *South Africa's Indians.*

21. H. E. Joosub, "The Future of the Indian Community," in *South African Dialogue,* ed. Nic Rhoodie (Johannesburg, 1972), pp. 426-427.

22. *A Survey of Race Relations in South Africa,* 1975 (Johannesburg: South African Institute of Race Relations, 1976), p. 163.

23. *South African Digest* (Pretoria), February 13, 1976.

24. *Indian Opinion* (Durban), June 8, 1907.

15
The Future of Ethnicity in Southern Africa

Wilf Nussey

The way history is evolving in the southern African subcontinent—the newest of the world's flashpoints—a head-on, cataclysmic collision seems inevitable between the fast-expanding extremes of white and black nationalism from the Zambezi River in the north to the Cape Peninsula in the south. It threatens simultaneously to plunge four states into a ghastly civil war, which would very quickly drag in at least six more, swamping some of them entirely and involving a total of more than 50 million people.

This situation has been gradually developing since the late 1950s, when the crumbling of colonialism and the advent of independence triggered winds of change in black Africa, while, at about the same time, radical white nationalism came to power in the south. The pace quickened when Rhodesia drew the line for the march of black independence in November 1965 by unilaterally declaring itself independent under white rule. It moved into high gear after April 1974, when Portugal's great empire abruptly collapsed—giving enormous practical and emotional impetus to black nationalism and, in direct proportion, strengthening the resolve of the diehards among southern Africa's whites.

The world, by and large, watched dispassionately and even cynically until late in 1975, when an event occurred that altered the character of this confrontation: from a purely regional affair it became one of worldwide significance, thus making it an important factor in the global ideological and strategic struggle between the communist East and the free West. This was the skillfully timed, dramatic entry of the Soviet Union into the Angolan civil war with heavy weapons and Cuban troops to back the Marxist MPLA movement against the guerrillas of its Unita and FNLA opponents. It was the first Big Power involvement in southern Africa, and in a way it was comparable to the prelude to the Vietnam debacle.

Since then, southern Africa has been exposed to unprecedented and anxious world scrutiny. There has been a considerable increase here of foreign media representatives, and there has been intense behind-the-scenes diplomatic maneuvering in Washington, London, Geneva, and elsewhere, including several African capitals. The accent in all this extraordinary attention is very much on how soon majority rule can be achieved and on the race war that almost everyone believes must intervene. But war is not, in fact, inevitable.

Largely unnoticed or purposely ignored, a fascinating new formula is emerging in southern Africa. It could not only defuse the racial bomb but also be nurtured into a fresh ideology enabling people of all creeds, colors, and cultures to achieve their aspirations while living together in complete equality, freedom, and peace. It is still a fragile philosophical flower, this thing, growing on southern Africa's dungheap of prejudice, emotion, and fear. It would wither instantly if left fully exposed to the blast of white or black extremism, but it is being determinedly sheltered by a growing number of people of all races who see no future for anybody under the political structures now on offer.

These are "one man, one vote"—majority rule by another name—or the universally condemned policy of apartheid, otherwise separate development, invented by colonialism and polished into diamond-hard discrimination by the ruling National Party of South Africa. In the prevailing southern African emotional climate, the former would undoubtedly lead to black rule, which in the rest of Africa, with rare exceptions, has become severe autocracy—sometimes benign but usually dictatorial and too often downright barbaric, and generally accompanied by decline in the economy and in the quality of life. The latter, apartheid, leads directly to conflict. In no way can blacks and browns and their aspirations be corralled territorially, socially, economically or in any other way (this is simply because they are black or brown) without an eventual explosion of frustration.

The new philosophy is not a panacea, a political pentothal to plunge everyone into happy, color-blind euphoria. Nor is it a middle-of-the-road attempt at compromise, which is notorious for failure in Africa. It is, rather, an entirely new direction. No firm label such as "apartheid" of "majority rule" has yet been coined for it, but it does have a tag, albeit a semantically awkward one: *ethnicity*.

The word is resented by those prickly to any suggestion of race definition, by those who fear it is another guise for discrimination. It is also resented by the segregationists, who see in it a Trojan horse of integration. In fact, ethnicity is certainly not discriminatory in the sense of denying or depriving people on the basis of their color. Nor does it go to the other extreme of forcing them to integrate. But it firmly grasps the nettle that the advocates of apartheid and majority rule refuse to even examine: that the society of southern Africa is made up of many different groups spanning almost the entire spectrum of humanity.

That this is so, and that this society contains enormous diversity in languages, traditions, cultures, and levels of development, by twentieth-century standards, are inescapable facts of southern African life. They are absolutely fundamental to any political philosophy that hopes to succeed. Equally fundamental is that this very diversity contains a potential richness of human art, science, energy, and cooperation unmatched anywhere else on the African continent, indeed, in very few other parts of the world.

Apartheid recognizes that there are differences among South Africa's many peoples. It is a philosophy rooted in the concept that these differences are so great that the various ethnic groups are incompatible and mutually harmful; therefore, they must by physically separated (as far as territorial and economic circumstances permit) and cooperate, as it were, at arm's length. In its practical application, however, apartheid has emerged as a means primarily to preserve white civilization against what its adherents regard as black barbarism and the debilitating influences of other cultures. In particular, it represents the climb to power, after generations of subjugation, of the Afrikaner "tribe"—as represented by the ruling National Party (though definitely not by all Afrikaners). But apartheid has failed because it crudely and cruelly defines people as superior or inferior by the arbitrary yardstick of color, with the whites clearly judged as superior. And the welter of laws that has grown up to segregate the others is sowing dragon's teeth of frustration and indignity.

Ethnicity also recognizes the differences, but it seeks only to define or distinguish people—*not* to separate them—by culture, tradition, and other ethnic characteristics that create the differences in their ways of life and aspirations. Color does not enter into it, any more than it does among the Welsh, Scots, Yorkshiremen, Cornishmen, and others who make up the polycultural society of Great Britain. Or among the multitude of hues that form the Brazilian nation. Or among the array of technologies, craftmanships, arts, and creeds of which America is composed.

At the same time, ethnicity places absolutely no restrictions on association among people, no barriers between them except those they wish to raise themselves for the protection of their own cultures and traditions—provided these do not impinge upon the rights of any other people. That all people are born equal is a fallacy maintained by philosophies seeking the unattainable holy grail of a totally homogeneous society, and it is quite understandably perpetuated by those who suffer discrimination. But ethnicity has the basic principle that all people should be born into equal opportunity and be able to pursue their ambitions with equal facilities and equal rights. Its only reservation is that no group or individual shall do so to the detriment of any other group or individual.

This simplified outline of the ethnicity idea might sound too idyllic to be practicable in southern Africa's present climate of overheated emotions. That it is a real, pragmatic policy being seriously pursued as the only viable solution to southern Africa's volatile problems has been revealed by several recent developments.

Foremost is the just-completed constitution for an interim government that will assume control in the controversial territory of South West Africa/Namibia about the middle of this year and lead it to full independence at the end of 1978. It was drafted by legal advisors to the various delegations at the independence conference held at the old Turnhalle in Windhoek; it has been approved with relatively minor alterations by the conference's constitutional committee; and it is expected to be the blueprint for the final independence constitution, which must be thrashed out in detail in the interim government's eighteen-month life-span.

The second significant development is the fourteen principles compiled by a committee appointed by South Africa's three opposition parties and headed by a retired Transvaal Supreme Court judge, Mr. Justice Kowie Marais. The principles, intended to form the basis for a united South African opposition party, are remarkably similar in concept to those adopted by the Turnhalle conference in Windhoek. Both are aimed at giving all groups and individuals their rightful places in the southern African sun while removing their natural fears of domination and persecution by other groups—and not necessarily by groups of other colors. Both the Turnhalle and Kowie Marais documents are, in essence, based on ethnicity.

Now this seedling has taken root in a most unexpected place: Rhodesia. It is almost hidden by the fogs of hysteria and enmity generated by the escalating guerrilla war there, but it emerges firmly in the decision of the Rhodesian Front government to abandon virtually all forms of discrimination based on race. In the process of forcing through legislation that emasculates the Land Tenure Act (Rhodesia's equivalent of South Africa's apartheid laws), Ian Smith and his supporters are also recognizing ethnicity. They are doing so by retaining the numerous Tribal Trust Lands reserved for the exclusive use of the country's many different African tribes, while at the same time preserving less than 1 percent of the entire country for whites, as opposed to nearly half the country previously.

The SWA/Namibia interim constitution is the clearest exposition of ethnicity. With wise and firm leadership from among the vast majority of South Africans of all races, who want no extremism of any kind, it could be used as a blueprint for a sane and peaceful South Africa. Although Rhodesia is already far along the road to chaos, which would benefit nobody except Moscow and Peking, it could just be the savior there, too.

Its keystone is a Bill of Rights to guarantee all fundamental human freedoms for everyone, irrespective of race, origin, sex, language, religion, or political persuasion—with the sole exception that it prohibits Marxist-Leninist parties as hostile to the state.

On top of this the constitution (which the South African Parliament is expected to ratify without demur) has a three-tier government structure that involves all the country's eleven ethnic groups while enabling each to preserve its identity.

At the top will be a sixty-member National Assembly, where all groups will have a modified proportional representation broadly related to population size, ranging from twelve members for the Ovambos with 396,000 people (46 percent of the population) to four for the Bushmen. The executive arm of this legislative body will be an eleven-member Council of Ministers, with all ethnic groups equally represented. Head of the council will be a twelfth man chosen by the National Assembly to be de facto prime minister.

The second tier of government will be representative authorities controlling the domestic affairs of the various ethnic groups and also serving as electoral colleges to select deputies for the National Assembly. However, if any ethnic council so chooses, it can leave its affairs in the hands of the National Assembly and act only as an electoral college.

The constitution has an ingenious safety valve: its empowers the National Assembly to create additional electoral colleges to represent "persons who cannot under the present arrangement be represented in the National Assembly." This clears the way for "open" councils for people who cannot identify with any one of the defined ethnic groups—such as those who reject ethnic representation on principle, or those who have reached a high academic level or otherwise advanced so far into modern-day society that they have shed their ethnic backgrounds and cultures. From this group could emerge political movements untrammelled by ethnic obligations, although the spirit of the constitution is that the ethnic representatives will do the same in matters of national interest.

The constitution contains another remarkable facility that powerfully underlines its inherent multiracialism. The National Assembly deputies of any ethnic group can appoint, as their member in the Council of Ministers, a person from any other group in the National Assembly. And if they do so, they are entitled to nominate from among themselves a full-time deputy minister to understudy the minister of their choice. The Namas, for instance, could choose a white as their minister, and the Bushmen an Ovambo. The permutations are considerable, and the system creates dramatic scope for cross-pollination of ideas and the growth of trust among all groups without the taint of racialism.

Within that broad framework, the equality of all people will be protected by the Bill of Rights, which is based heavily on West Germany's declaration of human rights. The only flaw (not a serious one in an interim government) is that the Bill will not be guaranteed by the independent Supreme Court as was first planned. Instead, the Turnhalle has opted for a special constitutional court to advise all levels of government on request. Its decisions will not be binding, but they will be made public; and draft legislation must be suspended while the court decides on its constitutionality (with emphasis on fundamental rights). In addition, the Supreme

Court, although it cannot declare laws invalid, will have common law review powers so it can nullify executive and administrative actions on constitutional and other grounds. Finally, an ombudsman will be appointed to deal with public complaints.

The Bill of Rights binds the legislature and executive to respect fundamental rights and human dignity. All people will be equal before the law without privilege or prejudice. Life, liberty, and physical person may be assailed only on the grounds of legal writ. Religion is free. The right to possess property is guaranteed, and dispossession is possible only in the public interest, on legal grounds, and with fair compensation. Free speech and freedom of the press are entrenched within certain general provisos, chiefly that they do not "assail the rights of others" or affect the state or public order. The individual's right to advance himself is protected, subject to similar provisos.

The constitution specifically states that if the legislature restricts any fundamental right, it must be in general and not for the individual. It goes even further—under no circumstances, it rules, may any fundamental right be assailed "in essence" or abolished. Hence if the legislature does apply restrictions, they must by nature be temporary.

In short, then, SWA/Namibia is to be a democratic, multiracial, free enterprise state, with no more than the reasonable restrictions found in democracies. But it has a novel addition: constitutionally entrenched protection not only for the individual, but also for ethnic groups.

The sagest analysis of this experiment in peaceful coexistence and of its weaknesses was made by Clive Cowley, Windhoek bureau chief of the Argus Africa News Service and for years the most authoritative commentator on the country's affairs. In a recent article, he wrote that the curiosity of the Turnhalle constitution is that the closest parallel lies in communist countries, "for the Soviet Union, Yugoslavia and Czechoslovakia are all constituted, on paper at least, as polyethnic federations." But, he said, the resemblance ended there because of the Turnhalle's ban on Marxism-Leninism. He goes on:

> The danger of a further, ominous parallel with the communist countries will lie in practice, rather than theory.
>
> With the ethnic groups all given a veto in the legislature and executive, the whites could block demands for change, to maintain superiority over other groups.
>
> If so, the similarity with the Soviet Union, where the Russians lord it over other populations through a better hold on a totalitarian machine, would reveal the constitution as a cynical pretence.
>
> It should not happen if realism prevails. The interdependence of ethnic groups is clearly recognised in the draft constitution.

Cowley points out that National Assembly decisions will need a majority vote from the deputies of every group, so that a single group could block legislation; and that because all are also represented in the Council of Ministers, the consensus principle applies there, too.

> The veto is therefore a double-edged sword. Black and brown deputies could use it as easily as the whites to gain benefits for their groups.
> Potentially the first-tier or national government is a framework for either conciliation or conflict. The use to which it is put will depend heavily on the human element.

Remarking that the long-dominant whites will enter ethnic government with all the advantages accrued from apartheid; with better education more money, and greater political experience; with a mass of discriminatory legislation still on the books; and with the country still beholden in the interim period to South Africa for defense and internal security, Cowley says:

> Still, the whites are unlikely to use their advantages, in the central government, to stop black and brown advancement.
> They would destroy themselves if they did.
> So the real struggle—or national reconciliation—will start with interim government.

There are still gaps in the Turnhalle constitution, and many finer details have yet to be worked out. But in its intent, its principles, and the spirit in which many races have cooperated to produce a formula for mutual survival, it is possibly the most remarkable document to come out of a continent renowned only for colonial hegemony and the dictatorial repression and institutionalized racism that have succeeded it.

It is already being attacked internationally by the textbook liberals, who cry that anything that makes ethnic distinctions is apartheid by another name and an evil scheme to perpetuate repression and exploitation. It is also being attacked by nationalists of all shades, who fear the formula will spoil their respective bids for total power.

But the sad, unavoidable truth of Africa is that if the overworked policy of majority rule were applied in the SWA/Namibia—or in South Africa or Rhodesia—it would very quickly lead to real repression and exploitation, probably via the bloody path of racial war or police dictatorship. As independent Africa has abundantly proved, racialism is by no means practiced only by whites against blacks. By far the bloodiest and most widespread race repression on this continent is black over black, tribe over tribe, and it was centuries before colonialism imposed some order—albeit at a fat profit and often brutally.

Independence and the replacement of the spear by the gun have allowed this menace to rise again. It is perhaps unfair to mention Uganda, where systematic genocide has become so much a part of routine administration that even Black Africa admits its severe embarrassment. But there are many other examples: Burundi, where the regime defeated its traditional tribal opposition simply by murdering them in the tens of thousands; Ethiopia, where political killings have become so commonplace since Haile Selassie's dethronement, it is difficult to keep track of the new leaders; the former Spanish Sahara, where indigenous tribes are fighting off efforts by Morrocco and Mauritania to grab their land; Angola, where civil war still rages on mainly tribal lines; the Sudan, where the Arab north and African south warred for seventeen years and trouble is still latent; Benin, where tribal feuding and poverty have caused seven changes of regime and many attempted coups in sixteen years.

There is ample evidence of totalitarianism. Only two states in Africa—Botswana and the Gambia—have genuinely democratic, multi-party, free-enterprise systems of government. Admittedly, some of the autocracies are benign, such as those of the Ivory Coast, Senegal, Mauritania, Zambia, Liberia, Kenya, Ghana, Swaziland (the world's last true monarchy) and a few others—although their efficiency varies widely. But others rank among the most vicious governments outside the communist bloc, governments such as the Central African Republic, Guinea and Equatorial Guinea, where life is hardly pleasanter than in Uganda but vastly less publicized.

If SWA/Namibia was tossed to the principle of majority rule, it would probably go totalitarian. It was plagued by war and enslavement among the Hereros, Ovambos, Damaras, and coloured (or mixed-blood) peoples before the Germans colonized it and in their own inimitable fashion brought lasting peace with a war of repression that nearly exterminated the Hereros and subdued the rest.

The diversity between the SWA/Namibia ethnic groups is astonishing. Yet because the country is so big and sparsely populated and because most of the groups have long lived in distinctly separate areas with foreign-enforced law and order, there has been little contact and friction among their cultures for generations, and race relations are unusually good now.

A simple examination of the country's peoples reveals the extraordinary breadth of the human spectrum and the riches it could produce for all—and also how volatile their mixture could be.

- *Ovambos.* Seven tribes, each with its own chief or headman. Devoutly Christian, dedicated to social gospel, they are natural traders with a shop for every 400 people. Show remarkable aptitude for operating earthmoving machinery and other heavy equipment in mines, such as the diamond operation at Oranjemund. Energetic and amiable, they form just under half the total South West African population.

- *Whites.* Most experienced politically, administratively, technologically, with highest education level. Basically Western and furthest advanced by twentieth-century standards. Mainly South African origin with strong German presence.
- *Coloureds.* Mixed-blood people similar to whites but less advanced. Vary from sophisticated to very poor and backward. Prominent in building and fishing industries.
- *Basters.* Racially akin to coloureds but historically distinct in origins and culture and tradition. Extremely insular, proud, and stubborn. Artisans and farmers, with very deep attachment to their own historically settled Baster territory.
- *Caprivians.* Two agriculturalist tribes with subsistence economy. Remote and long isolated by distance and desert. Languages unrelated to others in the country, except the Mbukushu of neighboring Kavango. Both Bantu tribes closely related to the Barotse of southeast Zambia.
- *Kavango.* Agriculturalists, fishermen, woodcarvers. Five tribes with a largely subsistence way of life. Only recently became involved elsewhere in any significant numbers, as migrant workers. Insular and distrustful of neighboring Ovambos.
- *Damaras.* Of somewhat mysterious Negro origin, they long ago lost tribal culture as slaves to Hereros and Namas, so they do not have to peel off layers of tradition before they adapt to modern life. Potential Ibos of southern Africa, they make good artisans. Their way of life is often fairly primitive. Their own language is lost, and they speak Nama.
- *Hereros.* Born cattlemen and warriors who dislike manual work and prefer jobs as clerks, drivers, and the like. Bantu with strong Hamitic influence, they are an extremely proud, politically sharp and alert people. They include two small subgroups in the remote Kaokoveld in northwest corner of the country: (1) the pastoral Himbas, who still wear the traditional leather apparel, use ocher and other ancient forms of adornment, and follow a wholly unsophisticated way of life far behind the twentieth century; and (2) the Tjimbas, who only a few years ago were discovered in the Kunene River region to be still using Stone Age techniques, such as flintstone weapons and skinning tools.
- *Namas.* Like the Bushmen, members of the Khoisan group of people, but their click language is unrelated to Bushmen tongues. Divided into about eleven subtribes, they are pastoralists with sheep and some cattle and are mainly in the arid south.
- *Tswanas.* An overflow from neighboring Botswana, they are of Sotho stock and are Bantu. Cattlemen like their brethren across the border, they are a mild, peace-loving people.

- *Bushmen.* Stone Age desert hunters and gatherers of wild foods, they are divided into unrelated bands. Many are still wild, but others work on white farms or as serfs to blacks. Whey they get training and opportunity, they display unusual ability with the internal combustion engine, reputedly because of their exceptional knowledge of animal anatomy. Addicted to tobacco.

It would take nothing short of a miracle to achieve homogeneity or successful majority rule among so many ethnic groups of such enormous disparity in so many respects.

Even democratic, nonracial Botswana tacitly admits the remoteness of the Bushmen from the modern social structure. It has created the Central Kalahari Game Reserve in the depths of that desert. While indeed a game reserve, it is also a safe shelter for the many wild Bushmen there, and the government has appointed a special official to safeguard the Bushmen's interests. The Bushmen's exceptional skill in surviving in a harsh wilderness where other people would die is in itself a form of civilization (insofar as that means turning one's environment to one's advantage), but it is eons away from twentieth-century politics and technology.

At about the time that the Turnhalle's legal advisers were drafting the SWA/Namibia constitution late last year, the same ethnicity concept emerged in South Africa. It came in the report by a steering committee, headed by Mr. Justice Kowie Marais and appointed by South Africa's three main official opposition parties—the United Party, the Progressive Reform Party, and the Democratic Party. The committee's function was to find a basis for the creation of a new, combined opposition party to challenge the monolithic National Party with a policy designed to meet the fast-changing, increasingly dangerous circumstances of southern Africa.

This is what the committee produced, as reported by the political correspondent of *The Star* of Johannesburg, South Africa's largest daily newspaper:

> The proposals, officially handed to the three party leaders at an hotel near Johannesburg last night (December 19), have been roundly welcomed by the leader of the Progressive Reform Party, Mr Colin Eglin, and somewhat surprisingly by Mr Theo Gerdener, leader of the Democratic Party.
>
> But Sir De Villiers Graaff [United Party leader] reserved his comments until he had time to "study the contents carefully."
>
> The steering committee . . . found that there was a growing realisation among the voters that South Africa was in a continuing crisis of survival and that the government's policies offered no solution.
>
> The 14 principles which the committee laid down as a recommendation for inclusion in the constitution of a new opposition party were:
>
> - That God is acknowledged as the ultimate authority in the destiny of all nations.

- All South Africans have an equal right to full citizenship and citizenship rights, either in a federation or confederation, which can only be realised in a country where there is no discrimination on the grounds of race, colour, religion or sex.
- Full citizenship is the basis for loyalty towards the State.
- Political rights must be shared by all South African citizens on an equitable and responsible basis.
- In a plural society like that of the Republic, a constitution and an entrenched Bill of Rights guaranteeing the rights of individuals and minorities are essential.
- All citizens have the right to an equal opportunity to share in the system of free economic enterprise.
- The party stands for a democratic system of government and rejects all totalitarian or authoritarian systems, such as communism.
- Except in the case of a duly declared state of emergency or war, every individual has the right to the protection of his life, liberty and property. In the protection of these rights, access to the judiciary must not be denied.
- The party guarantees the right of all our people to maintain their religious, language and cultural heritages.
- All educational systems must provide equal educational opportunities for citizens of all races.
- All inequitable forms of statutory or administrative discrimination on the grounds of race, colour, religion or sex, are unacceptable.
- The party accepts that certain geographic areas are being developed as economic and political growth points for certain sections of the population. Where the inhabitants of such-like areas freely elect to proceed with such development towards increasing self-determination, the party will respect their wishes in terms of its broad policy.
- The party will endeavour to have all the above principles as well as other guarantees, incorporated in a constitution for the Republic which will be drawn up after joint consultation and decision-making by the representatives of all the citizens of the country.
- As an inseparable part of the continent of Africa, the Republic accepts that peaceful relations with the states of Southern Africa, in the first instance, must be brought about and that technological, economic and political co-operation must be encouraged where possible by way of institutional arrangements.

With minor variations, these "Fourteen Principles," as they have become known, are identical to those inherent in the SWA/Namibia constitution.

That the three opposition parties could get together to compile them, on the initiative of Sir De Villiers Graaff, reveals an unexpected degree of common ground among them and raised high hopes among many South Africans of all races for an end to the divisions that have bogged down the competition against the National Party for so many years. All three opposition leaders agreed there was common ground, and their officials met to build a framework on the Marais Committee's foundations.

But finally, in February, this tentative unity collapsed in failure. The grassroot differences in beliefs, leadership, and structure were too great. Each party is going its own way, the United Party's mainly downhill toward what looks to be disintegration in the not too distant future.

The interpretation of ethnicity was a prime cause of the collapse. The United Party wants ethnic power sharing at all levels of government. The most conservative opposition party, it has long been fearful of integrated decision making. It also accepts the various "homelands" the government has created for various African tribes as a fait accompli—they are too well established now to be undone, but they are potential partners in a future South African federation. The Progressive Reform Party, on the other hand, seeks a common society in which power is shared mainly on the basis of educational and economic achievement, with ethnic distinction on a very much lower key.

Despite the failure, the "Fourteen Principles" have not been thrown overboard and remain as a tantalizing and provocative signpost to what could be. With continuing disruption and erosion within the United Party in particular, prospects appear good for a major reorientation of the official South African opposition. If this occurs, the "Fourteen Principles"—and ethnicity—will again come into prominence.

In an article in *The Sunday Tribune* (Durban) after the collapse, Kowie Marais succinctly pinpointed the pressures within South Africa:

> The failure . . . has grievously disappointed South Africa. This must especially be the case among those who are primarily concerned, the voteless have-nots—the more so when consensus seemed to be within the leaders' reach after the general adoption of the 14 Principles.
>
> Seldom has there been a more auspicious time for consolidating public opinion in opposition to policies. There is a groundswell of disillusionment throughout the country and in every section of society.

He states that there is a choice that no South African can avoid:

> The one alternative is confrontation and violence. The other is a dedication to seeking the cooperation of all South Africa.
>
> Those who are traditionally bound to distrust men of colour or who have been indoctrinated into believing that peaceful coexistence in a single society is impossible no matter what checks and balances and safeguards for the protection of minority rights might be devised, will, as a matter of course, choose the alternative of "shooting it out."
>
> Their opponents in the confrontation will be men who also believe that nothing can be gained without violence.
>
> That is the one option. The other is open to those who can be convinced that whatever the risks and difficulties involved, there is still, in 1977, sufficient goodwill in all sections of the population to establish a successful Turnhalle-type of joint decision-making; that all human beings, except the mentally defective, are innately inclined towards order and peace and that no

man in his senses would be prepared to risk life, or liberty, in order to exchange a truly fair deal for communism or a similarly evil system.

National Party reaction to this sudden surge of southern African ethnicity has been intriguing—and has exposed a strange ambivalence. The vigor with which Nationalists attacked the Kowie Marais report revealed a realization within even that hidebound party of the dramatic change taking place in the subcontinent, and also fear of the threat that an opposition based on the "Fourteen Principles" would pose (although it would take a political earthquake to unseat the government).

The party dragged out and dusted off the faithful old "Swart Gevaar" ("Black Menace") bogey it has used so effectively to stay in power for nearly thirty years. In Parliament, for example, the deputy minister of the interior, Louis le Grange, made the extraordinary statement that not only were white voters satisfied with the government, but that the majority of the voteless nonwhites were satisfied that the present dispensation offered them the maximum advantage. He did not enlighten a somewhat stunned public as to how the National Party tested nonwhite opinion. By accepting the "Fourteen Principles," he alleged, the opposition parties accepted black majority government.

It was a typical reaction. Yet, at the other extreme, the same party, the same people, do not merely permit but actively encourage the Turnhalle in Windhoek to adopt an ethnic-based constitution whose keystone is virtually identical to the "Fourteen Principles," a constitution that shares power right to the top and that will replace apartheid with multiracialism. When strong voices repeatedly urged that if the government could swallow all that for SWA/Namibia, it could swallow it for South Africa too, John Vorster, the prime minister, thunderingly rejected the very thought with utter scorn. South Africans can be excused for feeling bemused and frustrated—especially since Mr. Vorster is urging blacks and whites in turbulent Rhodesia to "find each other."

Many powerful and deep-seated emotions lie behind this dichotomy: the dogged determination of the Afrikaners represented by the National Party to retain their cultural identity and purity at all costs; their traditional calvinism; the addiction to power after thirty years of it; the bitterness and insularity that increase in direct proportion to the international censure heaped upon them; their love for a country that they, mainly, tamed and carved from the wilderness, and where they evolved as the world's newest people.

The Nationalists also have a pathetically keen yearning to be understood. Seldom has this been better expressed than in their reaction to an article I wrote recently on the emergence of the ethnicity concept and published in several South African opposition newspapers.

Die Beeld, the Johannesburg newspaper of the powerful Nationalist-controlled *Nasionale Pers* group, immediately ran a news report of the article and editorialized on it for two days.

> We find refreshing the acceptance of ethnicity as a solution to Southern
> Africa's racial problems, as suggested by a senior journalist, particularly
> because it comes from sources which have for so long and laboriously denied
> the basically correct philosophy of the South African Government on this
> matter.
> Nevertheless it is evident that there is still confusion about "apartheid"
> which he describes as a system which classifies people as inferior or superior.
> South Africa gave its final answer on this matter, for example, during the
> World Court case about South West Africa which we won. This policy is not
> based on inferiority or superiority, but otherness—or ethnicity.
> An understanding of this is essential to make possible an understanding of
> what is happening in South Africa today.

The following day *Die Beeld* said that "the key of the Nationalist
relationship policy is the black homelands, which are based absolutely on
the recognition of ethnicity." It further said the policy was best expressed
by Vorster when he said that discriminatory practices would be ended but
that differentiation would be maintained and expanded. Under this
process, some black homelands were independent or about to be, others
would soon be self-governing entities, and independent structures were
provided for the coloured and Asian communities. "On what other basis is
the new dispensation in South West Africa being developed?" *Die Beeld*
asks.

"Current Affairs," the anonymous radio program that spearheads
Nationalist propaganda on the state-controlled South African Broadcasting Corporation, commented sarcastically that Woodrow Wilson
promoted ethnicity close on sixty years ago and that in 1975 it was
preached by Philip Burnham of University College, London, and by
Nathan Glazer of Harvard. "Had it not been for the unyielding
commitment of Pretoria to ethnicity, there would have been no Turnhalle
and there would have been no meeting of the minds of the territory's eleven
ethnic communities," the ghost voice told listeners.

Curiously, United Party supporters said that ethnicity was what their
policy was all about—and so did a number of Progressive Reform Party
members.

All of this reflects the semantic speciousness that has for so long
befogged South African politics and turned the considerable common
ground among all sides into a morass. United Party policy is also based
firmly on racial distinction and largely on the yardstick of color, but it seeks
some form of federation or "confederation" embodying the black
homelands and with equal rights for all. "White leadership with justice,"
they call it. The Progressive Reform Party also recognizes that South
African society comprises many and varied ethnic groups but seeks a kind

of merit system pegged to academic and economic achievement to determine how power will be shared. Depending on how the merit measure is used, it could be a long time under this system, too, before blacks, in particular, gain any significant say in government. *Die Beeld* and "Current Affairs" were less than honest in their reactions to the examination of ethnicity. Anybody can see this with his own eyes by taking a $152, three-hour return trip from Johannesburg to Windhoek. It is all a question of the application of the fact of ethnic differences, a fact that all parties claim to recognize.

Under South Africa's mass of discriminatory legislation, the government bars blacks, coloureds, and Asians from virtually any area and activity it wishes; it admits only those it requires, such as workers, shoppers, and servants. National Party policy purports to aim for separate but equal facilities and amenities for the other races. But only the most blatant hypocrite would insist that much has been done to make them equal, that much is being done, or that the other races are allowed much scope to do it for themselves. South Africa is riddled with ridiculous (and costly) warts of discrimination—separate buses, separate taxis, separate beaches, separate toilets—the list seems endless. Admittedly, there has been marked (by South African standards) change in the past few years. All races are now free to use certain hotels, to eat and drink together under certain circumstances, and, at the more erudite levels of society, to communicate more easily. But the total effect is still very small and mainly cosmetic.

In SWA/Namibia, on the other hand, the dismantling of the entire discrimination machine has already begun. Multiracialism is on the way in as apartheid fades out. By the time full independence comes, nobody will have to suffer the slings and arrows of outrageous apartheid: being refused a drink or meal because his skin is not white, being prosecuted for sleeping with someone of another color, being unable to travel freely around his own country without a special permit within the limits of the system protecting ethnic groups and cultures, he will be able to send his children to any school he wishes and live where he likes, as long as he can afford it. So to equate apartheid in South Africa with ethnicity in SWA/Namibia, as the Nationalist media do, demands the mental gymnastics of a Lewis Carroll.

The *Star* of Johannesburg neatly summed up the position for South Africa in a February 14, 1977, editorial on attempts to realign the opposition:

> Whichever parties are involved, there can be only one valid foundation for a realigned opposition: a full commitment to hammering out an acceptable race policy with moderate black and brown South Africans.
> There is no other alternative to the Government's policy of imposed separate development. Thus the real tragedy of the current dispute within opposition ranks is the fact that it is very close to being irrelevant.

Particularly, within the ranks of the United Party, the shots are being called now by the "plattelanders" ["rural community"] who still see politics in an exclusively white context. They are out of touch with the new pressures on and within South Africa; they are also out of touch with urban voters in their own party.

White people are arguing about *white* interpretations of the Marais Committee's cardinal commitment to full citizenship and equal rights for all South Africans. Instead, they should be worried about how they can jointly reach a modus vivendi with the other races before there are no moderate leaders with whom to negotiate.

The Nationalists have thus far failed in their attempts to govern South Africa equitably on the basis of sharing the country on a racial basis. Except for offering a watered-down version of separate development (which must fail on almost every count), the opposition cannot offer anything except a sharing of political power in an undivided South Africa. And just as separate development cannot work without the support of the majority of black and brown people, so power sharing cannot work without that support.

South Africa needs a Turnhalle-type approach to its future. Black, brown and white leaders must negotiate a political settlement that takes South African realities into account. And nobody can go into a Turnhalle-type operation with a fixed and inflexible commitment. That is what the Government is trying to do. It will fail. So the opposition must prepare itself for a settlement based on *real* consultation, and those members of the opposition who cannot go along with a negotiated settlement will be swept into total irrelevance by the accelerated pace of events.

This is the real dividing line in white South African politics. It is the line between those who want to impose a solution on 21 million people and those who want to negotiate.

Any lingering suspicion that the Turnhalle constitution is a ploy to preserve white supremacy should be dispelled by the fact that the majority of the Turnhalle delegates are black and brown and include prominent leaders who for years risked prosecution by their vigorous opposition to apartheid and demands for independence. Whatever the South African government may have intended when it set up the Turnhalle, the conference is now its own master. Pretoria would wreck what remains of its own international credibility if it tried to negate the Turnhalle decisions.

The only cloud on the scene is that the South West African People's Organisation (Swapo) is not a party to its decisions. Swapo, which was recognized by the UN, almost the entire Third World and Eastern bloc, and some Western nations as virtually the Namibia government-in-exile long before the Turnhalle initiative began, refuses to have anything to do with it. It (Swapo) demands that South Africa hand over the territory to it through the United Nations.

But the cloud is not a large one. Swapo is increasingly being left out in the cold by the pace of the Turnhalle developments and could never have hoped to gain full independence by the end of 1978.

There are, moreover, two Swapos, though the movement's spokesmen disclaim this. One is the Swapo recognized within SWA/Namibia, whose supporters are seeing at first hand how the Turnhalle is doing what their leaders cannot.

The other is the external Swapo, which is essentially a guerrilla movement trying rather half-heartedly to carry its fight into the country from the north through a powerful barrier of South African security forces. However, the status of the external Swapo is being eroded. Many African countries were shocked by the Soviet-Cuban thrust into Angola and now treat Swapo with circumspection because of its rapid swing to the Russians and Cubans for help. They are made uneasy, too, by the split within the external Swapo, a split that climaxed in the detention of more than one thousand of its followers, including some of the hierarchy. The split was caused by a variety of factors. Chief among these were complaints from guerrillas of poor training and shortages of arms, food, and other supplies while the leaders lived well; accusations of corruption in the management of these supplies; and criticism of the leadership for repeatedly postponing a long overdue and crucial party congress.

Also eroding the Swapo image is the return to SWA/Namibia, since the Turnhalle began, of several of its former leading lights and other exiles. Swapo has branded them "sell-outs," but this is hardly true of such people as Mburumba Kerina, the man who coined the name "Namibia," Emil Appolus, and Jariretundu Kozonguizi, intellectual and head of the near dormant S. W. A. National Union.

The South African government flatly refuses to have anything to do with Swapo, external or internal, because Swapo chose the gun to settle the issue and because of its dependence on the communists. This is a mistake. Whatever Nujoma's guerrillas do abroad, Swapo undoubtedly has a large following (though it has never been tested by a head count or poll) inside SWA/Namibia, especially among the Ovambo people. At some time Swapo will have to be admitted to the new political scene as a party to take its chances against the rest in elections. By then, probably, its militancy and a large part of its original reason for existence will have been neutralized by the fact of multiracial independence.

SWA/Namibia is not the only part of southern Africa where ethnicity, as applied by the Turnhalle, could avoid conflict and enable peaceful coexistence. It is perhaps the ideal area because of the very clear cultural and other ethnic differences among most of its eleven groups. But it just might work even in war-torn Rhodesia, and that appears to be the direction in which Rhodesian politics are now moving.

It is commonly assumed that the Africans of Rhodesia are divided into only two tribes, the majority Shona and the minority Matabele, an offshoot of the powerful Zulu nation. In fact, though the Matabele are a proud and unitary tribe—who incidentally despise and used to raid the

Shonas for food, women, and cattle—the Shonas are not one but an entire collection of tribes. Shona is a generic term for all the tribes and clans loosely aligned linguistically and culturally—the Karangas, Manyikas, Korekores, Zezurus, Ndaus, and Kalangas, each of which has its own subgroups. They have already evolved a form of ethnicity, a mutuality that goes back many generations to when they would sink their differences to unite against such common enemies as the Matabele and, later, the new white rulers.

Support for the leaders in the turbulent, constantly changing melange of African nationalist movements varies considerably from tribe to tribe and accounts to a large degree for the number of leaders vying for power. But this ever-shifting kaleidoscope of tribes, parties, guerrilla armies and coalitions can be boiled down to three. They are the Zimbabwe African People's Union, the Zimbabwe African National Union, and the United African National Council.

The first, Zapu, is almost exclusively a Matabele movement. The second, Zanu, is Shona. Such is the competition between these two behind their thin facade of unity that many politically aware Africans fear their biggest problem is not attaining majority rule, but the civil war they believe must inevitably follow it. Both movements have guerrilla armies supplied by Russia or China or both with weapons and training. Zanu's army is much the larger and leans more to the Marxist side; its main bases for attack on Rhodesia are in the avowedly Marxist-Leninist state of Mozambique. Zapu is now building up its smaller guerrilla force—reputedly less to attack Rhodesia than to repel an attack by Zanu after independence.

Rhodesia's main hope for peaceful solution now lies in the UANC, which has grown out of the groundswell of black opposition to the settlement deal worked out in 1971 by Ian Smith and Alec Douglas Home, then the British premier. Zapu and Zanu briefly came under its umbrella thereafter, but they could not abide each other and went their own ways. However, the UANC (as it proved in 1971 and since) clearly has the greatest following among Rhodesian Africans, especially the huge population of youth. It has very few guerrillas, and it is led by Bishop Abel Muzorewa of America's United Methodist Church—a man whose prime concern is not personal power but avoiding war that would ruin the country, probably admit the communists, and go completely against the grain of his strong Christian ethics.

The probable scenario in Rhodesia now is a referendum called by Smith, ostensibly to prove that his conservative African supporters, such as the officially sanctioned tribal chiefs and headmen, have the greatest following. Bishop Muzorewa would enter the fray, and Smith would fervently hope that the bishop would win. Smith could then say: "Demonstrably the bishop has the approval of the people. He is the man with whom to negotiate a settlement."

As Deon du Plessis, Argus Africa News Service bureau chief in

Salisbury, points out, it would be a hard argument for the outside world to refute, especially if (as has been reported likely) the referendum was supervised by outside powers, for example, Britain and America. A new multiracial Rhodesia responsibly ruled by those thus proven to be dominant would be very quickly accepted by a Western world only too eager to get this monkey off its back. The guerrillas attacking from outside would have to either accept the referendum result and lay down their arms, or be reduced to the status of outlaws whom the West can help to suppress with a clear conscience.

Smith has already prepared the stage for this scenario by recently forcing through Parliament—by the narrowest of margins and with African support—the Land Tenure Amendment Bill. This has virtually blocked the fountain of discrimination in Rhodesia, the former Land Tenure Act, by ending the entrenched right of exclusive tenure in white farming areas. Now blacks can buy there, too.

But the numerous Tribal Trust Lands all over Rhodesia will remain exclusively for Africans. Covering a large part of the country, they are areas occupied for many generations by the same tribes, and their firm delineation by law put an end to intertribal competition for land. Therein is already a plank for ethnicity: the protection of minority groups, in this case including millions still so unsophisticated they would be unable to look after themselves in the face of skilled competition.

As Du Plessis says, "Rhodesia will never be the same again. Further changes can now be made by a simple majority in the House." These are likely to include accelerated dismantling of segregation (which in any case was never as severe as South Africa's) in hotels, schools, and other fields.

With the blacks' two main objections out of the way—segregation and their exclusion from real power—the next stage would be an interim government with time to devise an ethnicity constitution similar to SWA/Namibia's, one that safeguards all minorities, including the whites who want to stay, in a free-enterprise, democratic system.

Curiously, this is precisely what is being urged by twelve of Smith's members of Parliament who rebelled against the passage of the Land Tenure Amendment Bill. In a letter to the *Rhodesia-Herald* on March 14, one of them, Ted Sutton-Pryce, who until his defiance was deputy minister in the office of the prime minister, said the twelve wanted a Turnhalle-type constitution. He wrote that the group envisaged a national assembly with each community represented and with a second tier of government below that where each group "could retain responsibility in a meaningful sense of the administration of community interests and values." He compared the Swiss cantonal system with "the approach being taken in South West Africa" and noted that American Admiral Elmo Zumwalt found the Turnhalle "the most promising area of political development" in southern Africa.

Had the hard-line, totalitarian ideology of Marxism-Leninism not got a stranglehold first in Angola and Mozambique, ethnicity could have applied there, too. But the Portuguese withdrawal after April 25, 1974, was too precipitous, chaotic, and cynical for a carefully paced, democratic transition to independence. The present civil war in Angola by and large follows tribal lines; ideology is a lesser motive among most. Mozambique also has a great tribal disparity. The 2.5 million Macua-Lomue people of Zambezia and Mocambique provinces of northern Mozambique, for instance, opposed Frelimo from the outset of its guerrilla attack in 1964 and still do. But they are forced to toe the Frelimo rulers' line like everybody else, thus creating latent instability.

But it is in South Africa—vastly more developed than its neighbors, the fulcrum on which the fate of all southern Africa hinges—that the rewards of ethnicity could be greatest. Its material wealth and resources, and the diversity and sophistication of its communications, industry, commerce, and administration are renowned, making it more akin to a modern state in Europe than in Africa. Its human resources are greater still—about 18 million blacks of some ten different tribes, 4.2 million whites, 2.7 million coloured people, and 700,000 Asians.

Those statistics barely touch on their real richness. About nine million Africans live in the urban areas, and millions of these are so urbanized (some after several generations) that they have no links whatsoever with their tribal cultures and life-styles. Among them and those still in the traditional tribal areas are enough highly educated, sophisticated men of stature to form governments for every other state in Black Africa, it has been said. The cultural treasure houses of people such as the Zulu, Tswana, and Venda are now being supplemented by entirely new cultures evolving in such giant urban complexes as Johannesburg's Soweto, whose select suburbs and modern slums are the teeming, lively home of a million people. The whites—the driving force whose dynamism and skills match those of Europe and North America—are themselves a rich cultural blend of British, Dutch, Huguenot French, German, Portuguese, and others, from all of whom emerged the proud, tough Afrikaner nation. Add to this the spice of the Cape Malays, the Chinese, and the Asians—plus that ebullient product of unsegregated days, the coloured people—and the potential is limitless for a country of tremendous vivacity and energy in everything from industry to the arts.

Consider a multiracial South Africa in which all these people share power in a central government whose first concern is the survival of South Africa and whose second concern the protection of its many ethnic groups. Consider a South African state with traditional regions safeguarded by entrenched rights, but without discrimination in everyday life, and with great centers of activity (e. g., the Witwatersrand complex) open to all, equally. Consider cities where residential areas are not dictated by laws and police but by economic and social abilities, by the natural affinity people have for others of their own culture or quality.

This is a real possibility and not just a pipedream. But it will take time to build, and South Africa is fast running out of time. In the brief period since the Portuguese collapse, black and white nationalism have polarized dangerously. Emotions have been brought close to the boil by events such as the widespread urban upheaval here with its Black Power exploitation and its police overreaction; the growing bloodshed in Rhodesia; and the Big Power challenge flung by the Soviet Union. Increasingly, blacks and whites believe a crunch is inevitable and are siding with their own kind.

Polarization has not yet gone so far, however, that coexistence is irretrievable. Polarization is still a phenomenon of the extremes, and in between lies the great majority of the population, who fear the total disaster of war more than they do other races, who have seen the misery and devastation spelled out clearly in Angola, Mozambique, Rhodesia, the Congo, and elsewhere, and who see it worsening in Africa as Russia tightens the screw. These people, if they can act in concert to nullify the extremists of either side, hold the solution in their hands, and this is the message that Kowie Marais and others are vigorously trying to spread.

But the ruling National Party stonily blocks all efforts at interracial cooperation outside its own narrow and tortuous channels. Warnings and fresh ideas run off its granite walls like water. By using force instead of diplomacy in dealing with its internal troubles, it increases racial polarization. It plans still tougher control laws, such as the Newspaper Bill, under which it can effectively silence the press critics who have long been the country's only effective opposition, and can keep secret from its own voters what it is doing.

Still there is hope, albeit faint. It lies in a widening acceptance that the total integration of majority rule can no more be forced on southern African society than can apartheid. It lies in ethnicity.

The Turnhalle in Windhoek has proved that it is possible in the society to work out a potentially peaceful system by consensus. Nobody has come up yet with a better formula, and it has taken the Cinderella state of SWA/Namibia to do it.

Bibliography

Adams, R. C. "Kinship, Family Structure and the Family Life of Five Families in Bishop Lavis." Research Project Report, University of the Western Cape, 1975.

Alverson, Hoyt. "Labor Migrants in South African Industry." In *Migration and Anthropology*, edited by R. F. Spencer. Seattle: American Ethnological Society, 1970.

Ancilli, Sister. "African Sisters Congregations." Unpublished paper for the School of Oriental and African Studies Seminar. "Christianity in Independent Africa," London University, 1975.

Apter, David. *The Gold Coast in Transition*. New York: Atheneum, 1963.

Ardener, Edwin, ed. *Social Anthropology and Linguistics*. London: Association of Social Anthropologists Monograph 10, 1971.

Arens, William E. "Mto Wa Mbu: A Study of a Multi-Tribal Community in Rural Tanzania." Ph. D. dissertation, University of Virginia, 1970.

Awolubuluyi, A. Oladele. "Towards a National Language." *Ibadan* 23 (1966): 16-18.

Bagshawe, F. J. "Peoples of the Happy Valley (East Africa), Part III, The Sandawe." *Journal of the African Society* 24 (1924-25): 219-227.

Balandier, Georges. *Sociologie des Brazzavilles Noires*. Paris: Colin, 1955.

Bandoh, A. A. Unpublished manuscript, n. d.

Banghart, Peter. "The Effects of the Migrant Labourer on the Ovambo of South West Africa." Unpublished manuscript, 1971.

Banton, Michael. *Race Relations*. New York: Basic Books, 1967.

———. "The Restructuring of Social Relationships." In *Social Change in Modern Africa*, edited by Aiden Southall. London: Oxford University Press, 1961.

———. *West African City*. London: Oxford University Press, 1957.

Barnouw, Victor. *An Introduction to Anthropology*. Vol. 2, *Ethnology*. Homewood, Ill.: Dorsey Press, 1971.

Barth, Fredrik. *Models of Social Organization*. Occasional Papers no. 23. London: Royal Anthropological Institute, 1966.

Barth, Fredrik, ed. *Ethnic Groups and Boundaries: The Social Organization of Cultural Difference*. Boston: Little, Brown and Company, 1969. Originally published in Norway for Scandinavian University Books, Universitetsforlaget, Bergen, Oslo, Tromso, and in the United Kingdom by George Allen and Unwin, London, 1970.

303

Barth, Henry. *Travels and Discoveries in North and Central Africa.* 3 vols. New York: Harper and Bros., 1859.

Bastian, Adolf. "Der Volkergedanke." In *Kultur*, edited by C. A. Schmitz. Frankfurt: Akademische Verlaganstalt, 1963.

Beaglehole, E., et al. "A Statement by Experts on Race Problems." *International Social Science Bulletin* 2 (1950): 391-396.

Beals, Ralph L., and Hoijer, Harry. *An Introduction to Anthroplogy.* 4th ed. New York: Macmillan Co., 1972.

Bennet, John W., ed. *The New Ethnicity: Perspective from Ethnology.* New York: West Publishing Co., 1975.

Berger, Peter, and Luckmann, Thomas. *The Social Construction of Reality.* New York: Anchor, 1967.

Bernstein, Hilda. *For Their Triumphs and for Their Tears.* London: International Defence and Aid Fund, 1975.

Bidney, David. "The Concept of Value in Modern Anthropology." In *Anthropology Today*, edited by A. L. Kroeber. Chicago: University of Chicago Press, 1965.

Biener, Henry. *Tanzania: Party Transformation and Economic Development.* Princeton, N. J.: Princeton University Press, 1967.

Bley, Helmut. *South West Africa under German Rule 1884-1914.* London: Heineman, 1971.

Boas, Franz, ed. *General Anthropology.* New York: D. C. Heath and Co., 1938.

Bogardus, Emory S. "Social Distance and Its Practical Implications." *Sociology and Social Research* 22 (1938): 462-476.

_____ . "A Social Distance Scale." *Sociology and Social Research*, 17 (1933): 265-271.

Bosman, F. C. L. "Die Franse Stamverwantskap en Kulturele bydrae tot die Afrikaanse Volk." In *Kultuurgeskiedenis van die Afrikaner.* Edited by C. M. van den Heever and P. de V. Pienaar. Vol. 1. Cape Town: Nasionale Pers, 1945.

Bradlow, Edna. "Indians at the Cape During the Period of Responsible Government." In *South Africa's Indians: The Evolution of a Minority Community.* Edited by B. Pachai and Bala Pillay. Dalhousie-Longman African Series, forthcoming.

Brandel-Syrier, Mia. *Reeftown Elite.* New York: Africana Publishing Corp., 1971.

Braroe, Nils. *Indians and Whites.* Stanford, Calif.: Stanford University Press, 1975.

Brett, E. A. *African Attitudes.* Fact Paper no. 14. Johannesburg: South African Institute of Race Relations, 1963.

Brewer, Marilyn B., and Campbell, Donald T. *Ethnocentrism and Intergroup Attitudes.* New York: John Wiley and Sons, 1976.

Bruwer, J. P. "Die Matriliniere Order Van Die Kavango." Unpublishd manuscript, 1966.

Bryant, A. T. *The Zulu People as They Were before the White Man Came.* Pietermaritzburg: Shuter and Shooter, 1949.

Bulletin Quotidien d'Information (B. Q.). Distribué par le Service de l'Information de la Republic de Haute Volta. Ouagadougou.

Bulmer, M. "Sociological Models of Mining Communities." *The Sociological Review* 23 (1975): 61-92.

Chattopadhyaya, H. P. *Indians in South Africa. A Socio-economic Study.* Calcutta, 1970.

Coetzee, J. H. "Rasseverhoudinge in Suide-Afrika, 1652-1952." *Koers* 19 (1952).

Cohen, Abner. *Custom and Politics in Urban Archaeology.* Los Angeles: University of California Press, 1969.

Cohen, Abner, ed. *Urban Ethnicity.* London: Tavistock Publications, 1974.

Cohen, Ronald, and Middleton, John, eds. *From Tribe to Nation in Africa: Studies in Incorporation Processes.* Scranton, Pa.: Chandler Publishing Co., 1970.

Coleman, James S. "The Role of Tribal Associations in Nigeria." *Proceedings Annual Conference, West African Institute of Social and Economic Research.* Ibadan, 1952.

The Collected Works of Mahatma Gandhi. Ahmedabad, 1958.

Colson, Elizabeth. "Competence and Incompetence in the Context of Independence." *Current Anthropology* 8 (1969).

Cox, Oliver C. *Caste, Class and Race: A Study in Social Dynamics.* 1948. Reprint. New York: Modern Reader Paperbacks, 1970.

Dar es Salaam. National Archives of the United Republic of Tanzania. Ministry of Agriculture. Monthly Reports. Minutes of the Coordinating Committee, Arusha Region. April 18, 1964. Accession no. 305. File A/MR/R/AR.

Dennis, P. "The Role of the Drunk in an Oaxaca Village." *American Anthropologist* 77 (1975).

Despres, Leo, ed. *Ethnicity and Resource Competition in Plural Societies.* Chicago: Aldine, 1976.

Dierickx, Charles Wallace. "Magugu: Population and Land Use in a Resettlement Project in the Northern Province of Tanganyika," Ph.D. dissertation, Northwestern University, 1955.

Dim Delobson, A. A. *L'Empire du Mogho-Naba: Coutumes des Mossi de la Haute Volta.* Paris: Domat-Montchrestien, 1932.

Dobert, M. "Liberation and the Women of Guinea." *Africa Report*, October 1970.

Du Toit, Brian M. "Cooperative Institutions and Culture Change in South Africa." *Journal of Asian and African Studies* 4 (1969).

_____ . "Cultural Continuity and African Urbanization." In *Urban Anthropology.* Edited by Elizabeth M. Eddy. Southern Anthropological Society Proceedings, no. 2. Athens: University of Georgia Press, 1968.

_____ ."Religious Revivalism Among Urban Zulu." In *Man: Anthropological Essays.* Edited by E. J. de Jager. Cape Town: C. Struik (Pty), 1971.

_____ . "Strike or You're in Trouble." In *Urban Man in Southern Africa.* Edited by C. Kileff and W. C. Pendleton. Gwelo: Mambo Press, 1975.

Edel, May. "African Tribalism." *Political Science Quarterly* 80 (1965).

Edelstein, Melville L. *What Do Young Africans Think?* Johannesburg: South African Institute of Race Relations, 1972.

Epstein, A. L. *Politics in an Urban African Community.* Manchester: Manchester University Press, 1958.

Evans-Pritchard, E. E. *The Nuer.* Oxford: The Clarendon Press, 1940.

_____ . "The Nuer of the Southern Sudan." In *African Political Systems.* Edited by M. Fortes and E. E. Evans-Pritchard. London: Oxford University Press, 1950.

_____. "The Zande State." *Journal of the Royal Anthropological Institute* 93 (1963): 134-154.

Fiawoo, D. K. "The Influence of Contemporary Social Changes on the Magico-Religious Concepts and Organization of the Southern Ewe." Ph. D. dissertation, University of Edinburgh, 1959.

Fishman, Joshua. *Language and Nationalism*. Chicago: Newbury House, 1972.

Fishman, Joshua, Ferguson, Charles A., and Das Gupta, Jyotindra, eds. *Language Problems of Developing Nations*. Boston: Wiley, 1968.

Forde, Daryll C. "The Governmental Role of Associations among the Yako." *Africa* 31 (1961): 309-323.

Fortes, Meyer. *The Dynamics of Clanships among the Tallensi*. London: Oxford University Press, 1945.

Fortes, M., and Evans-Pritchard, E. E., eds. *African Political Systems*. London: Oxford University Press, 1940.

Francis, E. K. *Interethnic Relations*. New York: Elsevier, 1976.

_____. "The Nature of the Ethnic Group." *American Journal of Sociology* 52 (1947): 394-395.

Freeman-Greenville, G. S. P. *The Medieval History of the Coast of Tanganyika*. London: 1962.

Fried, Morton H. *The Evolution of Political Society: An Essay in Political Anthropology*. New York: Random House, 1967.

Gamst, Frederick C., and Norbeck, Edward, eds. *Ideas of Culture, Sources and Uses*. New York: Holt, Rinehart and Winston, 1976.

Gayton, A. H. "Areal Affiliations of California Folktales." *American Anthropologist* 37 (1935).

Geertz, Clifford. "Ideology as a Cultural System." In *Ideology and Discontent*. Edited by D. Apter. London: Free Press, 1964.

Geertz, Clifford, ed. *Old Societies and New States*. Glencoe, Ill.: Free Press, 1963.

Giliomee, Hermann. "The Development of the Afrikaner's Self-Concept." *Looking at the Afrikaner Today*. Edited by Hendrik W. van der Merwe. Cape Town: Tafelberg, 1975.

Glazer, Nathan. "The Universalization of Ethnicity." *Encounter*, February 1950.

Glazer, Nathan, and Moynihan, Daniel P. *Beyond the Melting Pot*. Cambridge, Mass.: Harvard University Press, 1963.

_____. *Ethnicity: Theory and Experience*. Cambridge, Mass.: Harvard University Press, 1975.

Glazer, Nathan and Moynihan, Daniel P., eds. *Ethnicity: Theory and Practice*. Boston: Harvard University Press, 1975.

Gluckman, Max. "Anthropological Problems Arising from the African Industrial Revolution." In *Social Change in Modern Africa*. London: Oxford University Press, 1961.

_____. *Custom and Conflict in Africa*. Oxford: Basil Blackwell, 1965.

_____. "Malinowski's Functional Analysis of Social Change." In *Social Change: The Colonial Situation*. Edited by Immanuel Wallerstein. New York: John Wiley, 1966.

Goffman, Erving. *Asylums*. New York: Anchor, 1961.

_____. *Relations in Public*. Harmondsworth: Penguin, 1972.

Goldblatt, I. *History of South West Africa*. Cape Town: Juta, 1971.

Gordon, Milton M. *Assimilation in American Life*. New York: Oxford University

Press, 1964.

Gordon, R. J. "Informal Labor Organization in Namibia." *South African Labour Bulletin*, in press.

Gray, Betty A. *Beyond the Serengeti Plains: Adventures of an Anthropologist's Wife in the East African Hinterland.* New York: Vantage Press, 1971.

Gray, Robert F. "A Sleeping Sickness Settlement." Newsletter for Institute of Current World Affairs, April 27, 1955. Mimeographed.

Greeley, Andrew M. *Ethnicity in the United States: A Preliminary Reconnaissance.* New York: John Wiley and Sons, 1974.

Greenberg, Joseph. *The Languages of Africa.* Bloomington: University of Indiana Press, 1966.

_____. "Urbanism, Migration and Language." In *Urbanization and Migration in Africa.* Edited by Hilda Kuper. Los Angeles: University of California Press, 1965.

Gregson, Sister Ann. "The African Sisterhoods." Unpublished paper for the School of Oriental and African Studies Seminar "Christianity in Independent Africa," London University, 1975.

Grové, D. "Sosiale Afstand tussen Kleurlinge en Indiërs en Pretoria." Master's thesis, University of Pretoria, 1967.

Gulliver, Philip H. "Introduction." In *Tradition and Transistion in East Africa.* Edited by Philip H. Gulliver. Berkeley: University of California Press, 1969.

_____. *Labour Migration in a Rural Economy.* Kampala: East African Institute of Social Research, 1955.

Gupta, Jyotindra Das. "Ethnicity, Language Demands, and National Development in India." In *Ethnicity: Theory and Practice.* Edited by Nathan Glazer and Daniel P. Moynihan. Boston: Harvard University Press, 1975.

Gutkind, Peter C. W., ed. *The Passing of Tribal Man in Africa.* Leiden: E. J. Brill, 1970.

Gutteridge, William. *The Military in African Politics.* London: Methuen and Co., 1969.

Haddon, A. C. *The Races of Man and their Distribution.* London: Milner and Co., 1909.

Hammond-Tooke, W. D. ed. *The Bantu-Speaking Peoples of South Africa.* London: Kegan Paul, 1974.

Handbooks of African Languages. London: International African Languages.

Harries-Jones, P. " 'Home-Boy' Ties and Political Organization in a Copperbelt Township." In *Social Networks in Urban Situations.* Edited by J. Clyde Mitchell. Manchester: Manchester University Press, 1969.

Hartley, E. L. *Problems in Prejudice.* New York: King's Crown Press, 1946.

Henderson, J. T., ed. *Speeches of the Late Right Honorable Harry Escombe.* Pietermaritzburg, n. d.

Herskovits, Melville J. *Dahomey: An Ancient West African Kingdom*, 2 vols. New York: J. J. Augustin, 1938.

Hodgkin, Thomas. *African Political Parties.* London: Penguin, 1961.

_____. *Nationalism in Colonial Africa.* New York: New York University Press, 1956.

Hunt, Chester L. and Walker, Lewis, eds. *Ethnic Dynamics.* Homewood, Ill.: Dorsey Press, 1974.

Huttenback, Robert A. *Racism and Empire: White Settlers and Colored Immi-*

grants in the Self-Governing Colonies, 1830-1910. New York, 1976.

Huxley, Julian S. and Haddon, A. C. *We Europeans: A Survey of "Racial" Problems.* 1935. Reprint. London: Penguin Books, 1939.

Ideal Woman (Nigeria). May 1974.

Indian Opinion. June 8, 1907.

Jacobson, David. "Friendship and Mobility in the Development of an Urban Elite African Social System." In *Urban Growth in Subsaharan Africa.* Edited by Josef Gugler. Kampala: Makerere Institute of Social Research, 1970.

Johnson, Frederick. *A Standard Swahili-English Dictionary.* Oxford: Oxford University Press, 1967.

Joosub, H. E. "The Future of the Indian Community in South Africa." In *South African Dialogue.* Edited by Nic Rhoodie. Johannesburg, 1976.

Kandovazu, E. *Die Oruuano-Beweging.* Karibib: Rynse Sending Drukkery, 1968.

Kelley, J. and Pendleton, Wade C. "Structure, Culture and Occupational Prestige: Data from Sub-Saharan Africa." *American Journal of Sociology,* in press.

Kimble, David. *A Political History of Ghana.* Oxford: Clarendon Press, 1963.

Kozonguizi, F. J. "Historical Background and Current Problems in South West Africa." In *Southern Africa in Transition.* Edited by J. Davis and J. Baker. New York: Praeger, 1966.

Kroeber, A. L., ed. *Anthropology Today.* Chicago: University of Chicago Press, 1952.

Kuper, H. *An African Aristocracy.* London: Oxford University Press, 1947.

Kuper, Leo. *An African Bourgeoisie.* New Haven: Yale University Press, 1965.

Laumann, Edward O. *Bonds of Pluralism.* New York: John Wiley and Sons, 1973.

Lerner, Daniel, and Robinson, Richard D. "Swords and Plowshares: The Turkish Army as a Modernizing Force." *World Politics* 12 (1960): 26-29.

Lever, Henry. *Ethnic Attitudes of Johannesburg Youth.* Johannesburg: Witwatersrand University Press, 1968.

Levy, Marion J., Jr. "Armed Force Organizations." In *The Military and Modernization.* Edited by Henry Bienen. Chicago and New York: Aldine Atherton, 1971.

Lewis, I. M., ed. *History and Social Anthropology.* London: Tavistock Publications, 1968.

Lewis, Ioan. *Perspectives in Social Anthropology.* Harmondsworth: Penguin, 1976.

Lieberson, S. *Social Stratification: Theory and Research.* Indianapolis, Ind.: Bobbs-Merrill Co., 1970.

Lindzey, Gardner, ed. *Handbook of Social Psychology.* Reading, Mass.: Addison-Wesley Publishing Co., 1954

Little, Kenneth. *African Women in Towns.* Cambridge: Cambridge University Press, 1975.

———. "A question of Matrimonial Strategy?" A Comparison Between Ghanaian and British University Students." *Journal of Comparative Family Studies* 7 (1976).

———. "Some Women's Strategies in Modern West African Marriage." Unpublished manuscript, n. d.

———. *Urbanization as a Social Process: An Essay on Movement and Change in Contemporary Africa.* London: Routledge and Kegan Paul, 1974.

———. *West African Urbanization.* Cambridge: Cambridge University Press, 1965.

Lowie, R. H. *The History of Ethnological Theory.* New York: Holt, Rinehart and

Winston, 1960.

_____ . *Social Organization*. New York: Rinehart & Co., 1948.

Lystadt, Robert. *The Ashanti*. New Brunswick, N. J.: Rutgers University Press, 1958.

Maasdorp, Gavin and Pillay, Wesen. "Indians in the Political Economy of South Africa." In *South Africa's Indians: the Evolution of a Minority Community*. Edited by B. Pachai and Bala Pillay. Dalhousie-Longman African Series, forthcoming.

MacAndrew, C., and Edgerton, R. *Drunken Comportment*. Chicago: Aldine, 1970.

MacCrone, I. D. *Race Attitudes in South Africa*. Johannesburg: Witwatersrand University Press, 1937.

Mafeje, Archie. "The Ideology of 'Tribalism.'" *Journal of Modern African Studies* 9 (1971).

Mair, Lucy. *An Introduction to Social Anthropology*. 2d ed. Oxford: Clarendon Press, 1972.

Maquet, Jacques. *The Premise of Inequality in Ruanda*. London: Oxford University Press, 1961.

Marwick, M. G. "Some Problems in the Sociology of Sorcery and Witchcraft." In *African Systems of Thought*. Edited by M. Fortes and G. Dieterlen. London: Oxford University Press, 1965 (b).

_____ . *Sorcery in its Social Setting: A Study of the Northern Rhodesian Cewa*. Manchester: Manchester University Press, 1965 (a).

Mayer, Phillip. *Townsmen or Tribesmen: Conservation and the Process of Urbanization in a South African City*. London: Oxford University Press, 1961.

Mazrui, Ali A. "Is the Nile Valley Emerging as a New Political System?" Paper presented at the Annual Social Science Conference of the Universities of Eastern Africa (USSC), Kampala, Makerere University, December 1971.

_____ ."Piety and Puritanism under A Military Theocracy: Uganda Soldiers as Apostolic Successors." In *Political-Military Systems: Comparative Perspectives*. Edited by Catherine M. Kelleher. Beverly Hills and London: Sage Publications, 1974.

_____ . "Some Sociopolitical Functions of English Literature in Africa." In *Language Problems of Developing Nations*. Edited by Joshua A. Fishman, Charles A. Ferguson, and Jyotirindra Das Gupta. Boston: Wiley, 1968.

Meillassoux, Claude. *Urbanization of an African Community*. Seattle: University of Washington Press, 1968.

Meltzer, B., Petras, J., and Reynolds, L. *Symbolic Interactionism*. Boston: Routledge and Kegan Paul, 1975.

Mercier, Paul. "On the Meaning of 'Tribalism' in Black Africa." In *Africa: Social Problems of Change and Conflict*. Edited by Pierre L. van den Berghe. San Francisco: Chandler Publishing Co., 1965.

Merle-Davis, J. *Modern Industry and the African*. London: Frank Cass, 1967.

Merton, Robert K.; Broom, Leonard; Cottrell, Leonard S., Jr. eds. *Sociology Today*. New York: Basic Books, 1959.

Mitchell, J. Clyde. *The Kalela Dance*. Manchester: Manchester University Press, 1956.

_____ . "Perceptions of Ethnicity and Ethnic Behaviour: An Empirical Exploration." In *Urban Ethnicity*. Edited by Abner Cohen. London: Tavistock Publications, 1974.

———. "Race, Class and Status in South Central Africa." In *Social Stratification*. Edited by A. Tuden and L. Plotnicov. New York: Free Press, 1970.

———. "Some Aspects of Tribal Social Distance." In *The Multitribal Society*. Edited by A. A. Dubb. Lusaka: Rhodes-Livingstone Institute, 1962.

———. "Theoretical Orientations in African Urban Studies." In *The Social Anthropology of Complex Societies*. Edited by Michael Banton. London: Tavistock, 1966.

———. "Tribalism and the Plural Society." *Middleton*, 1970.

———. "Tribe and Social Change in South Central Africa: A Situational Approach." In *The Passing of Tribal Man in Africa*. Edited by Peter C. W. Gutkind. Leiden: E. J. Brill, 1970.

Mitchell, J. Clyde, ed. *Social Networks in Urban Situations: Analysis of Personal Relationships in Central African Towns*. Manchester: Manchester University Press for the Institute of African Studies, University of Zambia, 1969.

Mortimer, Edward. *France and the Africans 1944-1960*. London: Faber and Faber, 1969.

Morton-Williams, Peter. "The Yoruba Ogboni Cult in Oyo." *Africa* 30 (1960): 362-374.

Moumouni, A. "Le Probleme Linguistique en Afrique Noire." *Partisans* 23 (1966): 37-48.

Mühlmann, Wilhelm. *Geschichte der Antropologie*. Frankfurt: Athenaum Verlag, 1968.

———. *Methodik det Volkerkunde*. Stuttgart: Enke, 1938.

———. *Rassen, Etnien, Kulturen*. Berlin: Hermann/Luchterhand Verlag, 1964.

Mujaju, Akiiki. "The Religio-Regional Factor in Uganda Politics." Paper presented at the Third International Congress of Africanists. Addis Ababa, December 1973.

Murphy, G., and Likert, R. *Public Opinion and the Individual*. New York: Harper, 1938.

Myrdal, Gunnar. *An American Dilemma: The Negro Problem and Modern Democracy*. New York: Harper & Brothers, 1944.

Nadel, S. F. *A Black Byzantium*. London: Oxford Unversity Press, 1942.

———. *The Nuba*. London: Oxford University Press, 1947.

———. "Witchcraft in Four African Societies." *American Anthropologist* 54 (1952): 18-29.

Natal Mercury, August 5, 1908.

Nelson, Nici. "Informal Sector Economic Activity in a Squatter Neighbourhood." Paper for "African Urban Culture Seminar." London School of Oriental and Asian Studies, February 11, 1976.

Newman, J. L. "Dimensions of Sandawe Diet." *Ecology of Food and Nutrition* 4 (1975): 33-39.

———. *The Ecological Basis for Subsistence Change among the Sandawe of Tanzania*. Washington, D. C: National Academy of Sciences, 1970.

———. "A Sandawe Settlement Geography." *The East African Geographical Review*, no. 7 (1969): 15-24.

Nyerere, J. K. *Socialism and Rural Development*. Dar es Salaam: Government Printer, 1967.

O'Barr, William M. "Multilingualism in a Rural Tanzanian Village." *Anthropological Linguistics* 13: 289-300.

Oberg, A. M. K. "The Kingdom of Ankole in Uganda." In *African Political Systems*. Edited by M. Fortes and E. E. Evans-Pritchard. London: Oxford University Press, 1940.

Obote, The Hon. Milton. "Language and National Identification." *East African Journal* 4 (1967): 3-6.

O'Toole, James. *Watts and Woodstock: Identity and Culture*. New York: Holt, Rinehart and Winston, 1973.

Pachai, B. "Aliens in the Political Hierarchy." In *South Africa's Indians: the Evolution of a Minority Community*. Edited by B. Pachai and Bala Pillay. Dalhousie-Longman African Series, forthcoming.

———. *The History of Indian Opinion*. Cape Town, 1963.

———. *The International Aspects of the South African Indian Question, 1860-1971*. Cape Town, 1971.

Paine, Robert. *Some Second Thoughts about Barth's Models*, Occasional Papers, no. 32. London: Royal Anthropological Institute, 1974.

Pendleton, Wade. "Ethnicity as a Factor in Occupational Prestige." In *ASSA Sociology Southern Africa 1973*. Durban: University of Natal, 1973.

———. "Herero Reactions: The Pre-Colonial Period, the German Colonial Period and the Period of South African Colonialism." In *African Responses to European Colonialism in Southern Africa*. Edited by David Chanaiwa. Northridge: California State University Foundation, 1977.

———. "Introduction." In *Urban Man in Southern Africa*. Edited by C. Kileff and W. C. Pendleton. Gwelo: Mambo Press, 1975.

———. *Katatura: A Place Where We Do Not Stay: The Social Structure and Social Relationships of People in an African Township In South West Africa*. San Diego, Ca.: San Diego State University Press, 1974.

———. "Social Categorization and Language Usage in Windhoek, South West Africa." In *Urban Man in Southern Africa*. Edited by C. Kileff and W. C. Pendleton. Gwelo: Mambo Press, 1975.

Penniman, T. K. *A Hundred Years of Anthropology*. 2d ed. London: Gerald Duckworth and Co., 1952.

Pettigrew, T. F. "Social Distance Attitudes of South African Students." *Social Forces* 38 (1960): 246-253.

Piddington, Ralph. *An Introduction to Social Anthropology*. Vol. 2. Edinburgh: Oliver and Boyd, 1957.

Pietermaritzburg. Natal Archives. Natal and Imperial Blue Books. Blue Book, 1870-1885, vol. 66. Commission Report, ch. 40.

Polak, H. S. L. *The Indians of South Africa: Helots within the Empire and How They Are Treated*. Madras, 1910.

Preston-Whyte, E. "The Adaptation of Rural-Born Female Domestic Servants to Town Life." In *Focus on Cities*. Edited by H. Watts. Durban: University of Natal Press, 1970.

The Race Concept. Paris: UNESCO Publications, 1952.

Radcliffe-Brown, Alfred R. "The Present Position of Anthropological Studies." In *Method in Social Anthropology*. Edited by M. N. Srinivas. Chicago: University of Chicago Press, 1958.

_____ . *Structure and Function in Primitive Society.* London: Cohen and West, 1952.

Radcliffe-Brown, Alfred R., and Forde, Daryll. *African Systems of Kinship and Marriage.* London: Oxford University Press, 1950.

Riesman, David. "Some Observations on Intellectual Freedom." *The American Scholar* 23 (1953).

Rivière, C. "La Promotion de la Femme Guineenne." *Cahiers d'Études Africaines,* 8 (1968).

Rosenblatt, P. C. "Origins and Effects of Group Ethnocentrism and Nationalism." *Journal of Conflict Resolution* 8 (1964).

Ross, Marc. *Grass Roots in an African City.* Cambridge, Mass.: The Massachusetts Institute of Technology Press, 1975.

_____ . "Measuring Ethnicity in Nairobi." In *Survey Research in Africa.* Edited by William M. O'Barr, David H. Spain and Mark A. Tessler. Evanston, Ill.: Northwestern University Press, 1973.

Rouch, Jean. "Migration on the Gold Coast." *Journal de la Société des Africanistes* 23 (1956).

_____ . *Les Songhay.* Paris: Presses Universitaires de France, 1954.

Rouch, Jean, and Bernus, E. "Note sure les Prostituées 'toutou' de Treichville et d'Adjamé," *Études Éburnéennes* 6 (1959).

Rudebeck, Lars. *Guinea-Bissau: A Study of Political Mobilization.* Uppsala: Scandinavian Institute of African Studies, 1974.

Sahlins, M. *Stone Age Economics.* Chicago: Aldine, 1972.

Samarin, William J. "Lingua Francas, with Special Reference to Africa." In *Study of the Role of Second Languages in Asia, Africa and Latin America.* Edited by Frank A. Rice. Washington: Center for Applied Linguistics, 1962.

_____ . "Self Annulling Prestige Factors among Speakers of a Creole Language." In *Sociolinguistics.* Edited by William Bright. The Hague: Mouton, 1968.

Sargent, S. Stansfeld. *Social Psychology: An Integrative Interpretation.* New York: Holt, Rinehart & Winston, 1976.

Scalapino, Robert A. "Which Route for Korea?" *Asian Survey* 2 (1962).

Schapera, I., ed. *The Bantu-Speaking Tribes of South Africa.* Cape Town: Maskew Miller Ltd., 1937.

Schermerhorn, R. A. *These Our People, Minorities in American Culture.* Boston: D. C. Heath and Co., 1949.

Schild Krout, Enid. "Ethnicity and Generational Differences Among Urban Immigrants in Ghana." In *Urban Ethnicity.* **Edited by Abner Cohen. London: Tavistock Publications, 1974.**

Schlosser, K. *Einegeborenen Kirchen in Süd-und-Südwestafrika.* **Kiel: Mülau, 1958.**

Seligman, C. G. "Human Types in Tropical Africa." *Discovery,* June 1936.

_____ . *Races of Africa.* 3rd ed. London: Oxford University Press, 1961.

Senghor, Leopold S. *Négritude et Humanisme.* Paris: Seuil, 1964.

Shack, William A. "Notes on Voluntary Associations and Urbanization in Africa, with special reference to Addis Ababa, Ethiopia." In "Migrants and Strangers," *African Urban Notes,* ser. B., no. 1. Edited by Niara Sudarkasa. (Winter 1974-75).

Shelley, J. Karen. "Township Teens: A Study of Socialization Processes Among

Urban Black School Girls in South Africa." M. A. thesis, University of Florida, 1975.

Shibutani, T., and Kwan, K. M. *Ethnic Stratification: A Comparative Approach.* London: The Macmillan Co., 1970.

Shirokogoroff, S. M. "Die Grundzuge der Theorie van Ethnos." In *Kultur.* Edited by C. A. Schmitz. Frankfurt: Akademische Verlaganstalt, 1963.

_____ . *The Psychomental Complex of the Tungus.* London: Kegan Paul, 1935.

Skinner, Elliott P. *African Urban Life: The Transformation of Ouagadougou.* Princeton, N. J.: Princeton University Press, 1974.

_____ . "Group Dynamics in the Politics of Changing Societies: The Problem of 'Tribal' Politics in Africa." American Ethnological Society. *Proceedings of 1967 Annual Spring Meeting.* Seattle, Wash.: University of Washington Press, 1967. pp. 170-185.

_____ . *The Mossi of the Upper Volta.* Stanford, Ca.: Stanford University Press, 1964.

_____ . "Strangers in West African Societies." *Africa* 33, (1963): 307-320.

_____ . "West African Economic Systems." In *Peoples and Cultures of Africa.* Edited by Elliott P. Skinner. New York: Doubleday, 1973.

Smith, Edward W. *African Ideas of God.* London: Edinburgh House Press, 1950.

Smith, M. G. "Institutional and Political Conditions of Pluralism." In *Pluralism in Africa.* Edited by Leo Kuper and M. G. Smith. Berkeley: University of California Press, 1971.

_____ . *The Plural Society in the British West Indies.* Berkeley: University of California Press, 1965.

Soja, E. W. "The Political Organization of Space." *Resource Paper No. 9.* Commission of College Geography, Association of American Geographers, 1971.

South African Digest. Pretoria, February 13, 1976.

Southall, Aidan W. *Alur Society.* Cambridge: W. Heffer and Sons, 1953.

_____ . "Forms of Ethnic Linkage." In *Town and Country in Central and East Africa.* Edited by David Parkin. London: Oxford University Press, 1961.

_____ . "The Illusion of Tribe." In *The Passing of Tribal Man in Africa.* Edited by Peter C. W. Gutkind. Leiden: E. J. Brill, 1970.

Southall, Aidan, ed. *Social Change in Modern Africa.* London: Oxford University Press, 1961.

Spoelstra, Bouke. *Die "Doppers" in Suid-Afrika, 1760-1899.* Cape Town, 1963.

Steady, Filomina. *Female Power in African Politics: The National Congress of Sierra Leone.* Munger Africana Library Notes, California Institute of Technology, 1975.

Stipp, John L.; Hollister, Warren; Dirrim, Alan W.; and Bauman, Harold L. *The Rise and Development of Western Civilization.* New York and London: John Wiley and Sons, 1969.

Sumner, William Graham. *Folkways.* New York: New American Library, 1960.

Sutton, J. E. G. "Archaeological Sites in Usandawe." *Azania* 3 (1968): 167-173.

Tabouret-Keller, A. "Sociological Factors of Language Maintenance and Language Shift: A Methodological Note Based on European and African Examples." In *Language Problems of Developing Nations.* Edited by Joshua A. Fishman, Charles A. Ferguson, and Jyotirindra Das Gupta. Boston: Wiley, 1968.

Temu, A. J. "The Rise and Triumph of Nationalism" In *A History of Tanzania.*

Edited by Isaria Kimambo and A. J. Temu. Nairobi: East African Publishing House, 1969.

Ten Raa, E. "Bush Foraging and Agricultural Development: A History of Sandawe Famines." *Tanzania Notes and Records*, No. 69 (1968), pp. 33-40.

_____ . "The Couth and the Uncouth: Ethnic, Social, and Linguistic Divisions among the Sandawe of Central Tanzania." *Anthropos* 65 (1970): 127-153.

_____ . "Dead Art and Living Society: A Study of Rock Paintings in a Social Context." *Mankind* 8 (1971): 42-58.

_____ . "The Moon as a Symbol of Life and Fertility." *Africa* 34 (1969): 24-53.

_____ . "Sandawe Prehistory and the Vernacular Tradition." *Azania* 4 (1969): 91-103.

Theal, George McCall. *History of South Africa*. Vol. 3. Cape Town: Struik, 1964.

Thieret, A. *L'enseignement du Français en Afrique. Le Senegal, Population, Langues, Programmes Scolaires*. Dakar: Centre de Linguistique Appliquée de Dakar, 1966.

Thomas, Elizabeth Marshall. *The Harmless People*. New York: Vintage Books, 1965.

Thompson, Leonard. "Historical Perspectives of Pluralism in Africa." In *Pluralism in Africa*. Edited by Leo Kuper and M. G. Smith. Berkeley: University of California Press, 1971.

Towles, Joseph. "Symbiosis and Opposition in Inter-Group Relations." Unpublished notes.

Trevor, J. C. "The Physical Characters of the Sandawe." *Journal of the Royal Anthropological Institute* 77 (1947): 61-78.

Trimingham, J. Spencer. *Islam in East Africa*. Oxford: Oxford University Press, 1964.

Turnbull, Colin M. *The Forest People: A Study of the Pygmies of the Congo*. New York: Simon and Schuster, 1962.

_____ . *The Structure of Human Populations*. London: Oxford University Press, 1972.

Toynbee, Arnold. *A Study of History*. Vol. 1-4 abridged by D. C. Somervell. London: Oxford University Press, 1948.

Ucendu, Victor C. *Igbo of Southeast Nigeria*. New York: Holt and Rinehart, 1965.

Uganda Argus, January 26, 1971.

Valentine, Charles. *Culture and Poverty: Critique and Counter-Proposals*. Chicago: University of Chicago Press, 1968.

Van Aswegen, "The Asians in the Orange Free State." In *South Africa's Indians: The Evolution of a Minority Community*. Edited by B. Pachai and Bala Pillay. Dalhousie-Longman African Series, forthcoming.

Van den Berghe, P. L. "Race Attitudes in Durban, South Africa." *Journal of Social Psychology* 57 (1962): 55-72.

_____ . *Race and Racism: A Comparative Perspective*. New York: John Wiley and Sons, 1967.

Van den Berghe, P. L., ed. *Race and Ethnicity in Africa*. Nairobi: East African Publishing House, 1975.

Van der Merwe, Hendrik W., ed. *Looking at the Afrikaner Today*. Cape Town: Tafelberg, 1975.

Verwoerd, E. "Die Vrou Se invloed op goeie rassebetrekkinge." *Die Taalgenoot,* November 1973.

Voipio, R. *Kontrak Soos die Owambo dit Sien.* Johannesburg: Christian Institute of South Africa, 1972.

Wallerstein, Immanuel. "Ethnicity and National Integration in West Africa." *Cahiers d'Études Africaines* 3, (1960).

_____. "Migration in West Africa: The Political Perspective." In *Urbanization and Migration in West Africa.* Edited by Hilda Kuper. Berkeley: University of California Press, 1965.

Walter, Paul A. F. *Race and Culture Relations.* New York: McGraw-Hill Book Co., 1952.

Wanji, Barri A. "The Nubi Community: An Islamic Social Structure in East Africa." *Sociology Working Paper* no. 115. Department of Sociology, Makerere University, 1971.

Warner, W. Lloyd, and Drole, Leo. *The Social System of American Ethnic Groups.* New Haven: Yale University Press, 1965.

Warr, P., and Wall, T. *Work and Well-Being.* Harmondsworth: Penguin, 1975.

Welch, Claude E., and Smith, Arthur, K. *Military Role and Rule: Perspectives on Civil-Military Relations.* North Scituate, Mass.: Duxbury Press, 1974.

Williams, Brett, and Gordon, Robert, eds. *Anthropological Studies of Total Institutions.* Champaign, Ill.: Stipes, 1977.

Wilson, Francis. *Migrant Labor in South Africa.* Johannesburg: Ravan Press, 1972.

Wilson, Monica. "Witch Beliefs and Social Structure." *American Journal of Sociology* 56 (1951): 307-313.

Wilson, Monica, and Mafeje, A. *Langa: A Study of Social Groups in an African Township.* Cape Town: Oxford University Press, 1963.

Winick, Charles. *Dictionary of Anthropology.* Paterson, N. J.: Littlefield, Adams and Co., 1964.

Wippur, Audrey. "The Politics of Sex: Some Strategies Employed by the Kenyan Power-Elite to Handle a Normative-Existential Discrepancy." *African Studies Review* 14 (1971).

World Radio and TV Handbook. Denmark: Hvidours.

About the Contributors

Miss R. C. Adams was until very recently a lecturer in social anthropology at the University of the Western Cape. Fieldwork for her master's degree concentrated on coloured family life, and she writes about her people with conviction. Recently married, she now resides in Namibia.

Johannes Hendrik Coetzee was born on April 23, 1916, in the province of the Orange Free State. Educated at rural schools, he took B.A. and M.A. degrees in sociology at the University College of the Orange Free State (constituent of the University of South Africa), D. Phil. at University of Potchefstroom. Changed to the study of ethnology at the latter. Attained a B.A. Honns. in ethnology. Took the chair in ethnology in 1956. Developed interest in race relations and ethnicity. Publications mostly in these fields in various articles in scientific journals.

Brian M. du Toit was born on March 2, 1935, in Bloemfontein, South Africa. He studied at the University of Pretoria, where he received his B.A. (1957) and M.A. (1961) degrees, and at the University of Oregon, where he received his Ph.D. (1963). He was a lecturer in social anthropology at the University of Stellenbosch and at the University of Cape Town, and currently is professor of anthropology at the University of Florida. Recent publications include *People of the Valley: Life in an Isolated Afrikaner Community in South Africa* (1974), *Akuna: A New Guinea Village Community* (1975), *Migration and Urbanization* (edited with Helen Safa, 1975), *Configurations of Cultural Continuity* (1976), *Content and Context in Zulu Folk-Narratives* (1976), and *Drugs, Rituals, and Altered States of Consciousness* (1977).

Robert Gordon was born in Windhoek, Namibia. He was educated at the universities of Stellenbosch and Illinois. He is now teaching at the University of Papua New Guinea. A monograph, entitled *Mines, Migrants & "Masters,"* which provides a detailed study of the mine on which his chapter is based, is being published.

Joseph V. Guillotte, III was born in New Orleans, Louisiana, on November 3, 1932. He received a B.A. in history from the University of Southwestern Louisiana and a Ph.D. from Tulane University in 1973. He is an assistant professor of anthropology at the University of New Orleans and has taught at Tulane University and Hamilton College. He has presented papers at various professional meetings on the topics of social change, ethnicity, language, and politics. He is currently investigating the connections between medicine and politics.

Ruth H. Landman, born 1926, Germany. Emigrated after Nazi takeover. B.A., Vassar; M.A., Ph.D., Yale. Research: German-Jewish childhood; Jewish drinking customs; Mexican immigrant acculturation; applied anthropology in Britain. Publications: Mexican-Americans' Problems with Government Programs. Report to the Intra-Mural Research Division, Soc. and Rehab. Services, HEW, 1968; Acculturation of Mexican Immigrants and Their Descendants in Southern California, Univ. Microfilms, 1970; Applied Anthropology in Britain, submitted, Human Organization.

Kenneth Little was born on September 19, 1908, and educated at Trinity College, Cambridge (M.A.), and London School of Economics (Ph.D.). He taught at the latter institution and for the past twenty-seven years has been attached to the University of Edinburgh, where he is professor of African urban studies. His publications on African material and lately on women are numerous, but some of the most important are *The Mende of Sierra Leone* (1951), *West African Urbanization: A Study of Voluntary Associations in Social Change* (1965), *Urbanization as a Social Process: An Essay in Movement and Change in Africa South of the Sahara* (1973), and *African Women in Towns: An Aspect of Africa's Social Revolution* (1975).

Ali A. Mazrui is professor of political science and associated with the Center for Afroamerican and African Studies, The University of Michigan, Ann Arbor. He was awarded the Ph.D. degree after studies at Manchester University, Columbia University, and Oxford University. He was on the faculty of Makerere University, Kampala, for ten years, where he served as head of the department of political science and dean of the faculty of social sciences. His books include *Towards a Pax Africana* (1967), *Cultural Engineering and Nation-Building in East Africa* (1972), *Soldiers and Kinsmen in Uganda* (1975), and, his latest book, *A World Federation of Cultures: An African Perspective*, (1976), as part of the World Order Models Project.

James L. Newman is associate professor of geography at Syracuse University. He lived among the Sandawe for over a year during 1965-1966 and, based on this research, published *The Ecological Basis for Subsistence Change among the Sandawe of Tanzania* (1970).

Wilf Nussey has spent twenty-two of his twenty-three years in journalism covering Africa, particularly east and southern Africa, for a wide range of publications. For the last thirteen years, he has been with the Argus Africa News Service, owned by the Argus Printing and Publishing Company, Limited, which controls thirteen newspapers; for the last ten years, he has been its editor. The function of this news service is to provide objective, in-depth coverage of African affairs to promote understanding and knowledge of the continent well beyond that provided by its routine news events, and for this purpose it maintains five bureaus and a network of correspondents. In this essay, Mr. Nussey has drawn not only from his own experience in SWA/Namibia, which goes back to 1961, but also on the reporting and analyses of *Clive Cowley*, his Windhoek bureau chief, and of *Deon du Plessis* and *Wilfred Mbanga*, bureau chief and a staffer in the Salisbury Bureau. He has also drawn from reports by political reporters and editorial writers of *The Star* of Johannesburg, South Africa's largest daily newspaper.

Bridglal Pachai was born in Ladysmith, South Africa, the descendant of an Indian indentured laborer. He was the first South African Indian to obtain a Ph. D. in that country. Pachai was awarded the doctorate by the University of Natal for his study published as *The South African Indian Question 1860-1971* (1971). He was professor of history at the University of Malawi and currently occupies the same position at Dalhousie University. Pachai has edited a number of volumes on the history of Malawi, and he and P. D. Pillay have recently completed an edited volume entitled *South Africa's Indians: The Evolution of a Minority Group.*

Wade C. Pendleton; born 1941 in Big Spring, Texas, U.S.A.; B.A. and Ph.D., University of California (Berkeley), 1970; professor of anthropology, San Diego State University, San Diego, California; primary research in urban anthropology in southern Africa, especially South West Africa (Namibia); major recent publications include: *The Peoples of South West Africa* (1977), *Urban Man in Southern Africa* (1975, jointly edited with Clive Kileff), and *Katatura: A Place Where We Do Not Stay* (1974).

Elliot P. Skinner was educated at Columbia University, where he received his Ph.D. and where he is currently professor and chairman of the Department of Anthropology. Between 1966 and 1969 he was United States ambassador to the Republic of Upper Volta, where he had conducted field work a year earlier. Among his publications are *The Mossi of Upper Volta* (1964), *A Glorious Age in Africa* (1965), and *African Urban Life: The Transformation of Ouagadougou* (1974).

Colin M. Turnbull was born in London and educated at Oxford University. Since 1951 he has had regular contact with pygmies, living among the Mbuti for periods in 1954 and 1957. He was associate curator of African ethnology at the American Museum of Natural History (New York) and subsequently taught at the Virginia Commonwealth University. Turnbull's publications include *The Forest People* (1961), *The Lonely African* (1962), *Wayward Servants: The Two Worlds of the African Pygmies* (1965), and *The Mountain People* (1972).